POLYMERIC LIQUID CRYSTALS

POLYMER SCIENCE AND TECHNOLOGY

Recent volumes in the series:

A Continuation Order Plan is available for this series. A continuation order will bring delivery of each new volume immediately upon publication. Volumes are billed only upon actual shipment. For further information please contact the publisher.

POLYMERIC LIQUID CRYSTALS

Edited by

ALEXANDRE BLUMSTEIN

University of Lowell
Lowell, Massachusetts

PLENUM PRESS • NEW YORK AND LONDON

Library of Congress Cataloging in Publication Data

Symposium on Polymeric Liquid Crystals (2nd: 1983: Washington, D.C.)
   Polymeric liquid crystals.

   (Polymer science and technology; v. 28)
   Includes bibliographical references and index.
   1. Liquid crystals—Congresses. 2. Polymers and polymerization—Con-
gresses. I. Blumstein, Alexandre. II. American Chemical Society. Division
of Polymer Chemistry. III. American Chemical Society. Meeting (1983:
Washington, D.C.) IV. Title.V. Series.
QD923.S96   1983   548′.9   84-25478
ISBN 0-306-41814-2

Proceedings of the Second Symposium on Polymeric Liquid Crystals, Division of
Polymer Chemistry, held as part of the American Chemical Society meeting,
August 28–31, 1983, in Washington, D.C.

©1985 Plenum Press, New York
A Division of Plenum Publishing Corporation
233 Spring Street, New York, N.Y. 10013

PREFACE

This book originated in the Proceedings of the Second Symposium on Polymeric Liquid Crystals held by the Division of Polymer Chemistry in the framework of the 1983 Fall Meeting of the American Chemical Society.

At the First Symposium in 1977, the literature in this field could be encompassed in a single volume. Today, that is no longer possible. The field of Polymeric Liquid Crystals grew, and continues to grow, at a very rapid pace. At present, we know of every major mesophase in its polymeric form and of polymeric glasses, elastomers and fluids in their liquid crystalline form. Every year, new polymeric mesophases are being discovered.

The aim of this book is to go beyond a compilation of papers presented at the 1983 ACS Fall Meeting. It is conceived as a learning tool for the benefit of the scientist interested in Polymeric Liquid Crystals.

The book is divided into three sections. The first section contains articles discussing synthetic, physico-chemical, structural and rheological aspects of Polymeric Liquid Crystals in their generality. A chapter on methods currently used in this field is also included. There are also chapters on theoretical and classification aspects of PLCs. These self-contained tutorial chapters provide an introduction to this field as well as to the specific papers given in the book. They provide an exhaustive coverage of literature on the subject from its inception to the present.

The second and third sections deal with thermotropic and lyotropic LCP systems. Papers are grouped together by structure of the polymer chain rather than by the nature of properties investigated. The reader can thus compare research on similar polymers approached from different perspectives.

Rigid and semi-rigid main chain systems, historically the first synthetic PLCs studied, are treated with emphasis on structural, morphological, and mechanical properties More recent flexible main chain systems are discussed, focusing mainly on structure-property relationships.

Various aspects of liquid crystallinity in flexible polymers with the mesogenic moiety in the side group are discussed in several articles. The focus here is on the liquid crystalline behavior in electric fields and possible applications of PLCs in electro-optical display and recording. The properties of a novel group of PLCs -- liquid crystalline elastomers -- obtained from mesogenic side group systems are also described.

In the last section of the book, lyotropic systems are treated. These concern derivatives of cellulose in various solvents as well as solutions of synthetic PLCs in low molecular mass liquid crystal solvents and poly-peptide solutions in water. The last article illustrates the tremendous variety of polymeric bio-mesogens encountered in living matter.

Thus, this book provides an exhaustive cross-section of the field from the historical, tutorial and present day "state of the art" perspectives.

I would like to thank Mrs. Sarah Goldman for her invaluable help with the preparation of this book. I would also like to thank the authors for their efforts and Plenum Press for their efficient cooperation.

Alexandre Blumstein
Lowell, Massachusetts
May, 1984

CONTENTS

THERMOTROPIC SYSTEMS:
Flexible Main Chain Polyesters

THERMOTROPIC SYSTEMS:
Polymeric Liquid Crystals with
Mesogenic Moieties in the Side Group

LYOTROPIC SYSTEMS:
Polypeptides

LYOTROPIC SYSTEMS:
Biomesogens

LIQUID CRYSTALLINE POLYMERS: PHENOMENOLOGICAL AND SYNTHETIC ASPECTS[1]

Anselm C. Griffin[a,b], Shailaja R. Vaidya[a] and
Marcus L. Steele[a,2]

Departments of Chemistry[a] and Polymer Science[b]
University of Southern Mississippi
Hattiesburg, MS 39406

INTRODUCTION

Since the discovery of liquid crystallinity by Reinitzer[3] in 1888 as he studied the melting behavior of cholesteryl benzoate, anisotropic structural ordering in fluid phases has been of considerable interest to chemists, physicists and other scientists. Polymers which exhibit liquid crystallinity either in solution (lyotropic) or in the neat state upon heating (thermotropic) have both theoretical and practical importance. Du Pont's Kevlar®, a high modulus polyamide fiber spun from a lyotropic solution, is a prime example of such an application. In order to profitably discuss the structure and properties of polymeric liquid crystals it is necessary to briefly describe and define several aspects of liquid crystallinity in small molecule (low molar mass) compounds.

Liquid crystallinity (mesomorphism) can be observed directly on heating from a solid phase (enantiotropic) or it can be seen only upon supercooling the isotropic liquid phase below the melting temperature (monotropic). In the former case the mesophase is thermodynamically stable and can be obtained on both heating and cooling cycles. In the latter case, the mesophase is metastable with respect to the solid and is seen in the cooling regime only. The liquid crystalline state is characterized by long range (as well as short range) orientationally ordered molecules. These molecules are usually rod- or lath- shaped and can exist in two major structural arrangements. These two types, the nematic and the smectic, are each characterized by parallelism of the major molecular axes and are shown in Figure 1. The nematic phase allows for translational mobility of constituent molecules; the smectic phase is composed of

1

NEMATIC

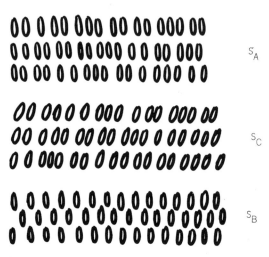

S_A

S_C

S_B

Figure 1.   Schematic of Molecular Organization in Nematic and Smectic Phases.

molecular layers in which translational mobility is minimal. There are a variety of smectic phases classified as S_A, S_B, S_C, etc., differing in i) the ordering (or lack thereof) of molecules within the same layer, ii) the tilt (or lack thereof) of the "average" molecular axis with respect to the layer plane and iii) the positional correlation of molecules in different layers. The cholesteric phase is locally similar to the nematic phase. However the constituent molecules are chiral giving rise to an asymmetric helical packing of molecular "sheets" due to the spontaneous twist resulting from molecular chirality.

Compounds forming small molecule thermotropic liquid crystals usually have the following molecular structural features:
- high length: breadth (axial) ratio
- rigid units such as 1,4-phenylene, 1,4-bicyclooctyl,

2

1,4-cyclohexyl, etc.

- rigid central linkages between rings such as -COO-, -CH=CH-, -N=NO-, -N=N-, etc.
- anisotropic molecular polarization.

There are exceptions to each generalization above, but the list can serve as a rough guide to predicting liquid crystalline (mesomorphic) potential of a given compound. Examples of thermotropic compounds are shown in Figure 2. In general the nematic phase is predominant for compounds having short flexible tails and the smectic phase is dominant for compounds with long tails. Both phases are often seen when tail length is of intermediate length. Figure 3 shows this behavior in schematic form. Lyotropic liquid crystals result from very specific interactions between an amphiphilic solute and solvent(s). Typical low molar mass lyotropic combinations are sodium dodecyl sulfate/water/1-alkanol and phospholipid/water.

Characterization of Liquid Crystals

In this section liquid crystal characterization methods commonly employed by synthetic/organic chemists will be presented. This list

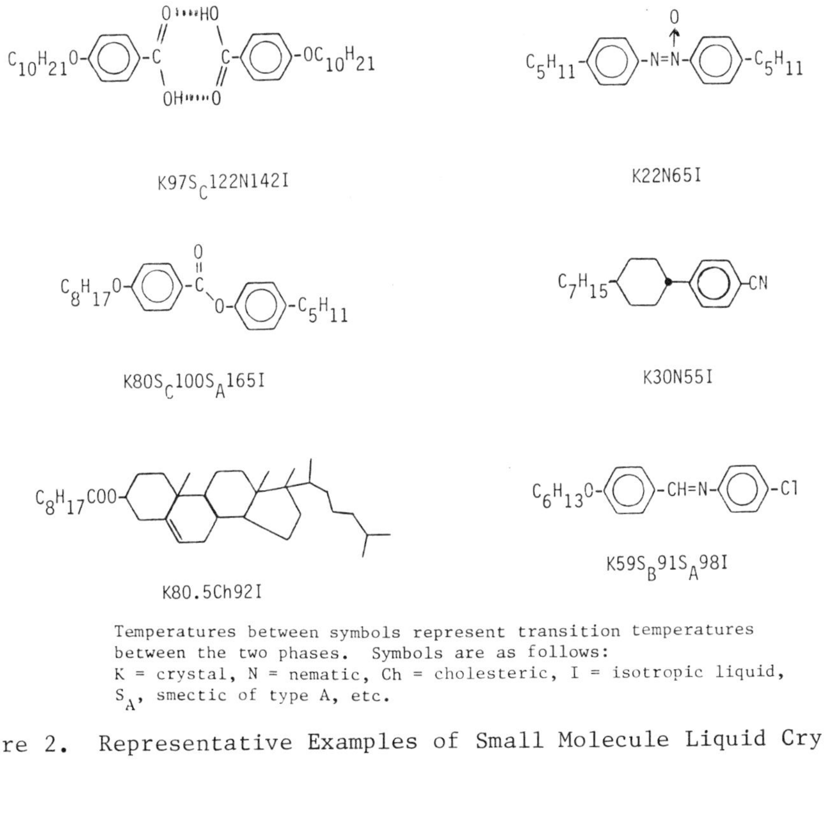
$C_{10}H_{21}O$
$OC_{10}H_{21}$
K97S$_C$122N142I
C_5H_{11}
-N=N-
C_5H_{11}
K22N65I
$C_8H_{17}O$
C_5H_{11}
K80S$_C$100S$_A$165I
C_7H_{15}
CN
K30N55I
$C_8H_{17}COO$
K80.5Ch92I
$C_6H_{13}O$
-CH=N-
Cl
K59S$_B$91S$_A$98I

Temperatures between symbols represent transition temperatures between the two phases. Symbols are as follows: K = crystal, N = nematic, Ch = cholesteric, I = isotropic liquid, S_A, smectic of type A, etc.

Figure 2. Representative Examples of Small Molecule Liquid Crystals.

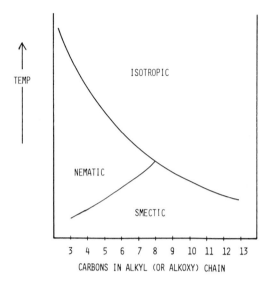

Figure 3.   General Trend in Phase Type Versus Number of Carbons in Tail.

is not exhaustive, but merely represents the first tack at identifying fundamental mesophase behavior. In a capillary tube the mesophase often appears to the naked eye as a cloudy, turbid fluid. A liquid crystalline sample sandwiched between glass plates when viewed through a polarizing light microscope is often highly birefringent. The details of these (often beautiful) optical textures are highly dependent upon mesophase type, sample thickness, and surface treatment of the glass among other factors. There has been, however, a considerable effort to use optical textures to differentiate/identify phase type. In fact Sackmann[4] and his school at Halle have produced a classification scheme for mesophases based on the criterion of complete miscibility of identical phases. Thus the mesophase of a new compound can be compared with known, standard mesophase types by miscibility using optical microscopy. There are available two excellent general references on phase identification by optical textures of liquid crystals[5,6].

Differential scanning calorimetry (DSC) is a valuable aid by which phase transition temperatures, transition heats, and transition entropies can be conveniently measured or calculated. This technique offers a direct and complimentary (to microscopy) evaluation of thermal behavior. Figure 4 shows the DSC curve for 4-octyloxybenzylidene-4'-chloroaniline[7] in which can be seen K_1-K_2, K_2-S_B, S_B-S_A, and S_A-I transitions. All transitions are enantiotropic and all are reversible. The most extensive supercooling occurs for the mesophase-solid transition, in this case the S_B-K_2 transition. Optical microscopy and/or x-ray diffraction is required to assign the specific mesophase type.

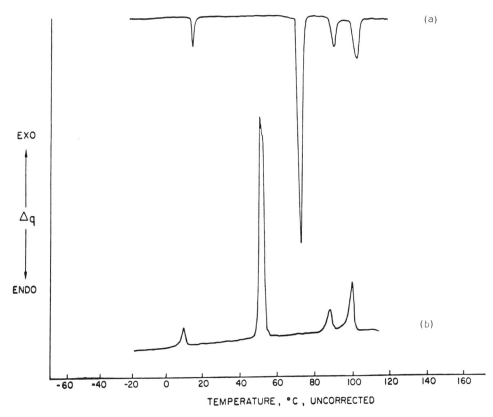

Figure 4. DSC Curves for 4-octyloxybenzylidene-4'-chloroaniline;
(a) is the heating curve and (b) is the cooling curve.

Thermal optical analysis (TOA), sometimes known as depolarized light
intensity (DLI), measures the intensity of light transmitted through
a sample. This method uses circularly polarized light and the transmitted intensity can be directly related to sample crystallinity[8].
To date little use has been made of this technique in the study of
liquid crystals for monitoring phase transitions and the intensity
of birefringence, but the potential is great. This is particularly
true for polymeric liquid crystals since the technique was originally developed to study polymers. Figure 5 shows a TOA curve for
4-octyloxybenzylidene-4'-trifluoromethylaniline[7]. The heating curve
shows a sluggish solid-solid transition K_1-K_2, at low temperature,
followed by relatively sharp K_2-S_B and S_B-I transitions. Upon cooling
the I-S_B transition is seen (with slight supercooling) followed at
much lower temperature by a S_B-K_1 transition after extensive supercooling of the S_B phase.

5

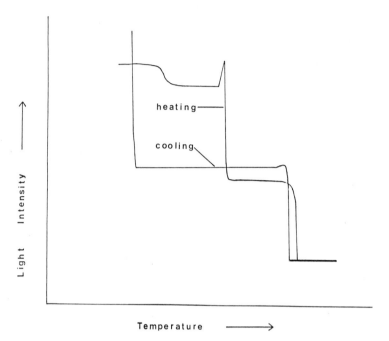

Figure 5. TOA curves for 4-octyloxybenzylidene-4'-trifluoromethyl-
           aniline.

X-ray diffraction is becoming increasingly more important in the
initial characterization of liquid crystals. Although single-crystal
determined atom maps and the analysis of diffraction by oriented
mesophases are still in the hands of experts in the field of x-ray
diffraction, the use of x-ray data from powder samples (often unoriented) is becoming ubiquitous. Nematic and smectic samples yield
comparatively simple diffraction patterns when powder samples are
run in the mesophase with no applied external orienting field. The
nematic phase exhibits a weak diffraction maximum around 4.5Å related
to the average distance separating these parallel rod-like molecules.
In the smectic phase an additional maximum appears at lower angles
(inner ring) which is related to the smectic layer thickness. From
this value the conformation and pairing of molecules in the smectic
phase can be inferred. A schematic diagram of an x-ray diffraction
photograph of an unoriented smectic phase is presented in Figure 6.

LYOTROPIC POLYMER SYSTEMS

The most extensively studied lyotropic polymer is poly-γ-benzyl-
L-glutamate (PBLG)[9]. This polymer exists in solution as a rigid
rod-like α-helix. A variety of solvents including dimethylformamide,

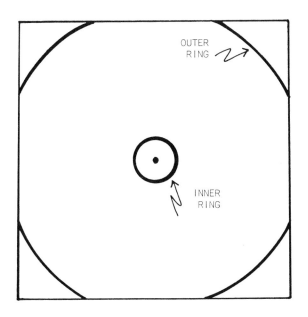

Figure 6. Generalized X-ray Powder Diffraction Photograph for a
Smectic Sample.

chloroform, meta-cresol, 1,4-dioxane and methylene chloride at

certain concentration and temperature ranges can selectively weaken
specific PBLG crystalline lattice sites resulting in lyotropic mesomorphism—a fluid yet still structurally ordered state. An intriguing
type of liquid crystallinity arises from the interaction of certain
block copolymers with solvents which have different affinities for
the different block types. The resulting lamellar structures are
quite closely related to smectic organization[10]. Other examples of
lyotropic polymers (many involving water as the solvent) are nucleic
acids (DNA, RNA), collagen fibrils, glycoproteins, cellulose derivatives and certain viruses, such as Tobacco Mosaic Virus, which have
a cylindrical supramacromolecular structure.

Synthetic polyamides such as du Pont's Kevlar® and Monsanto's

X-500® are examples of commercialized high-modulus, high-strength

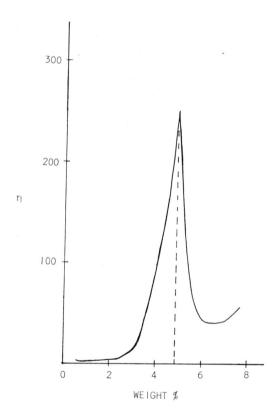

Kevlar®   X-500®

fibers spun from liquid crystalline solution. The characteristic
dramatic change in viscosity upon increasing concentration of the
rod-like solute is shown in Figure 7 in which the viscosity increases
rapidly until a critical concentration is reached and a sharp decrease in viscosity then accompanies the formation of the lyotropic
mesophase. Other examples of such polymers have been recently
collected and described[11]. They include aromatic, aliphatic, hetero-
cyclic, ring-substituted and hydrogen bonded materials.

Figure 7. Schematic Diagram of Viscosity versus Concentration Profile for a Lyotropic System Exhibiting the Characteristic
Viscosity Decrease upon Formation of the Mesophase.

THERMOTROPIC POLYMERS

In this category are included both side chain (comb) polymers and main chain polymers; the name implying the location of the meso-genic moiety.

Side Chain Polymers[12]

In these polymers the pendant liquid crystalline unit is con-nected to the polymer main chain (usually vinyl, methylvinyl or siloxane based) by a flexible spacer group, often methylene units. This flexible connection preserves the delicate interactions between the pendant liquid crystalline moieties by decoupling the main chain motion from that of the pendant group. Since the polymerizations are usually radical initiated the polymer is atactic and molecular weights are quite high ~100,000 daltons. Figure 8 shows in schematic form the nematic and smectic organization of a side chain polymer. Cholesteric polymers can be formed from chiral precursors.

Synthesis of side chain polymers is commonly performed by one of two routes -- an example[13] of route i) as described on the fol-lowing page is shown below:

$$HO(CH_2)_nCl \quad + \quad HO-\langle\bigcirc\rangle-CO_2H \xrightarrow{\text{NaOH}} HO(CH_2)_n-O-\langle\bigcirc\rangle-CO_2H$$

$$n=2,3,6$$

$$\underset{CH_2}{\overset{CH_3}{\underset{\diagdown}{\overset{\diagup}{C}}}}\underset{CO_2H}{} \quad \Bigg\downarrow H^+$$

$$\underset{CH_2}{\overset{CH_3}{\underset{\diagdown}{\overset{\diagup}{C}}}}COO-(CH_2)_n-O-\langle\bigcirc\rangle-CO_2H$$

$$\xrightarrow[\text{2) } HO-\langle\bigcirc\rangle-R]{\text{1) } SOCl_2}$$

$$CH_2{=}\underset{\diagdown}{\overset{CH_3}{\underset{}{C}}}COO(CH_2)_n-O-\langle\bigcirc\rangle-COO-\langle\bigcirc\rangle-R$$

$$\Bigg\downarrow$$

$$\left[\begin{array}{c}CH_3\\ |\\ CH_2-C\\ |\end{array}\right]_x \\ COO(CH_2)_n-O-\langle\bigcirc\rangle-COO-\langle\bigcirc\rangle-R$$

i)   vinyl polymerization of pendant moieties
     ii)   addition of pendant moieties to a preformed polymer[14]
Method i is much more common and an example using this approach is
given on the preceding page. Table 1 presents thermal data for
selected ones of the phenyl benzoate polymers described above as
well as for the corresponding monomers. Common linking groups
such as those described in the introductory section can be incorporated into side chain liquid crystalline polymers. It should
be mentioned that co-polymers can be made using different vinyl
monomers to tailor the end properties.

Main Chain Polymers[15]

Liquid crystalline polymers in which the mesogenic moiety is
incorporated into the polymer main chain are of two types:
     i)   rigid segments joined by flexible segments
     ii)   totally rigid segments (with very little or no chain
           flexibility)

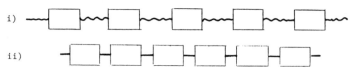

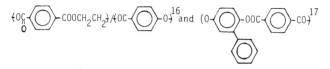

Examples of the latter class include homopolymer and copolymer
compositions such as

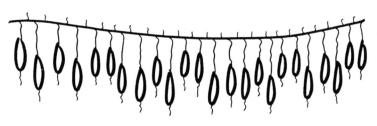

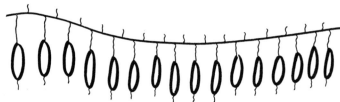

Figure 8.   Schematic of Side Chain Polymers: Nematic (top) and
            Smectic (bottom).

n	R	Phase Transition (°C)	n	R	Phase Transition (°C)
2	OCH_3	K69I	2	OCH_3	g101N121I
6	OCH_3	K47I	6	OCH_3	g95N105I
6	(phenyl)	K64S68N92I	6	(phenyl)	g132S164N184I
6	OC_6H_{13}	K47N53I	6	OC_6H_{13}	g60S115I

A number of observations can be made from this data:

i) The monomers have first order melting transitions whereas the polymers have only glass transitions into the mesomorphic state

ii) mesophase-isotropic temperatures are higher for the polymers

iii) The polymers are "more" mesogenic -- the first two entries of monomers are not liquid crystalline, whereas the related polymers are.

iv) For n=6, R=(phenyl), the monomer and polymer exhibit the same mesophase types - a common characteristic for many side chain polymers

v) For n=6, R=OC_6H_{13}, a reversal of phase type is observed; the monomer is nematic, the polymer is smectic.

Table 1.   Phase Transition Temperatures and Phase Types for Selected Side Chain Polymers and Corresponding Monomers.

This type of material often has quite high solid-mesophase temperatures and is insoluble in most organic solvents. These types of polymers are usually nematic, reasonably thermally stable and have Tg temperatures above 0°C. de Gennes[18] has predicted that for type i polymers smectic phases should be obtained if the flexible segments are of equal length allowing lamellar organization.
There have been several recent reports[15] of such smectic polymers although the majority of type i polymeric materials are nematic.
A schematic diagram of nematic and smectic organization in main chain polymers is shown in Figures 9 and 10.

Structure-property relations in small molecule liquid crystals have been the object of systematic investigation for some time and are at least qualitatively well understood[19]. The question naturally arises as to whether these relationships can be transferred to polymeric mesogens. At present the answer seems to be a tentative "yes", at least to the level of a first approximation. A few

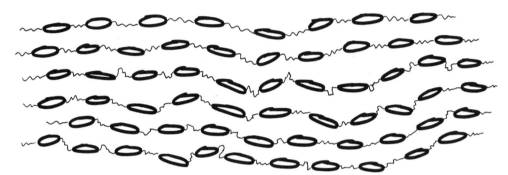

Figure 9. Nematic Organization in Main Chain Polymer.

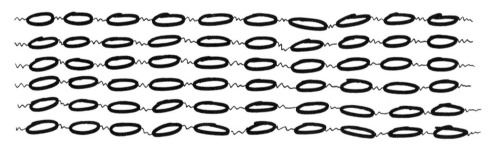

Figure 10. Smectic Organization in Main Chain Polymer.

examples illustrate this point--

1) Small molecule liquid crystals suffer a lowering of the
mesophase-isotropic transition temperature upon lateral
substitution, i.e. replacement of hydrogen on an aromatic
ring. As can be seen from the compounds below the same
trend is observed for both polymer and the chemically
analogous small molecule. It is known that for small
molecule mesogens the transition temperature, governed
by the relation $T_{NI} = \Delta H / \Delta S$, decreases upon lateral
substitution primarily due to an entropic effect. The
same holds true for polymer mesogens[20].

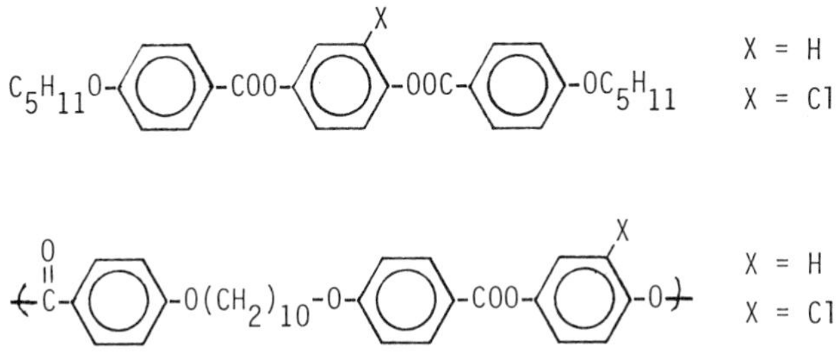

X = H
X = Cl

X = H
X = Cl

2) The amide linkage -CONH- while possessing the required
rigidity and chemical stability is by no means common to
liquid crystals. The usual explanation for the paucity of
liquid crystalline amides is that hydrogen bonding in the
solid state elevates the melting point above the stability
range of the mesophase. It has been recently shown that
through a reduction in intermolecular (interchain) hydrogen
bonding it is possible to produce both small molecule and
polymeric mesogens with enantiotropic liquid crystalline
phases. A lateral substitutent ortho to the amide nitrogen
produces the effect. The structures below are examples[21].

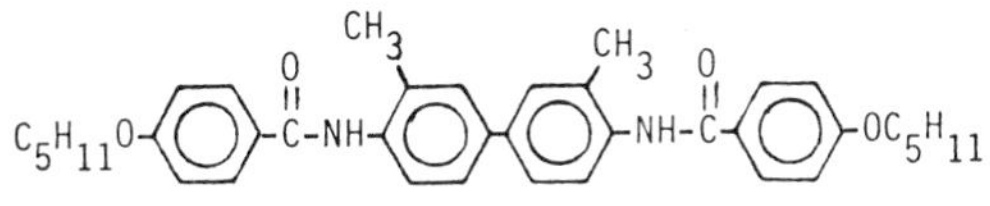

and

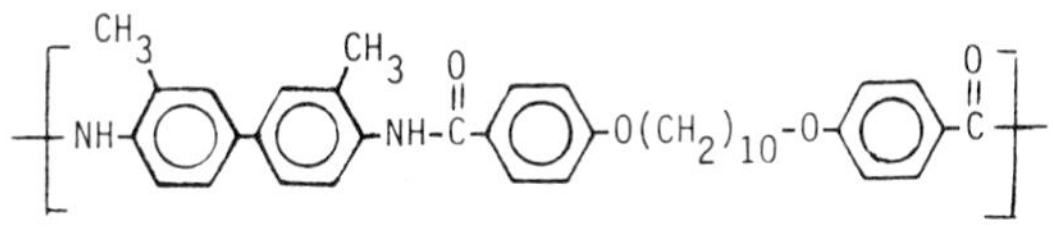

3) Roviello and Sirigu[22] in 1975 reported the first main chain (flexible rod)$_n$ polymer and soon thereafter the properties of the chemically analogous small molecule family[23]. This allows a comparison of phase types, phase sequences and transition temperatures and is valuable in interpreting the influences of chemical constitution on mesophase behavior. These compounds are shown below.

RCOO—⬡—C=N-N=C—⬡—OOCR   R= n-alkyl
            | |
           CH$_3$ CH$_3$

—[OC(CH$_2$)$_n$COO—⬡—C=N-N=C—⬡—O]—   n = 6, 8, 10   nematic
                        | |
                       CH$_3$ CH$_3$

Characterization of Main Chain Polymer Liquid Crystals. Thermal parameters (DSC) can be obtained for liquid crystalline polymers just as for their small molecule counterparts. The polymeric nature of these materials leads to exhibition of glass transitions, dramatic annealing effects (loss of some solid-solid transitions) and the comparative rarity of monotropic mesomorphism. In contrast to side chain polymers which form a mesophase via a glass transition, main chain polymers exhibit first order melting at the solid-mesophase transition although a glass transition may be seen at lower temperatures. Odd-even alternations in mesophase-isotropic temperatures and in ΔH and ΔS for this transition are commonly seen in small molecules as a function of the number of carbon atoms in the flexible segment[19]. The same is true with polymeric mesogens to an even greater extent, i.e. larger amplitude[24]. This alternation has been ascribed to regular variations in the packing of the alkylene chain and/or changes in the anisotropic "molecular" polarizability. Figure 11 depicts a typical DSC curve for a rigid-flexible polyester liquid crystal. The heating curve (bottom) shows a double melting phenomenon, characteristic of many such compounds, as well as a first order nematic-isotropic transition at higher temperature. Upon cooling (top) the mesophase-solid transition is seen to be much more prone to supercooling than the isotropic-mesophase transition. Polydispersity of polymeric mesogens broadens these transitions compared to their small molecule counterparts. Annealing effects are strongly evident in polymeric mesogens usually involving loss of some solid-solid transitions as well as changes in temperature, shape and area of endothermic/exothermic peaks. The longer relaxation times for polymers and the lack of positional "memory" of polymer segments

14

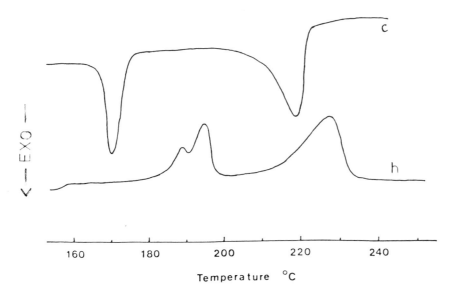

Figure 11. DSC curve for typical rigid-flexible polyester liquid crystals; h is heating curve; c is cooling curve.

result in this behavior. The mesophase structure can often be frozen into the solid by rapid cooling just as for side chain polymers, thereby permitting more convenient examination of this novel morphology. Optical microscopy is also used to identify polymeric mesophases. There is usually more difficulty in assigning mesophase type from optical texture for polymers due to the decreased definition of mesophase texture for polymeric mesogens. Surface orientation on the glass substrate is an important factor in texture development and polymers have fewer chain ends to facilitate such alignment. Nevertheless it is possible to use these textures alone for phase identification and also as the basis of miscibility studies of polymer/small molecule mixtures[25]. Side chain polymers have optical textures much more similar to those of small molecule liquid crystals. X-ray powder diffraction can also be used for mesophase characterization using both oriented and unoriented samples in a manner analogous to that for small molecule mesogens.

The primary goal of molecular weight determinations in mesogenic polymer systems is to establish whether the sample is indeed polymeric or simply oligomeric. Solution viscosity is perhaps the simplest informative technique for this purpose. However the lack of solubility of these rigid or semi-rigid rod-like polymers often presents a problem and aggressive solvents[26] such as H_2SO_4, CH_3SO_3H, phenol/tetrachloroethane must be employed. Due to the short-term

15

stability of some polymers (particularly the polyimines) in these powerful solvents, the sequential dilutions required to obtain intrinsic viscosities are not able to provide valid data and inherent viscosities, obtained from a single concentration, are the only solution viscosity measurement available. Other more definitive methods for molecular weight determination such as light scattering, osmometry and gel permeation chromatography are used but less commonly than are solution viscosities.

Chemical Structures and Synthesis. A variety of chemical structural units are available for use in the design and synthesis of liquid crystalline polymers. Flexible main chain polymers are depicted below in schematic form along with examples of chemical subunits. Rigid main chain polymers lack the flexible segment D.

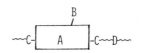

A	B	C	D
rigid core, often containing aromatic rings (p-substituted) joined by rigid linking groups such as: -N=N-, -COO-, CH=N-, -CH=CH-. ↓ O	lateral substituent on aromatic rings such as Cl, Br, CH₃, OCH₃, phenyl, n-alkyl	connecting units joining rigid core to flexible segment such as: ester, ether, amide, imine, urethane, carbonate	flexible spacer such as (CH₂)ₙ, siloxanes, chiral alkylene
Ring systems can be linearly substituted naphthyl, biphenyl, cyclohexyl, etc.			

A compilation of many specific examples has been recently published by Blumstein et al[27]. These polymers are virtually all prepared by condensation (step-growth) polymerization and as such are subject to the usual limitations on molecular weight for such polymers[28]. In practice these polymers are often synthesized by first constructing the monomers such that the flexible segment is incorporated into the monomer. The polymerization then proceeds to connect the requisite parts of the rigid core, route a. An alternative route to these polymers involves the reaction of rigid core units with the appropriately substituted flexible segments in the polymerization step (particularly useful for esters), route b.

The polymerization step is chosen appropriate to the monomers used and products produced. Solution polymerization, interfacial condensation, and ester interchange can be used to produce polyesters for example. References 27 and 29 should be consulted to locate specific reaction conditions and methods for a given polymer class.

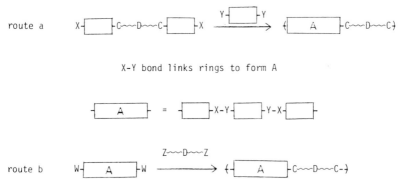

route a

X-Y bond links rings to form A

route b

W-Z bond forms linkage C

In an effort to bring about a detailed understanding of the relationship between small molecule liquid crystals and main chain polymeric liquid crystals, there has been recent work in the area of "oligomeric" liquid crystals. Figure 12 depicts schematically these relationships. These oligomers incorporating rigid-flexible segments can be either monodisperse species (such as dimers) or polydisperse oligomers. The results to date point to a continuum of liquid

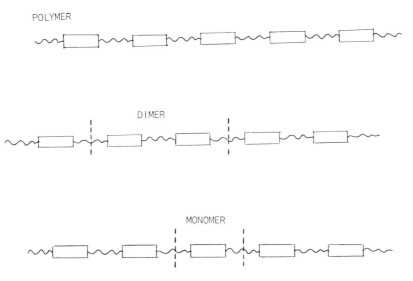

Figure 12. Schematic Diagram of Conceptual Relationship beween Monomeric, Oligomer (Dimer) and Polymer Liquid Crystals.

17

crystallinity from monomer to polymer with some "polymer" properties appearing at reasonably small DP values. Future work in this area should prove fruitful and of value to those involved in both small molecule and polymeric liquid crystals.

ACKNOWLEDGEMENT

We wish to thank the National Science Foundation for partial support of this work through a grant (DMR 8115703). Acknowledgement is also made to the Donors of the Petroleum Research Fund, administered by the American Chemical Society, for the partial support of this research.

REFERENCES

1. Part 7 of a series on Mesogenic Polymers. For part 7 see A. C. Griffin, T. R. Britt, R. S. L. Hung and M. L. Steele, Mol. Cryst. Liq. Cryst., in press.
2. Permanent address: Department of Chemistry, Henderson State University, Arkadelphia, AR 71923
3. F. Reinitzer, Monatsh. Chem., 9, 421 (1888).
4. H. Sackmann and D. Demus, Mol. Cryst. Liq. Cryst., 21, 239 (1973).
5. D. Demus and L. Richter, "Textures of Liquid Crystals", Verlag. Chemie, Weinheim, 1978.
6. D. Coates and G. W. Gray, The Microscope, 24, 117 (1976).
7. J. F. Johnson, G. J. Bertolini and A. C. Griffin, unpublished results.
8. E. M. Barrall, II and J. F. Johnson, Thermochem. Acta. 5:41 (1972).
9. E. T. Samulski in "Liquid Crystalline Order in Polymers", A. Blumstein, ed., Academic Press, NY, NY, 1978.
10. A. Skoulios in "Advances in Liquid Crystals," G. H. Brown, ed., Academic Press, NY, NY, 1975, p. 169.
11. P. W. Morgan, Macromolecules, 10, 1381 (1977).
12. A recent general review of this subject is given by H. Finkelmann in "Polymer Liquid Crystals", A. Ciferri, W. R. Krigbaum and R. B. Meyer, eds., Academic Press, NY, NY, 1982, p. 35.
13. H. Finkelmann, H. Ringsdorf, W. Siol and J. H. Wendorff in "Mesomorphic Order in Polymers", A. Blumstein, ed., ACS Symposium Series No. 74, 1978, p. 22.
14. C. M. Paleos, S. E. Filippakis and G. Margomenou-Leonidopoulou, J. Polym. Sci., Polym. Chem. Ed., 19, 1427 (1981).
15. A recent general review of this subject is given by A. Ciferri in "Polymer Liquid Crystals," A. Ciferri, W. R. Krigbaum and R. B. Meyer, eds., Academic Press, NY, NY, 1982, p. 63.
16. H. F. Kahfuss and W. J. Jackson, Jr., U.S. Patent 3778410, 1973.

17. C. R. Payet, Ger. Offen 2751653 (1978).
18. P. G. de Gennes, Faraday Discussions, 68, 96 (1980).
19. G. W. Gray, Phil. Trans. R. Soc. Lond. A 309, 77 (1983).
20. S. Antoun, R. W. Lenz, and J. -I. Jin, J. Polym. Sci., Polym. Chem. Ed., 19, 1901 (1981).
21. A. C. Griffin, T. R. Britt and G. A. Campbell, Mol. Cryst. Liq. Cryst. (Letters), 82, 145 (1982).
22. A. Roviello and A. Sirigu, J. Polym. Sci., Polym. Lett. Ed., 13, 455 (1975).
23. A. Roviello and A. Sirigu, Mol. Cryst. Liq. Cryst., 33, 19 (1976).
24. A. Blumstein and O. Thomas, Macromolecules, 15, 1264 (1982).
25. J. Billard, A. Blumstein and S. Vilasagar, Mol. Cryst. Liq. Cryst., 72, 163 (1982).
26. B. Millaud, A. Thierry, C. Strazielle and A. Skoulios, Mol. Cryst. Liq. Cryst. Lett., 49, 299 (1979).
27. A. Blumstein, J. Asrar and R. B. Blumstein in "Liquid Crystals and Ordered Fluids, Vol. 4", A. C. Griffin and J. F. Johnson, eds., Plenum Publishing Corp., NY, NY, 1984.
28. P. W. Morgan, "Condensation Polymers: By Interfacial and Solution Methods," Interscience Publishers, NY, NY, 1965.
29. J. -I. Jin, S. Antoun, S. Ober and R. W. Lenz, British Polymer Journal, 132 (1980).

IDENTIFICATION OF MESOPHASES EXHIBITED BY THERMOTROPIC LIQUID CRYSTALLINE POLYMERS

Claudine Noël

Laboratoire de Physicochimie Structurale et Macromolé-
culaire, ESPCI, 10 rue Vauquelin, 75231 Paris Cedex 05

INTRODUCTION

In 1975, Roviello and Sirigu [1], prepared new polyalkanoates
from p,p'-dihydroxy-α,α'-di-methylbenzalazine and appropriate acyl
chlorides. All the examined polymers melted to give fluid anisotropic phases whose textures and properties appeared quite similar
to those observed with conventional liquid crystals, hence the deno-
mination of "thermotropic liquid crystalline polymers". The discovery
of these new polymeric materials was a stimulating event in macromolecular science not only because of their potential as high
strength fibers but also because of their academic interest in the
theoretical scheme of structural order in fluid phases. In the past
few years a search for newer and newer liquid crystalline polymers
containing mesomorphic groups in the side chain or incorporated
in the main chain has begun on a much wider scale. Surprisingly
there have been few papers devoted to the study of their structures
as liquid crystalline phases. So far, the liquid crystals formed
by heating these polymers have been classified broadly as nematic,
smectic, cholesteric or twisted smectic. Identification of the type
of mesophase is thus an important step in the characterization of
these materials. The more definitive procedures used for classification of low molecular weight liquid crystals are [2-5] :

(a) Optical pattern or texture observations with a polarizing
microscope. There are, however, limitations and a complete classification of smectic phases by textures is not always possible.
Similar textures may be observed with two liquid crystal states
separated by a phase transition.

(b) The differential scanning calorimetry can be used to distinguish between thermotropic nematic and smectic phases by the magnitude of the enthalpy change accompanying the transition to the isotropic phase.

(c) Miscibility studies with known liquid crystals. Isomorphous mesophases are considered as equivalent and characterized by the same symbol.

(d) Possibility of inducing significant molecular orientations by either supporting surface treatments or external fields.

(e) X-ray investigations. Differences in the molecular long range order can be established by small angle X-ray diffraction. Classical X-ray methods allow to obtain short range order only.

Owing to high viscosity, broad molecular weight distribution, existence of poly-crystalline and amorphous material, the liquid crystalline nature of thermotropic polymers is usually established through a combination of these methods.

TEXTURE AND POLYMORPHISM OF LC POLYMERS

The term texture, as used by Friedel [6,7] designates the picture of a thin layer of liquid crystal observed by means of a microscope usually in linearly polarized light. The features of the various textures are caused by the existence of different kinds of defects. For a preliminary characterization, the optical patterns or textures are most useful. The microscopic observations offer the possibility of introducing some ideas on the structure of the liquid crystalline modifications.

In this paper, we are only dealing with the occurrence of different textures, without discussion of many details. Our purpose is to discuss the textures with the said system of liquid crystals.

Nematic Textures

Microscopic observations are sometimes misleading [8] owing to the difficulty with which thermotropic polymers give specific textures in the liquid crystalline state. This might be due to their multiphase nature and/or the high viscosities of the liquid crystalline melts [9,10]. The suggestion that long flexible spacers may hinder the formation of the texture has not been substantiated. Strzelecki and Liebert [11] observed that polyesters

$$\left[O - \langle O \rangle - COO - \langle O \rangle - OCO + CH_2 \frac{}{n} CO \right] \qquad I$$

develop in the melt a typical nematic texture only when n < 10

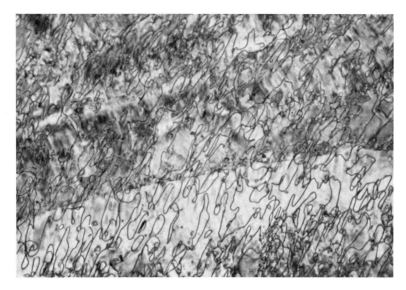

Fig. 1. Photomicrograph of nematic phase from copolyester, prepared
from a 50/25/25 (mol %) mixture of terephthalic acid, pyrocatechol and methylhydroquinone. Threaded texture (crossed
polarizers).

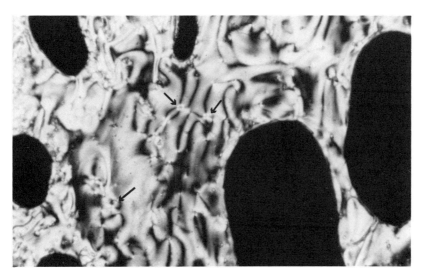

Fig. 2. Schlieren texture of copolyester formed by transesterification of poly(ethylene-1,2 - diphenoxyethane-p,p'-dicarbo-
xylate) with p-acetoxybenzoic acid. Recognizable singularities S = $\pm$ 1 (Crossed polarizers, 280°C, x 150). From
ref. 25.

but their viscosity at low shear rates was found to be strongly dependent on the number of methylene units. Lengthening the aliphatic chain from n = 5 to n = 12 increases the apparent viscosity from 11P to 332P [12]. However, the microscopic textures of several macromolecular phases observed under polarizing light microscope are very similar to those exhibited by low molecular weight compounds and can be described as threaded [11, 13-18] (Fig. 1) and/or Schlieren [19-25] (Fig. 2).

Usually, nematic liquids come from the isotropic ones in the form of the well-known droplets which after further cooling grow and coalesce to form large domains [26]. The schlieren textures display dark brushes which correspond to extinction positions of the mesophase. At certain points two dark brushes meet (Fig. 3), at others there are four brushes meeting (Fig. 2). These points indicate singularities i.e. disclinations in the structure [3]. The disclination strength is connected with the number of dark brushes meeting at one point :

$$|S| = \text{number of brushes}/4$$

The sign of S is positive when the brushes turn in the same direction as the rotated polarizers and negative when they turn in the

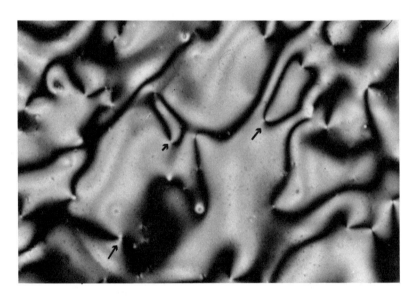

Fig. 3.Schlieren texture of copolyester formed by transesterification of poly(ethylene-1,2-diphenoxyethane-p,p'-dicarboxylate) with p-acetoxybenzoic acid. Recognizable singularities S = ± 1/2 (crossed polarizers, 260°C, x 200). From ref. 25.

24

opposite direction. Frank [27] has classified the types of disclinations that can be expected in a two dimensional ordered nematic liquid crystal and has given the trajectories of the directors in the region near a disclination (Fig. 4). From the observation of points at which only two dark brushes meet (the "strength" of the disclination S = ± 1/2) the mesophase can be identified unambiguously as a nematic phase since these singularities occur nowhere else. Indeed, in contrast with the nematic Schlieren textures in which points with two (S = ± 1/2) or four (S = ± 1) Schlieren are possible, the textures of smectics C, B and D display points with only four brushes [3,28].

Some optical observations of defects (disclination lines) in polymeric nematic phases exhibiting threaded textures showed that a massive organization of macromolecules occurs as the temperature is raised [14,18,19]. Immediately after melting of the specimen between two glass plates, a large number of threads are observed. When the temperature of the melt is increased, the number of threads diminishes. Simultaneously, the threads become loose and have a strong tendency to shrink in length. At high temperature most of them form closed loops (Fig. 5). They are unstable and after some time or by further heating they become smaller and smaller and to a great extent disappear. As a consequence, when the temperature is near T_{NI}, samples contain no threads and show homogeneous color across large areas between crossed polarizers. Similar microscopic observations have been reported by Millaud et al [29] for the lyotropic nematic phase of a poly(paraphenyleneterephthalamide). According to these

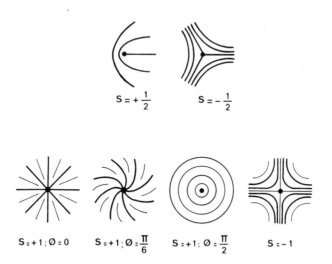

Fig. 4. Schematic diagram of molecular trajectories associated with disclinations of strength ± 1/2 or ± 1.

25

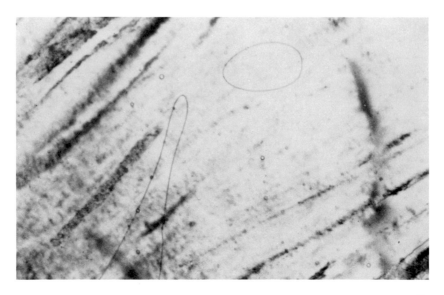

Fig. 5. Closed loop viewed between crossed polarizers at 250°C.
Copolyester prepared from a 50/25/25 (mol %) mixture of
terephthalic acid, pyrocatechol and methylhydroquinone.
From ref. 18.

authors most of the disclinations observed are of integer type and
more specifically equal to + 1. Elastic constants K_{11}, K_{22} and
K_{33} that oppose fundamental director modes of splay, twist and
bend (Fig. 6), are roughly the same for liquid crystals composed
of small molecules. On the contrary, according to de Gennes [30] the
bend elastic constant K_{33} is dominated by the rigidity of the macromolecular chains while K_{22} is mainly a function of the interaction
between neighbouring chains. The splay constant K_{11} is predicted to
increase either with the molecular length [31] or with the square of
the molecular length [30]. Millaud et al. [29] interpreted their results
in terms of high values of the bend constant K_{33} for polymers.
They proposed a pure radial splay model to describe the molecular
arrangement around the lines of strength S = + 1. In investigations
Mackley et al. [24] made on a copolyester prepared from 40 mol %
poly(ethylene terephthalate) and 60 mol % p-acetoxybenzoic acid, a
number of integer disclinations were also observed. However, according to Viney and Windle [23] who carried out some optical observations of defects in the same copolyester, if disclinations of
strength $\pm$ 1/2 are not seen below 340°C, they are more numerous
than those of strength $\pm$1 above this temperature. These observations
are at variance with results reported by Mackley et al. [24] but seem
to corroborate well with optical observations in polyester I
(n = 5) made by Kleman et al. [32]. One of the clearest conclusions
of their study is obtained from the predominance of |S| = 1/2 lines;

26

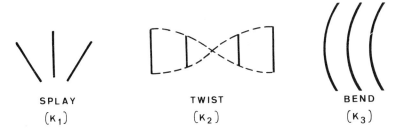

Fig. 6. The three types of elastic deformation in a nematic.

this implies that K_{11} and K_{33} are definitely larger than K_{22}, hence most of the geometry tends to avoid bend and splay. However K_{11} seems to be much larger than K_{33}. Differences between data reported in refs 29 and 32 might be due to the different type of polymer they have used : the rigidity of the macromolecular chain in poly (paraphenyleneterephthalamide) is higher than in polyester (note that the molecular weights in both of the above investigations are similar). Much study is needed in this area. For instance, systematic variations with either molecular weight or the number of methylene units in the flexible spacers can be looked for.

Thin layers of these mesophases exhibit other nematic characteristics such as intense movement within the melt and scintillation effects due to a directly observable Brownian motion. Homeotropic texture caused by a spontaneous orientation of the sample is also found [18, 24, 26]. This texture occurs when long axes of the mesogenic molecules are at right angle to the glass surfaces. With such liquid "single crystals" the field of view using crossed polarizers remains uniformly dark as the preparation is rotated. By slightly touching the cover glass of the preparation a flash-like brightness can be instantly produced, thus distinguishing between an homeotropic and an isotropic texture. The original dark field of view reappears after the reestablishment of a mechanically stable state.

A final remark will concern the "glassy liquid crystal". All the textures observed for both linear and comb-like polymers in the nematic state can easily be quenched and supercooled to room temperature [11, 18, 20, 23, 24, 33]. Both the homeotropic alignment and the planar one remain fixed in the glassy state on cooling [18, 34]. From the dichroic ratio of the bands in the IR spectra of the copolyester built from the following residues :

$$\left(HOOC-\langle O\rangle-OH\right)_{0.6} + \left(HO\text{-}CH_2\text{-}OH + HOOC-\langle O\rangle-OCH_2CH_2O-\langle O\rangle-COOH\right)_{0.4} \text{ II}$$

27

the transition of the nematic mesophase to a glass with nematic order upon quenching was confirmed quantitatively : in spite of the thermal shock, the macroscopic alignment remains throughout the whole process [25].

Cholesteric Textures

It is well known that the conventional nematic phases can be transformed to cholesteric ones by dissolving an optically active compound in them [15, 18, 21, 25]. For instance, Figure 7 shows that terephthalylidene-bis-4-((+)4* methylhexyloxy)aniline on addition to the polyazomethine

$$\left.+ N \bigcirc\hspace{-0.3em}O\hspace{-0.3em}\bigcirc - N = CH -\bigcirc\hspace{-0.3em}O\hspace{-0.3em}\bigcirc - CH \right=\hspace{-0.3em}|\hspace{2em} III$$
$$CH_3$$

in the nematic state causes the formation of the planar typical texture of cholesterics with "oily streaks" and moiré fringes.

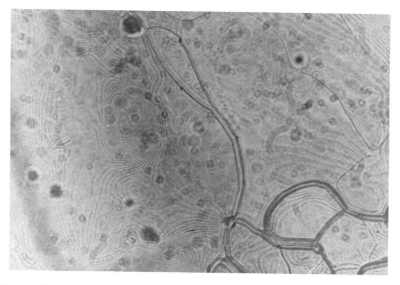

Fig. 7. Typical planar texture obtained at 260°C by the addition of an optically active compound to the polyazomethine III.

The same effect can be obtained by copolymerization or copolycondensation of a repeat unit, known to yield a thermotropic nematic homopolymer, with a chiral compound [21, 22, 35-41].

Finkelmann et al.[39] described the synthesis of what appears to be the first thermotropic cholesteric copolymers :

$$R_1 = + CH_2 +_6 O - \langle O \rangle - COO - \langle O \rangle - \langle O \rangle - OCH_3$$

$$R_2 = + CH_2 +_2 O - \langle O \rangle - COO - \langle O \rangle - CH=N-\overset{*}{C}H - \langle O \rangle$$
$$\hspace{10cm} |$$
$$\hspace{10cm} CH_3$$

$$\underline{IV}$$

Observations under a polarizing microscope indicate that above their glass transition temperature these copolymers display a cholesteric planar Grandjean texture. In this texture, cholesterics show reflection colours. The wavelength of the light at the centre of the reflection band λ_R is, for perpendicular incidence, equal to the length of the pitch P multiplied by a mean refractive index $\bar{n}$. Copolymers having $0.2 < x_2 < 0.3$ reflect in the visible region. It is noteworthy that this light reflecting texture can be frozen in the glassy state. Cholesteric films that may be commercially useful can be prepared in this way. The optical properties of these copolymers resemble those of conventional cholesteric phases :

- as the temperature increases, all the polymers show a blue shift of the reflection wavelength, the positive gradient $d(1/\lambda_R)dT^*$ (T^* = T measurement/T clearing point) increasing with increasing amount of chiral component in the polymer.

- the reciprocal reflection wavelength $(1/\lambda_R)$, which is a direct measure of the helical twisting, is a linear function of the mole fraction x_2 of the chiral monomer unit. The helical pitch of the cholesteric decreases with the increase in the content of the chiral monomer units.

Note that deviations from linearity was observed for copolymers :

$$R_1 = -O-\langle O \rangle-COO-\langle O \rangle-OCH_3$$

V

$$R_2 = -COO-$$

at high mole fractions of the chiral monomer [40]. It may be due to the difference in the chemical structure of the nematogenic and chiral moieties.

Slightly twisted cholesteric liquid crystals may be described by a director field on which the field of the twist axis is super-imposed [3]. Accordingly, in case of low MW molecules, the defects and the textures of cholesterics with low twist and nematics show many similarities. On the other hand, strongly twisted cholesterics can be considered to have a quasi-layered structure [3]. As a consequence, in low MW cholesterics with high twist, the defects and textures resemble the corresponding smectic textures, especially smectic A. These cholesterics can occur in two main textures of the non-planar type : the fan shaped textures and the polygonal textures, the latter existing usually in somewhat thicker preparations. Characteristic for these textures is the occurrence of ellipses and hyperbolae in confocal arrangement. Low MW cholesterics also appear in simple focal conic textures. Because of their somewhat disturbed appearance and less birefringence, the characteristic discontinuities are usually not as clearly visible as in the case of A modifications. In case of homopolymers prepared from chiral monomers, this makes it difficult to interpret the observed textures in terms of cholesterics or smectics. For example, Shibaev et al. [42] observed for comb-like homopolymers based on derivatives of cholesterol a confocal texture which seemed to suggest a cholesteric type of liquid crystals. However, the layered ordering in the arrangement of branches evidenced by X-ray scattering was indicative of a smectic texture. Furthermore, a first attempt to establish the cholesteric character of polyester synthesized from 4,4'-dihydroxy-α- methylstilbene and (+)-3-methyl adipic acid failed : fingerprint patterns or Grandjean planar texture with reflection colours were not seen. However, after examining a number of samples, a fan-like texture was observed [27]. This suggested a cholesteric with high twist, i.e. a pitch considerably smaller than visible wavelengths. By adding a small amount of either PAA, a low MW nematogen, or nematic polyester the cholesteric pitch was increased and a planar texture with typical "oily streaks" and bright colours was obtained.

Smectic Textures

The textures of polymeric smectic phases have not been investigated in detail. There occur several variants [15, 20, 43-47] which closely resemble the focal conic and fan shaped textures of low MW smectics A and C (Figs 8 and 9). These non planar textures consist of smectic layers

Fig. 8. Simple focal conic texture for a S_A homopolyester prepared from di-n-propyl-p-terphenyl-4,4" carboxylate and HO-CH(CH$_3$)-CH(CH$_3$)OH.

arranged in Dupin cyclides. The most usual defects in these structures are ellipses and hyperbolae in confocal relationship. Besides these defects, a number of irregularly shaped, different coloured regions are visible in smectic C [3, 28].

For homopolyesters prepared from di-n-propyl-p-terphenyl-4,4"-carboxylate and branched aliphatic diols, the existence of S_A phases was easily established from the simple focal conic or fan textures (Fig. 8) and the tendency to be homeotropic which indicates uniaxiality [48]. Oily streaks also occur. They appear as bright bands or ribbons which start from air bubbles in the dark homeotropic regions. On the other hand, homopolymer :

$$\left[\begin{array}{c} O \\ \parallel \\ C \end{array} - \hexagon - \hexagon - \hexagon - \begin{array}{c} O \\ \parallel \\ C \end{array} - O \left(CH_2 - CH_2 - O \right)_4 \right] \qquad \text{VI}$$

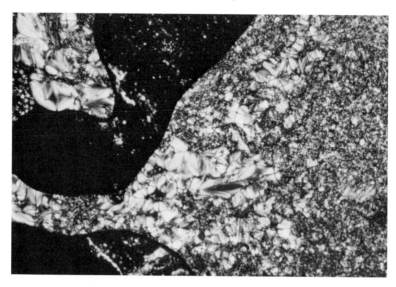

Fig. 9. Broken focal conic texture for a S_C homopolyester prepared from di-n-propyl-p-terphenyl-4,4" carboxylate and $HO(CH_2-CH_2-O)_4 H$. (Crossed polarizers, 240°C, x 200). From ref. 20.

exhibits a smectic C phase with broken focal conics (Fig. 9) [15, 20]. Chiral compounds on addition to this polyester cause the formation of the typical twisted smectic textures (Figs. 10 and 11)[15]. The striation can be explained by the periodicity of layers with equal twist angle. The separation of the stripes corresponds to the half-pitch of the smectic screw.

Transitions with the participation of liquid crystals may show characteristic phenomena. In this connection, for polyester of p,p' bibenzoic acid with hexamethylene diol, smectic liquid comes from the isotropic one in the form of bâtonnets which coalesce and grow into a focal-conic texture (Fig. 12) [20, 46]. Such observations are consistent with smectic A or smectic C mesophase but, usually the smectic C bâtonnets show regions with different interference colours. For a polyester prepared from di-n-propyl-p-terphenyl-4,4" carboxylate and tetramethylene glycol [20] a chevron texture (also called myelinic texture or striated texture) with typical transition bars was easily observed upon cooling from the nematic state (Fig. 13). Such a texture generally appears in the temperature range immediately below nematic/S_A or nematic/S_C transition. Bosio

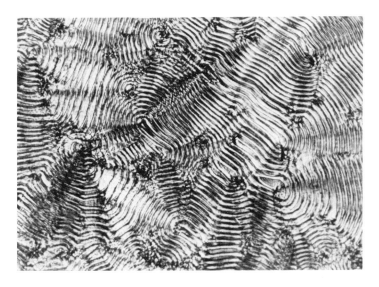

Fig. 10. Twisted smectic C texture obtained by the addition of
terephthalylidene-bis-4-((+)-4' methylhexyloxy) aniline
to the S_C phase of homopolyester VI. From ref. 15.

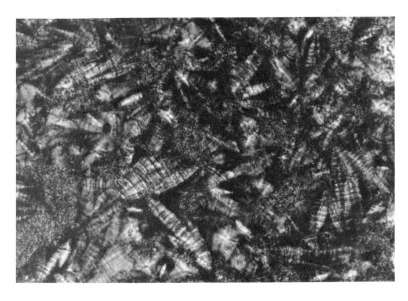

Fig. 11. Twisted smectic C texture obtained by the addition of
4'-(2-methylhexyloxy)biphenyl-4-carboxylic acid to the S_C
phase of homopolyester VI. From ref. 15.

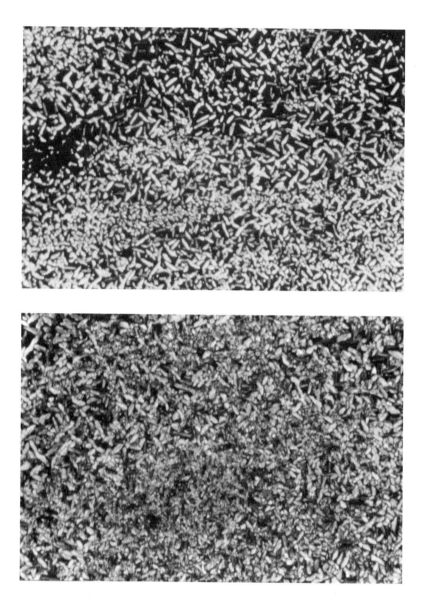

Fig. 12. Smectic bâtonnets observed at 229°C for a homopolyester of p,p'-bibenzoic acid with hexamethylene glycol. Crossed polarizers.

et al. [49] reported that some polyesters based on p-terphenyl exhibited N and S_A mesophases. From microscopic observations there is also some indication that these polyesters also form a more ordered smectic phase at lower temperatures. Indeed, upon cooling from the smectic A state, the fans were lightly chequered and some arcs ran laterally across them as in transient fan-shaped texture of smec-

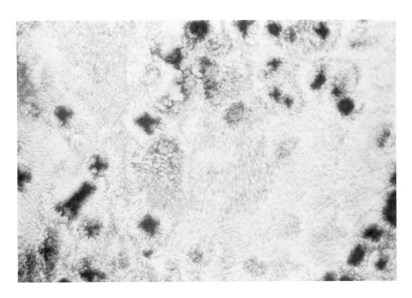

Fig. 13. Chevron texture with typical transition bars observed upon cooling from the nematic state for polyester prepared from di-n-propyl-p-terphenyl-4,4" carboxylate and tetramethylene glycol. Crossed polarizers. From ref. 20.

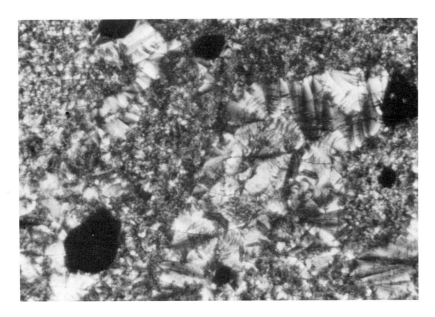

Fig. 14. Transient fan-shaped texture observed upon cooling from the smectic A state for polyester prepared from di-n-propyl-p-terphenyl-4,4" carboxylate and HO-CH(CH$_3$)-CH(CH$_3$)-OH. Crossed polarizers.

tic E (Fig. 14). By pressing on the coverslip with a fine steel needle it was possible to observe a reversible change in the texture. This observation rules out the possibility that the concentric arcs could be the result of strain cracks that may attend mesophase-solid transition.

DIFFERENTIAL SCANNING CALORIMETRY

As pointed out by Finkelmann in his recent review [50] the essential features of the phase behavior of side-chain polymers are relatively well established. The cooling curve of a low MW LC exhibiting one mesophase is characterized by a high temperature exotherm, indicating a first order transformation from the isotropic phase to the mesophase and a low-temperature exotherm indicating a first order transformation of the mesophase to the crystal (Fig. 15 α). On the other hand an amorphous conventional polymer exhibits a transition from the isotropic liquid to a glass (Fig. 15 β).For a liquid crystalline side chain polymer the DSC curve is different (Figs. 15 γ and δ). Compared to Fig. 15 α, a glass transition is observed instead of a low temperature crystallization. Compared to Fig. 15 β, the mesophase-isotropic phase transition is shifted to higher temperature. Thus, one observes a glass transition characteristic of the polymer backbone and a phase transformation from the mesophase to the isotropic phase due to the mesogenic side chains.

Complementary studies with a polarizing microscope reveal that the texture observed in the LC state of the polymer can be frozen without change in the glassy state.

The thermal behavior of liquid crystalline main chain polymers is more complicated. Depending on thermal history, different transitions may occur. The DSC curves of samples quenched from the melt may show the glass transition (positive ΔC_p), cold crystallization (exotherm), melting (endotherm)and the mesophase-isotropic liquid transition (endotherm)[Fig. 16].

- The mesophase to isotropic phase transition is little affected by changes in cooling and heating rates [1, 10, 20, 51]. Thus, usually, it is not possible to supercool the isotropic liquid. The glass transition is a transition of the solid mesophase.

- The entropy and enthalpy changes on disordering are larger than those for the low MW model compounds. This is an indication that at least a portion of the aliphatic chains participate in the ordered regions of the macromolecules [22, 51, 52].

- The melting transition shows a surprisingly small heat of melting which indicates that the crystal structure of polyester cannot have a high packing density or order [10].

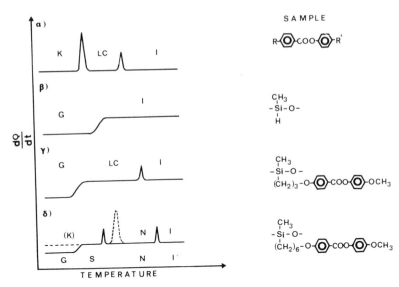

Fig. 15. DSC curves of (α) a low MW liquid crystal, (β) a conventional polymer and (γ), (δ) liquid crystalline side chain polymers. From ref. 50.

   - In contrast to the isotropic ↔ mesophase transition, complete supercooling to the glassy mesophase state is possible. The crystallization from the mesophase may be irreversible.

   Transitions from the solid to the nematic phase may be accompanied by several first order transitions [Fig. 17][10]. Multiple transitions may be due to :

   (i)   interconvertible forms of polymer, which differ only in degree of crystal size and perfection

   (ii)   fundamental differences in crystal morphology, for example folded-chain crystals and partially extended chain crystals

   (iii) "true" polymorphism

   (iiii)solid-mesophase and mesophase-mesophase transitions.
This makes it difficult to interpret the DSC curves and the nature of these transitions can be established only through a combination of optical observations and X-ray investigations. It is worth noting that usually large supercooling occurs for the crystal-mesophase transition in cooling cycles while mesophase-mesophase and mesophase-isotropic liquid transitions characteristically exhibit little or no supercooling.

37

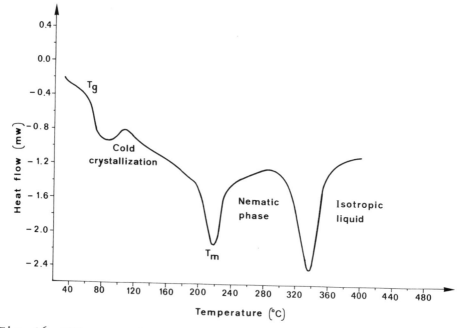

Fig. 16. DSC curve of copolyester formed by transesterification of poly(ethylene-1, 2 diphenoxyethane-p-p'-dicarboxylate) with p-acetoxybenzoic acid. From ref. 25.

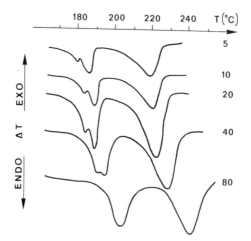

Fig. 17. DSC curves of polyester VII (x=5, y=10) at heating rates 5, 10, 20, 40 and 80 K/min. From ref. 10.

MISCIBILITY STUDIES

According to the rule of selective miscibility all liquid crystalline modifications which exhibit an uninterrupted series of mixed crystals in binary systems can be marked with the same symbol [2], [4], [28]. It means that those liquid crystalline modifications that have the same designation in no case exhibit an uninterrupted series of mixed liquid crystals with the liquid crystalline states of another designation. While uninterrupted miscibility established isomorphism, the converse is not true. In addition, two mesophases formed by the same compound but separated by a well defined phase transition are considered as different in their structures. In case of low MW liquid crystals, these two assumptions have been found to be consistent and have not led to contradictory results, although a large number of binary mixtures has been studied. For instance, Sackmann and Demus [28] have observed in only one case a heterogeneous region between two modifications of the same type but, in over 100 cases they have found heterogeneous regions between modifications of different types. Thus, the nature of a mesophase can be determined by establishing its isomorphy with a known mesophase of a reference compound, that is to say by establishing the isobaric temperature-composition phase diagram of a binary system composed of the polymer and a reference compound. DSC measurements and microscopic observations make it easier to choose the reference compound since they offer the possibility of determining the thermal stability range of the liquid crystalline phase to be identified and of introducing some ideas on its structure.

Use of the contact method [53] allows great rapidity in the assessment of the phase diagram. For this purpose, microscopic preparation is obtained by putting at high temperature, in preference in the isotropic state, a small amount of polymer between a glass slide and a cover slip. The reference compound is then introduced by capillarity in such a way that the polymer and the reference substance are in contact. Owing to the reference compound diffusion into the polymer, the composition continuously changes from one extremity of the preparation to the other one. After cooling, the preparation is examined under a polarizing microscope equipped with a heating stage. Note that well defined compositions can be examined for the phase diagram to insure a degree of accuracy for the composition coordinate.

Chronologically, the works that should be mentioned first are those of Millaud et al [54] and Noël et al [13], [15] who have determined the nature of the mesophase of a polyazomethine by establishing its isomorphy with low MW nematic Schiff's bases of similar structure. The binary diagrams are similar to the typical phase diagrams encountered for mixtures of two low MW nematogens : an eutectic is seen in the solid-nematic curve and there is a continuous pathway from the nematic phase of reference compound to the mesophase of polyazomethine. However, only a small part of the nematic-isotropic curve could be obtained due to the decomposition of the polyazomethine well below its clearing point. Polyesters of general structure :

$$\left[\begin{array}{c} O \\ \parallel \\ C \end{array} - \bigcirc - O \left(CH_2 \right)_x O - \bigcirc - \begin{array}{c} O \\ \parallel \\ C \end{array} - O - \bigcirc - O \left(CH_2 \right)_y O - \bigcirc - O \right] \quad \text{VII}$$

have also proven to be miscible with p-phenylene-bis(4-methoxyben-zoate) and terephthal-bis(4-n-butylaniline) in the nematic state [55,56] . Figure 18 shows that the nematic-isotropic curve is nearly linear, exhibiting almost ideal behavior. An eutectic is observed in the solid-nematic curve at a composition of approximately 77 % polyester by weight. The eutectic composition and temperature can be calculated for a binary system using the Schröder-van Laar equation [57,58] . However, the calculated values do not agree with those obtained experimentally (Fig. 18). This discrepancy may be due either to difficulties in packing long macromolecular chains with small molecules or to the polydispersity of the polyester. Similar phase diagrams have been reported for mixtures of nematic copolyester and low MW nematogen [18,25] . Recently, Billard et al [59] have identified, by means of the contact method, the nematic and cholesteric mesophases exhibited by linear polyesters based on 4,4'-azoxybenzene and 4,4'-azoxy 2,2'-methylbenzene. These authors observed that two enantiomers form perfect solid, cholesteric and liquid solutions : the solid-cholesteric and cholesteric-isotropic curves are reduced to horizontal straightlines. This type of diagram has been already encountered for mixtures of two low MW enantiomers [60] . Finally, for linear polyesters prepared from di-n-propyl-p-terphenyl-4,4" carboxylate and aliphatic diols, the assignment of N, S_A and/or S_C phases essentially founded on texture observations [20] was further confirmed by miscibility studies [15,48,49] . Unlimited and limited miscibility of smectic modifications are illustrated in Fig. 19.

The applicability of the rule of selective miscibility has been also examined for mixtures of nematic side chain polymers with conventional nematogens [61-65] . In contrast to linear polymers, unlimited miscibility seems to occur only if the chemical structure of the polymer-bound side groups is similar to that of the low MW molecule. Unlimited miscibility is illustrated in Fig. 20. As the polymer investigated does not crystallize, no eutectic is seen : the melting point of the reference compound is depressed as the polymer concentration is increased and the glass transition of the polymer is lowered as the mole fraction of the conventional nematogen is increased (plasticizing effect). As a consequence, at certain compositions the thermal stability range of the nematic phase is noticea-

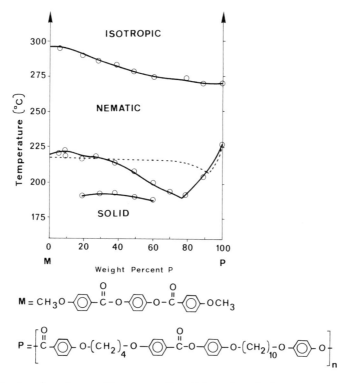

Fig. 18. Isobaric phase diagram of a LC main chain polyester and a
small molecule nematogen. From ref. 55.

bly enlarged towards lower temperatures. On the other hand, mixtures consisting of compounds with different chemical structures,
show phase separation in the liquid crystalline state. Thus, with
the same polymer as in the above investigation, the replacement of :

$$H_{13}C_6O - \bigcirc - COO - \bigcirc - O-CH_2-CH=CH_2$$

by

$$H_{13}C_6O - \bigcirc - COO - \bigcirc - OC_8H_{17}$$

leads to limited miscibility. This points out that these systems
are much more sensitive to the details of the molecular structure
than the mixtures containing linear polymers. Much study is needed
to define to what extent miscibility is changed by a structural
modification.

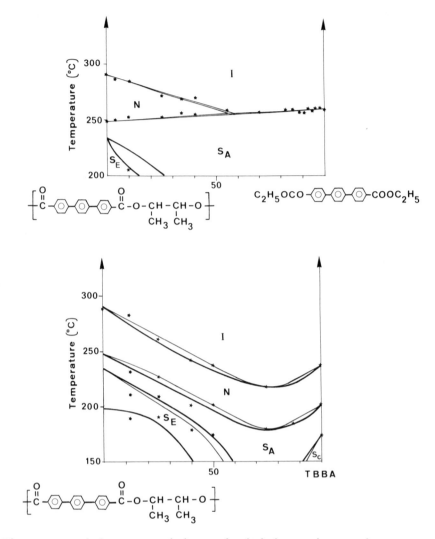

Fig. 19. Unlimited and limited miscibility. Binary mixtures of a
LC main chain polyester and low MW liquid crystals.
From ref. 49.

ALIGNMENT OF POLYMERIC NEMATIC LIQUID CRYSTALS

The molecular alignment within a classic nematic mesophase is
described schematically in Fig. 21. The main feature of molecular
organization is the orientational order of the molecular long axes.
The distribution of the molecular centres of gravity, however, is

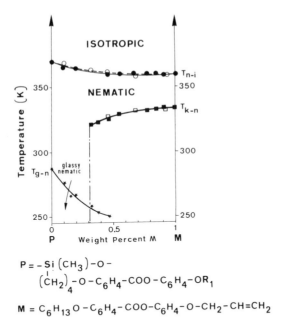

$P = -Si\,(CH_3)-O-$
$\quad\quad (CH_2)_4-O-C_6H_4-COO-C_6H_4-OR_1$

$M = C_6H_{13}O-C_6H_4-COO-C_6H_4-O-CH_2-CH=CH_2$

Fig. 20. Isobaric phase diagram of a LC side chain polymer and a small molecule nematogen . From ref. 64.

Fig. 21. Molecular alignment within a conventional nematic phase.

without long range order as in normal amorphous isotropic liquids. The alignment of molecules within a nematic mesophase is not precisely parallel because of thermal fluctuations of the individual molecular long axes. The average direction of these long axes defines the director $\vec{n}$ which may be treated as a vector, both directions of which $+\vec{n}$ and $-\vec{n}$ are equivalent.

Usually, the molecular organization in the whole sample volume is far from being as the idealized structure in Fig. 21. The specimen is split up into domains, the orientation of which varies. However, suitable treatment of the slides or glass surfaces [66] between which the nematic liquid is observed or the application of external field will result in an uniform molecular alignment and give a liquid "single crystal" instead of a liquid" crystalline powder". This means that the sample is macroscopically anisotropic with a unique direction of preference for the whole sample volume.

The degree of parallel alignment can be described by the order parameter S :

$$S = \frac{< 3 \cos^2\theta >- 1}{2}$$

where θ is the angle between the individual molecular long axis and the director $\vec{n}$. For a completely random isotropic phase S = 0; for perfect alignment of molecules parallel to an external reference, such as the direction of an applied magnetic or electric field, S = 1, and, for perpendicular alignment S = 1/2.

Alignment of Polymeric Nematic on Treated Surfaces

A report by Noël et al [18] indicates that an homeotropic alignment can be achieved by simple treatment of glass slides with boiling chromic sulfuric acid, acetone and methanol (sequentially interspersed with water rinses) and rinsing with hot distilled water. Films thus prepared appear completely dark when viewed vertically between crossed polarizers. They show no birefringence, which implies that optical axes of the molecules lie in the observation direction, perpendicular to the supporting surfaces. On the contrary glass surfaces which have been made anisotropic by directed oblique evaporation of silicon monoxide lead to planar samples in which the director lies parallel to the glass surfaces. Well aligned planar specimens were also obtained by using freshly cleaved mica surfaces [25]. Some samples were prepared after having coated the glass plates with a solution of polyimide in N-methylpyrrolidone, heating at 150°C, rubbing in a fixed direction (à la Châtelain), and heating again at 220°C [32]. Under these conditions the polymer I with n = 5 aligns along the direction of rubbing quite easily. However, Krigbaum et al [8,67] and Corazza et al [36] reported contradictory results for a poly(ethylene terephthalate-co-1,4 benzoate) containing 60 mol % p-oxybenzoyl units, a polyester synthesized from 4,4'-dihydroxy-α-methylstilbene and adipic acid and a polyester prepared from 4,4'-isopropenylene diphenol [68] and adipic acid. For these polymers, the Châtelain technique has proved to be barely efficacious in producing samples with uniform uniaxial alignment of the molecules. When well aligned planar samples are viewed from the top between crossed polarizers, birefringence is dependent on the sample orientation with respect to the incoming linear polarized light. Four positions of extinction are found.

Observations of the conoscopic images formed in a monochromatic beam converging in the sample are needed to check the uniformity of alignment (Figs 22 and 23). From conoscopic images, evidence has been found for the optically positive uniaxial behavior of

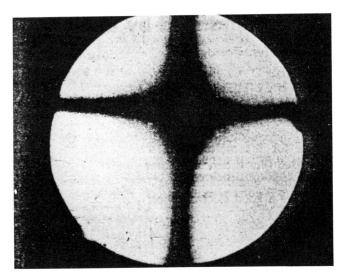

Fig. 22. Conoscopic interference pattern given by homeotropic texture of nematic phase of copolyester prepared from a 50/25/25 (mole %) mixture of terephthalic acid, pyrocatechol and methylhydroquinone.

linear copolyester prepared from a 50/25/25 (mole %) mixture of terephthalic acid, pyrocatechol and methylhydroquinone [8].

With the aid of conoscopic observations, Finkelmann [69] has also proved the positive uniaxial character of comb-like polymers:

$$-CH_2-\underset{\underset{COO}{|}}{C}(CH_3)- \quad (CH_2)_6 \; O -\langle O \rangle - COO -\langle O \rangle - CH_3 \qquad \text{VIII}$$

and

$$-\underset{\underset{(CH_2)_3}{|}}{Si}(CH_3)-O- \quad O -\langle O \rangle - COO -\langle O \rangle - OCH_3 \qquad \text{IX}$$

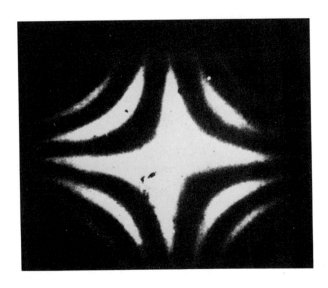

Fig. 23. Conoscopic interference pattern given by planar texture
of nematic phase of copolyester prepared from a
50/25/25 (mole %) mixture of terephthalic acid, pyrocatechol and methylhydroquinone.

which exhibit textures of conventional nematic phases. It is interesting to note that this author reported contradictory results
with polymers :

$$- CH_2 - \underset{\underset{COO \underbrace{+CH_2+}_{2} O -\langle O\rangle- COO -\langle O\rangle- OCH_3}{|}}{C} (CH_3+) \qquad \textbf{X}$$

and

$$- CH_2 - \underset{\underset{COO -\langle O\rangle- COO -\langle O\rangle}{|}}{CH} - \qquad \textbf{XI}$$

which were found optically negative. It is unclear whether this
was due to an erroneous mesophase identification owing to the lack
of typical texture or to the chemical structure of polymers. Indeed
a possible explanation for this specific behavior might be a disturbing effect of the main chain which has to be expected when the
polymer backbone and the side chain motions are little or not decoupled by a flexible spacer.

Coupling With External Fields

Macromolecules capable of forming mesophases are characterized

46

by asymmetrical shapes and, for this reason, would be expected to have anisotropic magnetic susceptibilities and dielectric constants. There should be a tendency for such macromolecular chains to become aligned when placed in a magnetic field, an electric field or a viscous flow. This area of polymer liquid crystals was recently reviewed by Krigbaum [70]. We shall concern ourselves with the effects of a magnetic field since a few reports appeared after this review was published.

In low MW nematic liquids, the application of a magnetic field H leads to uniform alignment of the molecules owing to the diamagnetic anisotropy of the volume elements and the free mobility of the molecules. The anisotropy of the magnetic susceptibility $\Delta\chi$ is the difference between the susceptibilities parallel and perpendicular to the main axis of the uniaxial medium (the director $\vec{n}$) :

$$\Delta\chi = \chi_{\|n} - \chi_{\perp n}$$

Concerning "rod-like" liquid crystal systems, for positive magnetic anisotropy and high magnetic fields, the long molecular axis will tend to align with the applied field H so that $\chi_{\|n}$ and $\chi_{\perp n}$ can be regarded as $\chi_{\|H}$, the susceptibility parallel to the magnetic field, and $\chi_{\perp H}$, the corresponding value perpendicular to the field, respectively. Under these conditions, measurements of $\chi_{\|H}$ and χ, the isotropic susceptibility, allow determinations of the anisotropy of the magnetic susceptibility:

$$\Delta\chi = \chi_{\|H} - \chi_{\perp H} = 3/2(\chi_{\|H} - \bar{\chi})$$

Noël et al [71] investigated ordering effects of a magnetic field on a copolyester synthesized from terephthalic acid and an equimolecular mixture of bisacetates of methylhydroquinone and pyrocatechol. Results summarized in Fig. 24 show that as long as the magnetic field is too weak to cause complete orientation in the direction of the field, the measured magnetic susceptibility depends on the magnetic field strength and decreases (in absolute value) with increasing field strength. This is followed by a period of rapidly increasing alignment which appears as a sharp decrease of the magnetic susceptibility. Then, the magnetic susceptibility approaches a nearly constant value. Practically complete orientation for copolyester in the nematic state was approximated at 6 KG. It is of interest to note that above the threshold field orientation time was less than 2 min for a sample having weight average molecular weight 220 000 and apparent viscosity 300 Pa. sec at a shear rate of 32 sec.$^{-1}$. The anisotropy of the magnetic susceptibility was evaluated as $1.4.10^{-7}$ emu cgs g^{-1}, a slightly higher $\Delta\chi$ than that of the low MW liquid crystals ($\Delta\chi \simeq 10^{-7}$)[4,72]. Such a difference may be attributed to a greater rigidity of the macromolecular chains as compared to the conventional liquid crystals which contain two free terminal groups, usually alkyl groups.

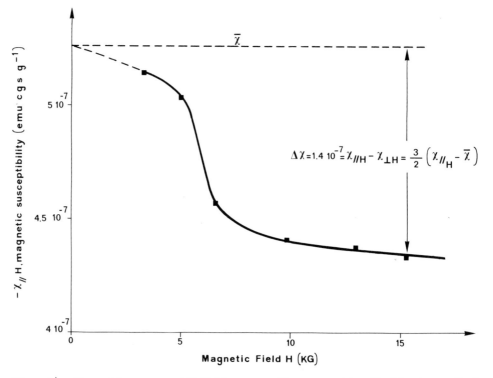

Fig. 24. Magnetic susceptibility χ vs the magnetic field strength H.
280°C. From ref. 71.

Further experimental studies have more recently been carried out
by Hardouin et al. [12] on polyesters I with $5 \leqslant n \leqslant 12$. The magnetic
field strength required for nematic orientation was found to be strongly dependent on the number of the methylene groups n. Lengthening
the aliphatic chain from n = 5 to n = 9 increased the threshold
field strength from 4 KG to 13 KG for a 1-min orientation time. At
high fields, complete orientation was obtained only for the lower
homologues of the series having the less viscous nematic phases.
Thus, the anisotropy $\Delta\chi$ of polyesters C_5 and C_6 approached a nearly
constant saturation value of $1.1.10^{-7}$ emu cgs g^{-1} which corresponds
to the diamagnetic anisotropy usually exhibited by conventional
liquid crystals in the nematic phase. The polymer with n = 9 reached
a plateau value of $0.5.10^{-7}$ emu cgs g^{-1}. For polyester with n = 12,
i.e. the nematic phase of highest viscosity, no orientation effect
was evidenced after 200 min. in the experimental magnetic field
range $0 \rightarrow 15$ KG.

48

The suggestion that the high viscosity of the melt may hinder the molecular alignment has not been substantiated. For polyester C_5 Liebert et al [73] reported that the tendency for macromolecular chains to become aligned when placed in a magnetic field decreased when the molecular weight increased. However, as above-mentioned, Noël et al [71] observed complete orientation for a copolyester having reasonably high molecular weight and viscosity. Maret and Blumstein [74-76] have submitted a number of linear polyesters to the test of orientation in strong magnetic field. A simple experiment was to slowly cool the samples under a constant high magnetic field and to check the resulting orientation by X-rays at room temperature. They observed that some polyesters forming a well-identified nematic phase orient easily in a field of 10 - 20 T irrespective of intrinsic viscosity while other polyesters having similar viscosities or molecular weights cannot be oriented macroscopically. According to the authors, the lack of orientation in the latter polymers can be due either to the fact that neither of these polymers gives a "true" nematic phase or to a destruction of the orientation by a random crystallization of the sample on cooling. It is interesting to observe that cholesteric copolyesters synthesized from 4,4'-dihydroxyazoxybenzene and mixtures of (+) -3-methyladipic acid and dodecanedïoic acid do not orient. Systematic variations with molecular weight and careful analysis in certain homologous series should give information on the influence of the intrinsic flexibility of the chains, the apparent viscosity of the melt and certain basic microscopic processes such as entanglement effects on the tendency for macromolecular chains to become aligned when placed in a magnetic field.

Finally, the movement of a polymeric nematic liquid in a rotating magnetic field was used for measurements of the twist viscosity γ_1 [72]. It is of great interest to note that, contrary to elementary liquid crystals, linear polyesters investigated gave no evidence for a viscous behavior. They behave as solids even at low rotation frequency of the magnetic field.

The effect of a magnetic field on comb-like polymers was first investigated by Casagrande et al [77]. These authors began with the determination of the diamagnetic anisotropy χ_a by using the Faraday method. Then, they determined the splay elastic constant K_{11} and the twist viscosity coefficient γ_1 from the dynamics of a Fredericks transition. By appropriate treatment of the supporting surfaces, the samples were aligned so as to obtain a planar texture. A magnetic field $H > H_c$ was applied perpendicular to the glass slides. (The critical field H_c, at which the onset of realignment occurred, is given by

$$H_c = (\pi/d)(K_{11}/\chi_a)^{1/2}$$

where d is the sample thickness). Then, at time t = 0, the field was rapidly reduced to $H_f < H_c$ and the planar texture reappeared after the reestablishment of a stable state. In the limit of weak distortions it has been shown for conventional nematics that the relaxation time τ is given by :

$$\tau^{-1} H_f = (2 \chi_a/\gamma_1)(H_c^2 - H_f^2)$$

the ratios γ_1/χ_a and K_{11}/χ_a were obtained from measurements of τ at different values of H_f. Casagrande et al [77] found that the diamagnetic anisotropy, the threshold field and the splay elastic constant K_{11} of the comb-like polymer :

$$\left\{ Si \left(CH_3 \right) - O \right\}_{95}$$
$$\left(CH_2 \right)_6 O - \langle O \rangle - COO - \langle O \rangle - OCH_3$$

XII

were similar to those obtained for the model compound :

$$H_{13}C_6O - \langle O \rangle - COO - \langle O \rangle - OCH_2 - CH = CH_2$$

In contrast to the previously described linear polymers, only the mesogenic side chains are responsible for the liquid crystalline order. Under these conditions, one can understand that the behavior of this mesogenic side chain polymer is very similar to that of the conventional low molecular weight nematics. However, some differences also exist between side chain liquid crystal polymers and low molecular weight liquid crystals. By comparing the values of the relaxation times of the model compound and of the polymer investigated, Casagrande et al [77] noted that polymerization slows down the dynamics of the Fredericks transition. The twist viscosity coefficient γ_1 of the polymer is several orders of magnitude higher than that of the model compound.

Order Parameters

The degree of order $S = 1/2 < 3 \cos^2 \theta - 1 >$, which gives the average deviation of the orientation of molecular axes, can be determined experimentally by various methods. It can be found from the principal refractive indices [50, 69, 78]. Other possibilities [79-81] are the use of IR dichroism [25], nuclear resonance spectra [82], electron spin resonance or X-ray scattering [25,73].

The first reliable results on the temperature dependence of the birefringence were reported by H. Finkelmann [69]. Results obtained on comb-like polymer VIII which exhibited a typical nematic texture, showed that the dependence of Δn upon temperature was similar to that observed for low MW liquid crystals. More recently[83], using the extrapolation method proposed by Haller et al [83], Finkelmann et al [50] determined the order parameter S of cholesteric copolymers :

$$CH_3-\overset{\displaystyle |}{\underset{\displaystyle |}{Si}}-(CH_2)_m-O-\underset{\displaystyle \bigcirc}{}-COO-\underset{\displaystyle \bigcirc}{}-OCH_3$$
$$\overset{\displaystyle |}{O}$$
$$CH_3-\overset{\displaystyle |}{\underset{\displaystyle |}{Si}}-(CH_2)_3-O-\underset{\displaystyle \bigcirc}{}-COO-Chol$$
$$\overset{\displaystyle |}{O}$$
$$\overset{\displaystyle |}{}$$

XIII

The mean refractive indices of the cholesteric structure, $n_{e,n}\star$ and $n_{o,n}\star$ were converted into the refractive indices of the corresponding untwisted nematic structure :

$$n_{o,n} = n_{e,n}\star$$

$$n_{e,n} = (2\,n_{o,n}^2\star - n_{e,n}^2\star)^{1/2}$$

Results thus obtained illustrate a number of features that were also observed in work by Finkelmann and Rehage [78], Wassmer et al [82] and Piskunov et al [79] on comb-like homopolymers. The first feature is the effect of the temperature. Similar temperature dependences are observed if macromolecular and low molecular weight liquid crystals are compared. However, there is a slight decrease of S on changing from model compounds to polymers. The second feature is the effect of the length of the flexible spacer. Shortening the aliphatic chain does not change the order parameter S. This experimental result clearly indicates that, although the distance of the mesogenic group to the main chain varies, there is no difference of orientation of the principal axis of the mesogenic groups with respect to the director. A decrease of the spacer length is only associated with more hindered rotation of the mesogenic groups around their long molecular axes [50].

A quite different situation is observed with the linear polyesters. From the FTIR polarized spectra of an oriented nematic copolyester II the temperature dependence of the order parameter was determined [25]. It is apparent from Fig. 25 that S decreases steadily from 0.55 to 0.45 as the temperature is raised from 250°C (~ 32° above the melting point) up to 285°C. The values of S calculated from the X-ray diagrams are in relatively good agreement with infrared data. Liebert et al [73] reported S = 0.54 for polyester I (n = 5) having molecular weight $\overline{M}_w$ = 25 000. These results seemed

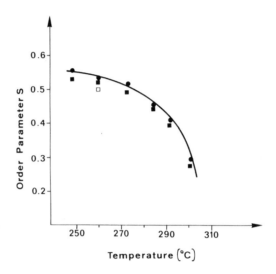

Fig. 25. Temperature dependence of orientational order parameter S
for copolyester II. From ref. 25.

to indicate that a polymeric nematic phase in which the mesogenic
groups are in the main chain can have order parameters approaching
those of low MW nematics. However, Volino et al [80,81] deduced from
NMR spectra of the polymer (Mn = 4 000) :

$$\left\{O-\bigcirc{\overset{CH_3}{}}-N=N-\bigcirc{\overset{CH_3}{}}-OCO-\left(CH_2\right)_{10}-CO\right\}$$ **XIV**

unusually high values for the order parameter S, ranging from 0.69
to 0.84. From a detailed analysis of the NMR spectra these authors
concluded that the flexible spacers aligned in the magnetic field
with a degree of order comparable to that of the mesogenic elements
and adopted a rather extended conformation. More detailed and specific structural studies by different techniques are required for
the complete clarification of the microscopic structure of these
polymers.

X-RAY DIFFRACTION PATTERNS

Detailed reviews [84-91] have appeared laying stress upon the liquid crystalline order in polymers with mesogenic side groups (see also 41-45 and 92-98). Thus, we shall confine our attention mainly to structural studies performed on polymers with mesogenic moieties and flexible spacers in the main chain.

X-Ray Diffraction Patterns for Powder Samples

For powder samples, the well-known Debye-Scherrer technique is used. This method gives all the reticular spacings but no information about the spacial orientation of these planes. Powder data are useful for determining relative intensities where it may be preferable to study a good powder specimen rather than a poor monodomain.

Until recently, in addition to nematics, only two types of smectics were known for low MW liquid crystals, on the one hand four smectic phases which exhibit 3 D-order : the smectic E and smectic B (the difference between them being the order inside a layer) and their tilted modifications the smectic H and smectic G, on the other hand the smectic A and the tilted smectic C phases which are true two-dimensional liquids with 1 D density waves along the director of molecules. Recently, the S_F and S_I phases were discovered. They appear as new stages between the three dimensionally ordered S_B and S_G and the "2 D liquid" S_C phases [99]. They are made of stacked uncorrelated layers and exhibit either a topological 2 D order in S_I or a short range order within a layer in S_F (hexatic phase). Hence, the mesophases can be divided into three groups according to the characteristics of their X-ray diffraction patterns at large diffraction angles [99-101]. By large diffraction angles we mean angles of the order of $20°$ which correspond to distance of approximately 4.5 A for the usual wavelengths ($CuK_\alpha = 1.54$ A and $CoK_\alpha = 1.79$ A). Small diffraction angles are taken to be of the order of $3°$ and correspond to distances of about 30 Å. The first group is composed of the disordered phases (N, S_A and S_C) which give only one diffuse, broad diffraction maximum at wide angles. In the second group, we find the S_E, S_B, S_H and S_G phases whose diffraction patterns show one or few sharp Bragg reflections instead of a diffuse ring. The S_F and S_I phases are intermediate between these two groups.

Data on unoriented specimens can be very useful for initial characterization of samples where it is usually possible to distinguish unambiguously between a number of phases (e.g. S_E and S_B, S_A and N) but this is not always so (e.g. S_A and S_C may be confused). X-ray patterns of polymers with mesogenic moieties and flexible spacers in the main chain are often too diffuse to be of help for identification of mesophases, especially in the case of unoriented

nematics or cholesterics. Nevertheless, the literature reports
X-ray diffraction patterns which are compatible with a nematic
structure [18, 19, 33, 37, 38, 102-105]. They present at wide angles
a diffuse halo arising from the intermolecular spacings perpendicular to the long axes of the molecules. Such a diffuse halo indicates a lack of periodic lateral order. Average intermolecular spacings can be obtained by the rather arbitrary use of the formula
$d = 1.2 \lambda/2 \sin \theta$ based on arguments of cylindrical symmetry. In
all cases, the average intermolecular spacing lies in the range
4-6 Å which corresponds approximately to the average width of the
molecules but is certainly smaller than the diameter of a freely
rotating molecule. Nematic patterns may present, at low diffraction angles, a diffuse ring corresponding to distances which are
usually close to the maximum molecular lengths measured on molecular models. Such a diffuse ring indicates that there is no order
in the direction of the molecular long axes.

In contrast, other X-ray patterns obtained from linear polymers in the liquid crystalline state are characteristic of a
disordered lamellar structure [47, 49, 103 c, 105, 106]. They present at wide angles a diffuse ring associated with the unstructured
nature of the layers and, at small angles, a sharp ring (with
sometimes its second order) corresponding to the lamellar thickness. With unoriented samples it was impossible to draw an infe-
rence on the nature of these disordered smectic liquid crystals.
However, in some cases optic observations and/or miscibility tests
have removed this indetermination. Ober et al [47] reported that
polyesters

XV

exhibited smectic A or smectic C phases with x = 10, 12
or 9. Occurrence of smectic A mesophases was reported by Bosio
et al [49] for polyesters

XVI

where R = $(CH_2)_x$ or branched alkyl segments. On the other hand,
a smectic C mesophase was evidenced for polyester VI.

Finally, there is some indication that polymers with mesogenic
moieties and flexible spacers in the main chain may also form smectic phases which exhibit 3 D-order. The X-ray diffraction patterns
of poly (p,p'-benzylidene aminophenylsebacate) and poly(4,4'-di-
phenylsebacate) in the smectic state were characterized by a sharp
central ring and a sharp outer ring which indicates a periodic
order within the lamellae. Such results are typical features of

the diffraction patterns of S_B and its tilted modification [102]. Such an assignment is, however in conflict with that made by Van Luyen and Strzelecki [107] who reported the poly(4,4'-diphenylsebacate) to be nematic. X-ray patterns of oriented specimens will be necessary to substantiate the nature of this mesophase. Bosio et al [49] reported from texture observations that the lower homologues in series XVI may exhibit a 3-dimensional ordered smectic E phase in addition to 2 D-liquid smectic A and nematic phases. At low temperature, the inner sharp diffraction ring and the four sharp outer rings are consistent with the formation of such a highly ordered mesophase.

X-Ray Diffraction Patterns for Oriented Samples

It is obvious that monodomain samples provide much more information than powder samples but they are not always available. For nematics, monodomains can be obtained by orientating a powder sample with a magnetic field. An alternative procedure, useful for preparing monodomains of the more ordered (and more viscous) smectic phases, is by careful melting of either a single crystal or an oriented fiber.

Some examples of polymeric nematics thus oriented can be given [18, 22, 25, 73, 76, 103d, 108]. As observed for low MW liquid crystals [100, 101] the diffraction patterns of all the aligned specimens investigated are characteristic of materials having a definite texture axis. In Figure 26 we see typical X-ray diffraction patterns. The anisotropy is clearly shown and there are two symmetry directions . One is parallel to the director (the long molecular axis) and can be described as the meridional section of reciprocal space in the plane of the film. The other direction is perpendicular to the director, this is the equatorial section of reciprocal space in the plane of the film. The dominant feature is that the outer diffuse halo evidenced for powder samples is split into two crescents symmetrical about the equatorial plane. For all the materials investigated, these features are qualitatively similar, but differ quantitatively mainly in the length of the arc which reflects the degree of parallel alignment. The scattering about the meridional direction differs qualitatively between the different types of materials showing that the molecular packing in the nematic phase varies significantly with the type of polymer. The dominant features are usually arcs or short bars which are due to intramolecular scattering. Their positions correspond to repeat distances of the order of the repeat unit lengths. They appear distinctly only when films are over exposed since their intensity is smaller than that of the strong equatorial crescents [18, 22, 25, 73, 76].

Blumstein et al [22, 76] reported for polyester prepared from 4,4'-hydroxy (2,2'-methyl) azoxybenzene and dodecanedioic acid a

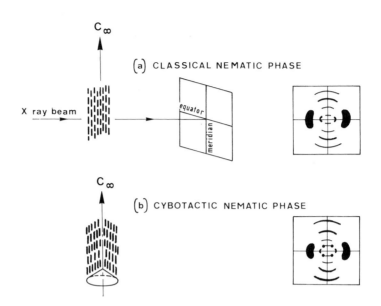

Fig. 26 – Schematic X-ray patterns of the main types of oriented
nematic mesophases.

X-ray diffraction pattern which shows that the first of the meridional arcs splits up into four sharp spots (Fig. 26 b). The diffraction pattern of the nematic then shows the development of
enhanced order characteristic of the smectic C phase. This has[109]
been already observed for some 4,4'-di-n-alkoxyazoxybenzenes
and also for bis-(4'-n octyloxybenzylidene)-2-chloro-1,4-phenylene
diamine[110]. This phenomenon is incompatible with the classical
concept of the nematic phase and De Vries[110] suggested an additional order of molecules within cybotactic groups. The simplest[22, 76]
way to account for the pattern obtained by Blumstein et al
is then "to have a bundle of mesogenic moieties arranged into
strata in which the plane of the layer is tilted with respect to
the mesogen by an angle approximately equal to the azimuthal angle
of the sharp spots".

On the other hand, the X-ray diffraction patterns obtained for
copolyester synthesized from terephthalic acid and an equimolecular mixture of bis acetates of methylhydroquinone and pyrocatechol
showed not only two pronounced crescents but also two diffuse spots
at smaller angles along the equator[18]. Such diffuse spots have[111]
been already reported for certain conventional nematic systems
but without any justification. Taking into account that similar
diffraction features have been observed for helical structures
[112, 113], Noël et al[18], in their analysis of the X-ray patterns

of copolyester, considered the possibility of such arrangements of the chains but without long-range order. The two diffuse spots along the equatorial line would be expected for roughly parallel chain bundles (typically 10-12 Å diameter) in the form of two-four stranded ropes. The polymer chains would be otherwise poorly packed together due to random monomer sequences. They are spaced an average distance 5.0-5.2 Å apart but are not otherwise correlated. The periodicity along the chains and also the pitches of the helices would give rise to the diffuse disks along the meridian.

Monodomains of the "2 D-liquid" smectic phase of the polyester VI were obtained by careful melting of oriented fibers [49]. Schematic diffraction pattern is shown in figure 27 b : the two sharp small angle reflections on meridian

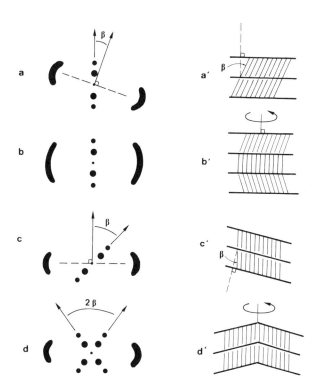

Fig. 27 - Schematic X-ray patterns of the main types of oriented smectic C phases. From ref. 114.

have to be attributed to regular smectic layer distance while the diffuse arcs observed along the equator are due to the short range positional order of the chains within the smectic layers. The relative position of the small angle reflections with respect to the diffuse crescents yields X-ray patterns similar to those of the class smectic A [100]. However, the smectic C nature of the mesophase was determined by establishing its isomorphy with the smectic C phases of standard materials [75]. On account of these results, the authors admitted that the polyester VI exhibited a smectic C configuration which is azimuthally disordered with the layers orientationally ordered (Fig. 27 b').

Finally, results of structural investigations on random copolymers were recently reported. X-ray methods were used to investigate the physical structure of high strength melt-spun liquid crystalline aromatic copolyester fibers prepared from p-hydroxybenzoic acid (B), 2,6-dihydroxynaphthalene (N) and terephthalic acid (P) [115]. The X-ray data obtained for fibers prepared from different B/N/P ratios showed variations in the positions and intensities of the maxima. Excellent agreement was obtained between the positions of these observed maxima and those predicted for an aperiodic array of points where each point represented a monomer in a random chain, separated from adjacent points by the appropriate monomer lengths.

Wide angle X-ray scattering from an aligned sample of copolyester prepared from 40 mol % polyethylene terephthalate and 60 mol % p.acetoxybenzoic acid was reported together with an analysis of the derived cylindrical distribution function (CDF) [116]. From the data in the form of an s-weighted reduced intensity function si (s,α), the immediate conclusion which might be drawn was that the material was well oriented. In fact, there was a high local orientational correlation of chains but no longitudinal (as opposed to orientational) correlation of adjacent chains. Comparison of the meridional spacings in the CDF with those of model chains, together with a consideration of the scattered intensity distribution along the meridian suggested that the distribution of the chemical components of the copolymer was random.

REFERENCES

1. A. Roviello and A. Sirigu, J. Polym. Sci. Polym. Lett. Ed. 13:455 (1975)
2. G.W. Gray and P.A. Winsor, "Liquid Crystals and Plastic Crystals", Horwood, Chichester, England (1974)
3. D. Demus and L. Richter, "Textures of Liquid Crystals", Verlag Chemie, Weinheim (1978)
4. H. Kelker and R. Hatz, "Handbook of Liquid Crystal", Verlag Chemie, Weinheim (1980)
5. G.R. Luckhurst and G.W. Gray, "The Molecular Physics of Liquid Crystals". Academic Press, New-York (1979)

6. G. Friedel, Z. Kristallogr., 79:26 (1931)
7. G. Friedel and E. Friedel, J. Physique Radium (VII)2: 133 (1931)
8. W.R. Krigbaum, H.J. Lader and A. Ciferri, Macromolecules, 13:554 (1980)
9. W.R. Krigbaum, A. Ciferri, J. Asrar, H. Toriumi and J. Preston, Mol. Cryst. Liq. Cryst., 76:142 (1981)
10. A.C. Griffin and S.J. Havens, J. Polym. Sci. Polym. Phys. Ed., 19:951 (1981)
11. L. Strzelecki and L. Liebert, Eur. Polym. J., 17: 1271 (1981)
12. F. Hardouin, M.F. Achard, H. Gasparoux, L. Liebert and L. Strzelecki, J. Polym. Sci. Polym. Phys. Ed., 20: 975 (1982)
13. C. Noël and J. Billard, Mol. Cryst. Liq. Cryst. (Lett.), 41:269 (1978)
14. B. Millaud, A. Thierry, C. Strazielle and A. Skoulios, Mol. Cryst. Liq. Cryst. (Lett.), 49:299 (1979)
15. B. Fayolle, C. Noël and J. Billard, J. de Physique, 40:C3-485 (1979)
16. L. Strzelecki and D. Van Luyen, Eur. Polym. J., 16:299 (1980)
17. B.W. Jo, J.I. Lin and R.W. Lenz, Eur. Polym. J., 18:233 (1982)
18. C. Noël, J. Billard, L. Bosio, C. Friedrich, F. Lauprêtre and C. Strazielle, Polymer, 25:263 (1984)
19. S. Antoun, R.W. Lenz and J.I. Jin, J. Polym. Sci. Polym. Chem. Ed., 19:1901 (1981)
20. P. Meurisse, C. Noël, L. Monnerie and B. Fayolle, Brit. Polym. J., 13:55 (1981)
21. W.R. Krigbaum, A. Ciferri, J. Asrar, H. Toriumi and J. Preston, Mol. Cryst. Liq. Cryst. 76:79 (1981)
22. A. Blumstein, S. Vilasagar, S. Ponrathnam, S.B. Clough and R.B. Blumstein, J. Polym. Sci. Polym. Phys. Ed., 20:877 (1982)
23. C. Viney and A.H. Windle, J. of Materials Science, 17:2661 (1982)
24. M.R. Mackley, F. Pinaud and G. Siekmann, Polymer, 22:437 (1981)
25. C. Noël, F. Lauprêtre, C. Friedrich, B. Fayolle and L. Bosio, Polymer, in press.
26. G. Galli, E. Chiellini, C.K. Ober and R.W. Lenz, Makromol. Chem., 183:2693 (1982)
27. F.C. Frank, Discuss. Faraday Soc., 19(1958)
28. H. Sackmann and D. Demus, Mol. Cryst. Liq. Cryst., 21:239 (1973)
29. B. Millaud, A. Thierry and A. Skoulios, J. Physique, 39:1109 (1978)
30. P.G. de Gennes, Mol. Cryst. Liq. Cryst. (Lett.), 34:177 (1977)
31. R.B. Meyer, Macroscopic Phenomena in Nematic Polymers, in "Polymer Liquid Crystals", A. Ciferri, W.R. Krigbaum and R.B. Meyer eds., Academic Press, New-York (1983)
32. M. Kleman, L. Liebert and L. Strzelecki, Polymer, 24:295 (1983)
33. R.W. Lenz and J.I. Jin, Macromolecules, 14:1405 (1981)
34. R.V. Talroze, S.G. Kostromin, V.P. Shibaev and N.A. Platé, Makromol. Chem., Rapid Commun., 2: 305 (1981)

35. D. Van Luyen, L. Liebert and L. Strzelecki, Eur. Polym. J., 16:307 (1980)
36. P. Corazza, M.L. Sartirana and B. Valenti, Makromol. Chem., 183: 2847 (1982)
37. S. Vilasagar and A. Blumstein, Mol. Cryst. Liq. Cryst. (Lett.), 56:263 (1980)
38. A. Blumstein and S. Vilasagar, Mol. Cryst. Liq. Cryst. (Lett.), 72:1 (1981)
39. H. Finkelmann, J. Koldehoff and H. Ringsdorf, (a) Angew. Chem., 90:992 (1978), (b) Angew. Chem. Int. Ed. Engl., 17:935 (1978)
40. H. Finkelmann and G. Rehage, Makromol. Chem. Rapid Commun., 1:733 (1980)
41. N. A. Platé and V.P. Shibaev, J. Polym. Sci. Polym. Symp., 67:1 (1980)
42. V.P. Shibaev, N.A. Platé and YA.S. Freidzon, J. Polym. Sci. Polym. Chem. Ed., 17:1655 (1979)
43. B. Hahn, J.H. Wendorff, M. Portugall and H. Ringsdorf, Colloid and Polymer Sci., 259:875 (1981)
44. V.P. Shibaev, S.G. Kostromin and N.A. Platé, Eur. Polym. J., 18:651 (1982)
45. S.G. Kostromin, V.V. Sinitzyn, R.V. Talroze, V.P. Shibaev and N.A. Platé, Makromol. Chem. Rapid Commun. 3:809 (1982)
46. W.R. Krigbaum, J. Asrar, H. Toriumi, A. Ciferri and J. Preston, J. Polym. Sci. Polym. Lett., 20:109 (1982)
47. C.K. Ober, J.I. Jin and R.W. Lenz, Makromol. Chem. Rapid. Commun., 4:49 (1983)
48. C. Noël, P. Meurisse, C. Friedrich and L. Bosio, to be published.
49. L. Bosio, B. Fayolle, C. Friedrich, L. Lauprêtre, P. Meurisse, C. Noël and J. Virlet, "DSC, Miscibility and X-ray Studies of the Thermotropic Liquid Crystalline Polyesters with Aromatic Moieties and Flexible Spacers in the Main Chain", in "Liquid Crystals and Ordered Fluids", A. Griffin and A. Johnson eds., Plenum Press, New-York, vol. 4, 401 (1984)
50. H. Finkelmann, "Synthesis, Structure, and Properties of Liquid Crystalline Side Chain Polymers", in "Polymeric Liquid Crystals", A. Ciferri, W.R. Krigbaum and R.B. Meyer eds, Academic Press, New-York (1982)
51. J. Grebowicz and B. Wunderlich, J. Polym. Sci. Polym. Phys. Ed., 21:141 (1983)
52. A. Blumstein and O. Thomas, Macromolecules, 15:1264 (1982)
53. L. Kofler and A. Kofler, in "Thermomikromethoden", Verlag Chemie, Weinheim, (1954)
54. B. Millaud, A. Thierry and A. Skoulios, Mol. Cryst. Liq. Cryst. (Lett.), 41:263 (1978)
55. A.C. Griffin and S.J. Havens, J. Polym. Sci. Polym. Lett., 18:259 (1980)
56. A.C. Griffin and S.J. Havens, Mol. Cryst. Liq. Cryst. (Lett.), 49:239 (1979)

57. I.Z. Shröder, Z. Phys. Chem., 11:449 (1893)
58. J.J. Van Laar, Z. Phys. Chem., 63:216 (1908)
59. J. Billard, A. Blumstein and S. Vilasagar, Mol. Cryst. Liq.
    Cryst. (Lett.), 72:163 (1982)
60. M. Leclercq, J. Billard and J. Jacques, Mol. Cryst. Liq. Cryst.,
    8:367 (1969)
61. K. Nyitrai, F. Cser, M. Lengyel, E. Seyfried and Gy. Hardy,
    Eur. Polym. J., 13:673 (1977)
62. F. Cser, K. Nyitrai, G. Hardy, J. Menczel and J. Varga,
    J. Polym. Sci. Polym. Symp. 69:91 (1981)
63. H. Ringsdorf, H.W. Schmidt and A. Schneller, Makromol. Chem.
    Rapid Commun. 3:745 (1982)
64. H. Finkelmann, H.J. Kock and G. Rehage, Mol. Cryst. Liq. Cryst.
    89:23 (1982)
65. C. Casagrande, M. Veyssie and H. Finkelmann, J. Phys. Lett. 43:
    L-675 (1982)
66. E. Guyon and W. Urbach in " 4 BBC Symp. Nonem. El. Disp.",
    Kmetz and Von Willisen eds., Plenum Press, New-York (1976)
67. W.R. Krigbaum, C.E. Grantham and H. Toriumi, Macromolecules,
    15:592 (1982)
68. P. Châtelain, Acta Crystallogr., 1:315 (1948)
69. H. Finkelmann, "Thermotropic Liquid Crystalline Polymers", in
    "Liquid Crystals of one-and two-dimensional order",
    W. Helfrich and G. Heppke eds, Springer-Verlag, Berlin (1980)
70. W.R. Krigbaum, "The Effects of External Fields on Polymeric
    Nematic and Cholesteric Mesophases" in "Polymer Liquid
    Crystals", A. Ciferri, W.R. Krigbaum and R.B. Meyer eds,
    Academic Press, New-York 1982
71. C. Noël, L. Monnerie, M.F. Achard, F. Hardouin, G. Sigaud and
    H. Gasparoux, Polymer, 22:578 (1981)
72. F. Hardouin, M.F. Achard, G. Sigaud and H. Gasparoux, Mol.
    Cryst. Liq. Cryst. 39:241 (1977)
    M.F. Achard, F. Hardouin, G. Sigaud and H. Gasparoux, J. Chem.
    Phys., 65:1387 (1976)
73. L. Liebert, L. Strzelecki, D. Vanluyen and A.M. Levelut, Eur.
    Polym. J., 17:71(1981)
74. G. Maret, A. Blumstein and S. Vilasagar, Polym. Prepr. Am. Chem.
    Soc. Div. Polym. Chem., 22:246 (1981)
75. A. Blumstein, S. Vilasagar, S. Ponrathnam , S.B. Clough and
    R.B. Blumstein, J. Polym. Sci. Polym. Phys. Ed., 20:877
    (1982)
76. G. Maret and A. Blumstein, Mol. Cryst. Liq. Cryst., 88:295
    (1982)
77. C. Casagrande, M. Veyssié, C. Weill and H. Finkelmann,
    Mol. Cryst. Liq. Cryst. (Lett.), 92:49 (1983)
78. H. Finkelmann and G. Rehage, Makromol. Chem. Rapid Commun.,
    3:859 (1982)
79. M.V. Piskunov, S.G. Kostromin, L.B. Stroganov, V.P. Shibaev
    and N.A. Platé, Makromol. Chem. Rapid Commun., 3:443 (1982)

80. F. Volino, A.F. Martins, R.B. Blumstein and A. Blumstein, C.R. Acad. Sci., 292:II-829 (1981); J. Phys. (Lett.) Paris, 42: L-305 (1981)

81. A.F. Martins, J.B. Ferreira, F. Volino, A. Blumstein and R.B. Blumstein, Macromolecules, 16:279 (1983)

82. K.H. Wassmer, E. Ohmes, G. Kothe, M. Portugall and H. Ringsdorf, Makromol. Chem. Rapid Commun., 3:281 (1982)

83. J. Haller, H.A. Huggins, H.R. Lilienthal and T.R. Mc Guire, J. Phys. Chem., 77:950 (1973)

84. J.H. Wendorff, in "Liquid Crystalline Order in Polymers", A. Blumstein ed., Academic Press, New-York, (1978)

85. J.H. Wendorff, H. Finkelmann and H. Ringsdorf, Am. Chem. Soc. Polym. Prepr., 74:12 (1978)

86. S.B. Clough, A. Blumstein and A. de Vries, Am. Chem. Soc. Polym. Prepr., 74:1 (1978)

87. A. Blumstein and E.C. Hsu, in "Liquid Crystalline Order in Polymers", A. Blumstein ed., Academic Press, New-York, (1978)

88. V.P. Shibaev and N.A. Platé, Polymer Sciences USSR, 1605 (1977)

89. V.V.Tsukruk, V.V. Shilov and Yu S. Lipatov, Makromol. Chem., 183: 2009 (1982), Eur. Polym. J., 19:199 (1983)

90. V.V. Tsukruk, V.V. Shilov, I.I. Konstantinov, Yu.S. Lipatov and Yu.B. Amerik, Eur. Polym. J., 18:1015 (1982)

91. V. Frosini, G. Levita, D. Lupinacci and P.L. Magagnini, Mol. Cryst. Liq. Cryst., 66:21 (1981)

92. P.L. Magagnini, Makromol. Chem. Suppl., 4:223 (1981)

93. B.A. Newman, V. Frosini and P.L. Magagnini, Am. Chem. Soc. Polym. Prepr., 74:71 (1978)

94. Y. Osada and A. Blumstein, J. Polym. Sci. Polym. Lett. Ed., 15:761 (1977)

95. A. Blumstein, Y. Osada, S.B. Clough, E.C. Hsu and R.B. Blumstein, Am. Chem. Soc. Polym. Prepr., 74:56 (1978)

96. A. Blumstein, R.B. Blumstein, S.B. Clough and E.C. Hsu, Macro-molecules, 8:73 (1975)

97. V.P. Shibaev, R.V. Talroze, F.I. Karakhanova and N.A. Platé, J. Polym. Sci. Polym. Chem. Ed. 17:1671 (1979)

98. C.M. Paleos, G. Margomenou-Léonidopoulou, S.E. Filippakis and A. Malliaris, J. Polym. Sci. Polym. Chem. Ed., 20:2267 (1982)

99. J.J. Benattar, F. Moussa and M. Lambert, J. de Chim. Phys., 80:99 (1983)

100. A.J. Leadbetter, in "The Molecular Physics of Liquid Crystals", G.R. Luckhurst and G.W. Gray eds., Academic Press, New-York, (1979)

101. J. Doucet, in "The Molecular Physics of Liquid Crystals", G.R. Luckhurst and G.W. Gray, eds, Academic Press, New-York, (1979)

102. A. Blumstein, K.N. Sivaramakrishnan, R.B. Blumstein and S.B. Clough, Polymer, 23:47 (1982)

103. A. Roviello and A. Sirigu, (a) Eur. Polym. J., 15:61 (1979),
     (b) Makromol. Chem., 181:1799 (1980), (c) Gazz. Chim. Ital.,
     110:403 (1980), (d) Makromol. Chem., 183:895 (1982)
104. P. Iannelli, A. Roviello and A. Sirigu, Eur. Polym. J., 18:745,
     753 (1982)
105. V. Frosini and A. Marchetti, Makromol. Chem. Rapid. Commun.,
     3:795 (1982)
106. A. Thierry, A. Skoulios, G. Lang and S. Forestier, Mol. Cryst.
     Liq. Cryst. (Lett.), 41:125 (1978)
107. D. Van Luyen and L. Strzelecki, Eur. Polym. J., 16:303 (1980)
108. A. Roviello and A. Sirigu, Makromol. Chem., 183:409 (1982)
109. I.G. Chistyakov and W.M. Chaikowsky in "Liquid Crystals 2",
     G.H. Brown, ed., Gordon and Breach, London, (1969)
110. A. de Vries, Mol. Cryst. Liq. Cryst., 10:31, 219, (1970)
111. K. Usha Deniz, A.S. Paranjpe, V. Amirthalingam and K.V. Mura-
     lidharan in "Liquid crystals", S. Chandrasekhar, ed.,
     Heyden, London (1980)
112. B.K. Vainshtein in "Diffraction of X-rays By Chain Molecules",
     Elsevier, Amsterdam, (1966)
113. R.S. Bear and H.F. Hugo, Ann. N.Y. Acad. Sci., 53:627 (1951)
114. C. Noël, C. Friedrich, L. Bosio and C. Strazielle, Polymer, in
     press.
115. J. Blackwell and G. Gutierrez, Polymer, 23:671 (1982), Polymer
     24:937 (1983)
116. G.R. Mitchell and A.H. Windle, Polymer, 23:1269 (1982).

SOME PHYSICO-CHEMICAL ASPECTS OF POLYMERIC LIQUID CRYSTALS

Edward T. Samulski

Department of Chemistry & Institute of Materials Science
University of Connecticut U-136
Storrs, CT 06268 U S A

INTRODUCTORY REMARKS

Apart from their obvious macroscopic differences that follow from their respective molecular dynamical differences, ordinary liquids are readily distinguished from solids on a microscopic scale by the absence of long-range translational and orientational molecular order. Nevertheless, one frequently refers to "structure" in liquids: short-range, near-neighbor spatial and angular correlations which, in the absence of specific intermolecular interactions (e.g. dipolar, hydrogen bonding, etc.) facilitate the "packing" of molecules at liquid densities.(1) Such correlations may be somewhat more exaggerated in liquids composed of anisometric molecules with high aspect ratios (rod- or disclike molecules), but even in such fluids the correlation lengths rarely exceed a very few (two or three) molecular dimensions.

Liquid phases of conventional, flexible polymer chains share with ordinary liquids the characteristic absence of long-range translational and orientational order. This may be surprising at first as these chains are a contiguous string of "submolecules" or monomers whose translational and orientational independence would appear to be intrinsically curtailed by the covalent linking of the monomers together to form the polymer chain. However, these intrachain constraints have a limited range: isomerization about the chemical bonds comprising the chain rapidly attenuates correlations along the chain (i.e., orientational and translational correlations become insignificant over a few monomer units), and intermolecular correlations approach those in low molar mass liquids. Consequently, experimental techniques which are sensitive to long-range correlations show negligible differences between melts or solutions of flexible polymers and the fluid phases of ordinary liquids, i.e., in

the condensed phases of non-crystalline polymers the intertwined
"random flight" chains behave ideally. This was anticipated in the
late 1940s by Flory and unequivocally demonstrated a quarter of a
century later with neutron scattering experiments.(2) Hovever,
polymer liquids do exhibit substantial macroscopic differences when
contrasted with low molecular mass liquids. The transport properties of polymers, in particular, are strongly dependent on the polymer chain's length.(3)

In this paper I will focus on how these similarities (and differences) between ordinary liquids and polymeric fluids may be
transposed to liquid crystalline phases of low molar mass mesogens
and their polymeric analogues. This general theme is prefaced by a
brief review of the two general classes of polymeric liquid crystals
that is intended for researchers unfamiliar with the development of
the field. The literature citations are not comprehensive but rather are designed to enable one to assess current references to the
topics considered here. The elementary physico-chemical aspects of
liquid crystals — applicable to polymeric liquid crystals as well
— are exhaustively summarized in Kelker and Hatz's Handbook of Liquid Crystals. Blumstein has edited a text, Liquid Crystalline Order
in Polymers, that reviews some of the earlier work on PLCs.(5) More
recent work is summarized in Polymer Liquid Crystals edited by Cifferi, Krigbaum and Meyer.(6)

THERMOTROPIC POLYMERIC LIQUID CRYSTALS

Our understanding of the liquid crystalline phases of polymer
melts is largely derivative of studies of low molar mass thermotropic mesogens (hereafter referred to as "monomeric" liquid crystals
(MLCs) irrespective of whether or not the mesogen has functionality
suited to polymerization). Polymeric liquid crystals (PLCs) like
MLCs are characterized by fluidity and, of course, long-range orientational order and the various degrees of translational order (e.g.,
viscous nematic PLCs may also exhibit some long-range, intramolecular translational ordering along the chain that is derived from the
connectivity of mesogenic units to form the polymer chain). If we
restrict the discussion to polymers derived from prolate, anisometric MLCs, i.e., the conventional mesogenic molecules consisting of
a rigid "core" (conjugated aromatic rings or linear aliphatic cyclic
frameworks), the history of the development of thermotropic PLCs is
transparent.

Thermotropic PLC's evolved in academic research laboratories by
incorporating known monomeric liquid crystals into polymer chains.
From such studies two types of PLC's have been developed: 1) sidechain polymers with variable flexibility in the main chain, and 2)
semi-flexible linear polymers. In the former the monomeric mesogen
appears as a pendant sidechain attached to the main chain by a flex-

ible spacer (usually methylene units; Figure 1). These sidechain polymers were initially a by-product of attempts in the mid-1960s to conduct synthetic chemistry using liquid crystals as solvents. Researchers anticipated that they would be able to observe changes in the kinetics and/or stereospecificity of chemical reactions of low molar mass molecules carried out in the anisotropic solvents. Generally speaking, the findings were not dramatically different from similar studies conducted in isotropic solvents; primarily because the phenomena investigated were controlled by short-range molecular interactions. In the nematic solvents such short-range interactions are quite similar to those operative in isotropic liquids. Subsequently, however, the influences that ordered phases might induce in macromolecular stereochemistry together with the ready availability of monomeric mesogens with polymerizable functional groups prompted the study of polymerization reactions in the liquid crystal phases of mesomorphic monomers. The ensuing polymers (often solid) retained a petrified version of the liquid crystal texture that existed prior to polymerization.

In a very low resolution picture of polymeric sidechain liquid crystals we might view them as monomeric liquid crystals with an impurity solute -- the flexible main chain of the polymer. In contrast with the observations in MLCs, this impurity does not destabilize the mesophase, but may increase the range of mesophase formation. Blumstein,(7) Finkelman,(8) Ringsdorf(9) and others have studied the flexibility of the main chain, the length of the spacer chain of the mesogenic sidechain (i.e., the degree to which the sidechain and mainchain are coupled), the chemical character of the mesogen unit itself (chiral, achiral, polar, etc.), and have obtained polymers that exhibit all of the common liquid textures: nematic, cholesteric (twisted nematic), and smectic phases. More recently synthesized sidechain polymers melt into fluid liquid crystal phases reversibly and oriented textures may be locked into solid glassy state by polymerizing and/or quenching liquid crystalline polymer melts in the presence of applied external electric (magnetic) fields.(10)

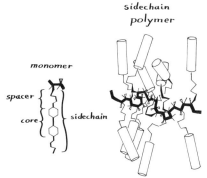

Fig. 1. A schematic representation of a sidechain polymeric liquid crystal.

Continuing this line of reasoning, in 1975 the synthetic chemists Roviello and Sirigu(11) reported the synthesis of linear thermotropic polymers derived by covalently linking mesogens end-to-end
(Figure 2). In these materials the linking spacer chain is not a
superfluous appendage but is integral to the polymer and dictates
the flexibility of the polymer backbone. While these polymers exhibit properties identical to those of their monomeric counterparts,
they also retain macroscopic features unique to ordinary polymer
melts. For example, their time scale for reorientation in response
to external stimuli is comparable to that of viscous isotropic polymer melts, i.e., minutes, sometimes hours are required to achieve
macroscopic order with electric (magnetic) fields.(12) Additionally
there may be interesting microscopic features unique to linear PLCs.
For example, the local short-range "structure" and long-range orientational order in the linear PLCs appears somewhat better developed
than that in MLCs.(13) The ubiquitous role of the spacer will be
considered in this context later.

Investigations on linear chain PLCs have paralleled those on
side-chain polymers (exploring the role of spacer length on the liquid crystal phase type, phase stability, etc.). Today there are examples of linear thermotropic polymers with stable liquid crystal
phases spanning a range of more than 300 degrees. Concurrently
there is a substantial industrial research effort focused on finding
melt-processable linear PLCs that exhibit unique physical properties
in the solid state. The patent literature reflects the scope of this
effort; see for example the citations concluding Calundann and
Jaffe's review of anisotropic polymers.(14) Current research includes screening of alloys (blends of rigid chains with one or more
flexible polymers), copolymers (incorporating rigid monomers into a
chain of conventional flexible monomers), and novel semiflexible
polymers synthesized explicitly to exhibit stable liquid crystal
phases.

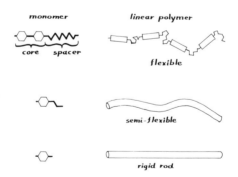

Fig. 2 Schematic representations of linear polymeric liquid crystals having primary structures with different inherent flexibility.

I conclude the introduction to thermotropic PLCs by calling attention to the polymeric analogues to discotic liquid crystals -- melts of oblate shaped mesogens.(15) Pitch is comprised of planar, fused aromatic molecules and it exhibits a birefringent mesophase during the course of heat polymerization. While such material is rather ill-characterized, it has all of the characteristics of a discotic mesophase and does play an important role in the fabrication of ultra-high strength carbonaceous solids.(16)

LYOTROPIC POLYMERIC LIQUID CRYSTALS

First I want to draw attention to polymers wherein specific solute-solvent interactions do promote mesophase formation. There are two general classes of polymer-solvent systems wherein instances of mesomorphism result from intermolecular interactions that are closely related to those that stabilize the amphiphilic MLC lyotropics: 1)Sidechain polymers with amphiphilic sidechains appended to the polymer backbone may form lyotropic PLCs in water. 2) Block copolymers exhibit organized, gel-like phases when one block is preferentially solvated. The former PLCs are intimately related to lyotropic MLCs, however, compared to the monomer amphiphiles, new phases and often phase stability is influenced by the attachment of the amphiphilic core to the chain. Finkelmann and co-workers have recently considered such amphiphilic sidechain PLCs.(17)

In the swollen block copolymeric materials, the insoluble block of the polymer aggregates into microphase separated domains that in turn are dispersed in well-defined, spatially periodic supramolecular structures.(18) While the morphologies of these structures are sometimes indistinguishable from textures assumed by thermotropic (and lyotropic) MLCs, these block copolymer phases are frequently more solid-like in their physical properties. Multiple interconnections between microphase separated domains tend to establish static, 3-dimensional, translationally ordered topologies in the materials. These aspects of the mesogenic block copolymers can be emphasized if, for example, multi-block copolymers -- chains with several distinct monomer compositions along its contour length -- are swollen.(19) Reference 20 provides a current entry to the literature of mesomorphic block copolymers.

The investigations of thermotropic PLCs commenced almost two decades after studies of mesophases employing rodlike polymers(21) and virus particles(22) in solution were initiated. Like their MLC analogues (aqueous solutions of amphiphilic molecules) the polymeric lyotropic liquid crystals which form in solutions of rigid, rodlike polymers (helical biopolymers, and more recently, semiflexible aromatic amides) constitute a distinct class of liquid crystals. Unlike the MLC analogues, however, the lyotropic PLCs are not necessarily stabilized by specific interactions between the polymer chain and the solvent. Rather, mesophase formation is driven by excluded volume effects (i.e., "hard core" repulsive interactions among dense configurations of rodlike particles).

Although instances of lyotropic PLCs predate studies of thermotropic PLCs, as they involved solutions of comparatively esoteric species -- virus particles and helical polypeptides -- studies of these liquid crystals were isolated to a few laboratories. Nevertheless, observations on these lyotropic PLCs did stimulate the first convincing theoretical rationalizations of spontaneously ordered fluid phases (see below). Much of the early experimental work was devoted to characterizing the texture of polypeptide solutions.(23) The chiral polypeptides (helical rods) generate a cholesteric structure in the solution; the cholesteric pitch is strongly dependent on polymer concentration, dielectric properties of the solvent, and polymer molecular weight. Variable pitch (<1 - 100 μm) may be stabilized and locked into the solid state by (for example) evaporating the solvent in the presence of a nonvolatile plasticizer.(24)

More recently, lyotropic PLCs were catapulted into the forefront of polymer research when ultra-high strength polymer fibers were commercialized by spinning mesomorphic solutions of poly(p-aromatic amides), e.g., Du Pont's Kevlar. I have recently reviewed these aspects of PLCs.(25)

MODELING THE NEMATIC-ISOTROPIC TRANSITION

In the late 1940s and early 1950s, observations of liquid crystal phases in solutions of the rodlike Tobacco Mosaic Virus and the helical synthetic polypeptides stimulated two distinct theoretical approaches aimed at describing the "isotropic-nematic" phase transition. Onsager used the viral expansion and predicted the onset of long-range orientational correlations in a gas of impenetrable rods above a critical volume fraction of rods.(26) Flory extended the lattice model successful in describing solutions of flexible polymer chains to the case of rigid rodlike chains and also predicted long-range orientational correlations at high chain concentrations.(27) In both theories the aspect ratio of the rod (the shape anisotropy of the solute) determines the critical volume fraction of rods which, if exceeded, results in a spontaneously ordered phase of quasiparallel rods (a nematic lyotropic phase).

Generally speaking, the virial expansion model, asymptotically exact for very large aspect ratios, is applicable only when the volume fraction of rods is very small (second virial approximation), whereas the lattice model may be more reliable at rod concentrations in typical lyotropic phases. The implications of both models are still being explored today. Inclusion of a distribution of rod lengths, rod flexibility (i.e., modeling semiflexible chains composed of freely jointed linear sequence of connected rodlike segments), and the addition of orientation-dependent attractive interactions are currently under investigation in these two models. A hybridization of thermal versions of the virial and the lattice models to describe PLCs has been reviewed.(28) The recent extension of the lattice model to predict phase equilibria in multicomponent nematics is indicative of the continued interest in this model.(29)

Alternate approaches to modeling semi-flexible PLC phases are now appearing.(30,31) Coarsely, these statistical theories address the influence of the intrinsic flexibility of the chain on the stability of ordered fluid phases. Such modeling may help define the prerequisites for PLC formation suggesting how non-mesogenic polymers may be modified to exhibit mesomorphism and thereby help identify new classes of PLCs.

EXPERIMENTAL CHARACTERIZATION OF NEMATIC ORDER

Essentially all of the techniques developed to characterize MLCs can be applied to PLCs with the realization that phenomena that are dependent on reorientation processes in the liquid crystal must be considered on considerably longer time scales in a PLC. Underlying most of the physical measurements performed on liquid crystals is the relationship between the observed anisotropic properties of the mesophase and orientational order of the mesogen. For uniaxial nematic phases and idealized low molar mass mesogens (cylindrical molecules), this relationship is embodied in the following equation:

$$A_{\parallel} - A_{\perp} = (A_{\ell} - A_{t})S_{z} \qquad (1)$$

The anisotropy of the macroscopic observable physical attributes, $A_{\parallel} - A_{\perp}$ (extracted from measurements along and normal to the nematic director), is proportional to the nematic order parameter S_{z}. The proportionality constant is the anisotropy of the corresponding molecular attribute ($A_{\ell} - A_{t}$). When the longitudinal and transverse molecular attributes A_{ℓ} and A_{t}, respectively, are accessible, the nematic order parameter and its temperature dependence may be determined. Properties such as the molar refractivity, polarizability, magnetic (dielectric) susceptibility, etc., are typical quantities that have been studied to infer the nematic order parameter.

While this general approach has been frequently exploited in MLCs, in fact, complex mesogen symmetries (less than cylindrical) and internal molecular flexibility (rotation about chemical bonds), renders eqn(1) only a coarse indicator of long-range orientational order in the liquid crystal. This is compounded in PLCs of the type shown in Figures 1 and 2 as the idealization of the mesogenic monomers is further complicated by their connectivity in the polymer. In sidechain and linear PLCs, the nematic order is frequently arbitrarily decomposed into two contributions termed the "core" and the "spacer" order parameters. This practice is reinforced by measurements employing experimental techniques which exhibit unequal sensitivities to the core and spacer contributions. (For example, certain x-ray diffraction azimuthal intensity distributions may be dominated by scattering from the core orientational correlations in the mesophase.) NMR labeling techniques (see below) are ideally suited to monitor submolecular behavior; in MLCs the core and the terminal chain can be examined independently. However, it is not common in such MLC materials to speak of "core" and "chain" order parameters separately. Moreover, it is the nematic order of the "average" mesogen in the homogeneous phase that is critical to testing molecular field models.

In both MLCs and PLCs, the appropriate internal averaging over the two distinct fragments of the monomer (core and spacer) must be carried out to derive the longitudinal and transverse attributes of the "average" monomer. Then the idealized quantity $(\overline{A}_\ell - \overline{A}_t)$ may be used in eqn(1) to yield the "nematic order parameter". The importance of using such a composite average attribute of the monomer (core + spacer) to determine the actual nematic order can be readily demonstrated. Consider, for example, the case of magnetic susceptibility measurements wherein aliphatic spacers and aromatic cores have diamagnetic anisotropies of opposite sign. In a homologous series of a linear PLC, it is possible for the observed anisotropy of the phase $(\chi_{\parallel} - \chi_{\perp})$ to decrease with increasing spacer length yet the phase maintain constant nematic order; the apparent decrease is observed because $(\overline{\chi}_\ell - \overline{\chi}_t)$ decreases as the spacer length increases.

An example of the use of this type of internal averaging has recently been reported by Sigaud et al.(32) However, it should be emphasized that despite such averaging of the monomer attribute, there still remain inherent limitations to the use of eqn(1). These limitations originate in approximations utilized to define longitudinal and transverse "average" mesogen properties. In particular, as the covalent attachment of the mesogenic core to the polymer chain may very well exaggerate biaxial molecular motions, the assumption implicit in eqn(1), namely a non-biaxial order tensor ($S_x - S_y = 0$), and the presumed identification of its principal frame (the "art" of defining the symmetry axis or "long molecular axis" in a monomer), further limits the use of eqn(1) in quantitative tests of theory.

As most of the physico-chemical characterization of PLCs may be directly transferred from studies of MLCs and as most of these techniques are considered in the separate contributions to this volume, I single out only one technique; this section is concluded by drawing attention to the recent application of deuterium nuclear magnetic resonance (NMR) to the characterization of the spacer component

of a deuterium labeled linear thermotropic PLC.(33) The labeling --
replacing protons with deuterons -- is an innocuous perturbation of
the mesogen and explicit information from different labeled sites in
the molecule is obtained.

Deuterium Nuclear Magnetic Resonance

In the last decade deuterium NMR (DMR) has emerged as a powerful technique for characterizing alkyl chain flexibility in liquid
crystals. DMR is sensitive to molecular motion because of the
coupling between the nuclear quadrupole moment of the deuteron and
the local electric field gradient colinear with the C-D bond vector.
Liquid crystals are ideally suited to DMR investigations. The rapid
anisotropic molecular motion in these fluids averages the C-D bond
orientation along the liquid crystal director and the residual quadrupolar interaction appears as resolved quadrupolar splittings in
the DMR spectra of labeled molecules. Hence, the residual quadrupolar interactions quantitatively determined with DMR measure the efficacy of the molecular motion averaging the C-D bond vector in liquid crystals.

In the case of labeled, non-rigid molecules, if the relevant
molecular motion may be separated into intra- and extramolecular
contributions, DMR gives inferences about molecular flexibility. To
this end, a considerable DMR effort has been focused on studying
labeled alkyl chains covalently attached to nematic MLCs. Generally, observations show a decrease in the magnitude of the methylene
quadrupolar splittings $\Delta \nu_i$ with increasing distance of the CD_2 unit
(i) from the comparatively rigid and well-oriented mesogen core.(34)

It is clear that this pattern of spacer quadrupolar splittings
will differ for the labeled spacer chain in a linear PLC as the
chain is attached at both of its ends to oriented cores. Figure 3
shows a DMR spectrum of an oriented nematic PLC, a poly(ester-linked
spacer + core) with ten CD units in the labeled spacer. The $\Delta \nu_i$
fall into two groups of unequal intensities. This immediately implies that the all- trans conformation is not the only accessible
spacer conformation, i.e., differential motional averaging (isomerization) occurs along the spacer. Above T_g such motions may be assumed to be rapid on the DMR time scale although the reorientations
may be somewhat restricted in the mesophase relative to an isotropic
liquid. Hence, in the fluid PLC, a coarse interpretation of the ef-
fectiveness of the averaging of the C-D bonds may be derived using
an equilibrium statistical mechanical average over all of the conformations. Following the idealizations underlying the use of eqn
(1), the analogous expression for the observed quadrupolar splittings in an aligned nematic is

$$\Delta \nu = qS_z \langle P_z (\cos\theta) \rangle \qquad (2)$$

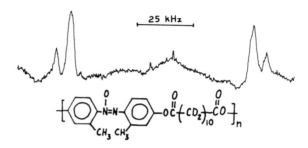

Fig. 3. Deuterium NMR spectrum of the nematic phase of a linear PLC with a labeled spacer chain.

where q is the quadrupolar coupling constant ($2q/3 = 168$ kHz), and θ is the angle that the C-D bond vector in the ith segment makes with the symmetry axis of the mesogen; $< >$ indicates a time average (equivalent to a statistical mechanical average in the rapid motion limit) wherein extra- and intramolecular motion is decoupled.(33,34)

If S_a can be determined independently, this simple exposition enables one to extract coarse information about the internal mobility of labeled sites in the spacer of the PLC (the average orientation of the C-D bond relative to the "symmetry axis", i.e., $<P(\cos\theta)>$, from experimental values of the quadrupolar splittings. The analysis is more complex when the relevant motions are slow.(35) Nevertheless, the distribution of C-D bond orientations and details about the nature of the motion can be ascertained. Via such a DMR analysis, the extent of decoupling of the motions of the mesogenic core from that of the mainchain in sidechain PLCs has been addressed.(36)

The analysis of the spectrum in Fig. 3 suggests that there is increased spacer orientational order relative to that found in chains attached to MLCs. However, the comparatively larger magnitudes of the observed quadrupolar splittings are primarily due to the relatively large S in the PLC. Extreme values of the nematic order at the nematic-isotropic transition in the polymer mesophases are not unexpected. The attachment of the spacer between two cores yields an important difference between a MLC nematic and the PLC phase. Covalently linking mesogens into a linear chain amplifies the dynamic cooperativity within the spacer and results in increased nematic order for certain spacer chain lengths, i.e., in the polymer, orientational order at one core can be propagated intramolecularly through the spacer chain to successive cores enhancing the local order. However, as the thermodynamic measurements indicate,(37,38) the efficacy of such enhanced ordering in linear PLCs appears to be strongly dependent on the number of bonds (even or odd) in the spacer. This point is further discussed in the following section.

SPACER CHAIN BOND CORRELATIONS

In this section an attempt to characterize the spacer chain's propensity to propagate orientational correlations is prefaced with a brief introduction to the established methodology for computing conformationally averaged properties of flexible (finite length) chains. Pre-eminent among these methods is the rotational isomeric state (RIS) model, a single chain model based solely on intramolecular interactions. In the model bond rotation potentials determine the local energetics of chain conformers in the limit of a small number of judiciously chosen discrete values of the dihedral angle variables. Strictly speaking, the RIS model by itself is only applicable to an isolated chain (intermolecular interactions of condensed phases are absent). However, it has been parameterized by Flory and coworkers to give excellent agreement with experimental data; configurationally averaged properties of chains such as molecular dimensions, dipole moments, etc., as a function of both chain length and temperature are nicely described.(39) The RIS description of the chain component of liquid crystal molecules is used for the intramonomer averaging discussed above, and this description is the starting point for molecular-field models of flexible mesogens. I consider it here as a useful route to phenomenological descriptions of PLCs.

The following two parameters are traditionally used to characterize the flexibility of an isolated alkyl chain: 1) the rms end-to-end vector $\langle r^2 \rangle$, a measure of chain deminsions; 2) the RIS probabilities $p_{t;i}$, a measure of the frequency of occurance of the trans state at a particular C-C bond. Both parameters involve an equilibrium statistical mechanical average (conformational average) over all of the conformations $\{\phi\}$ of the chain. They may be computed with a high degree of accuracy using the RIS scheme.

The average of any function of the conformation $f\{o\}$ is given by

$$\langle f\{\phi\} \rangle = \sum_{\{\phi\}} f\{\phi\}\exp(-U\{\phi\}/RT)/Z = \sum_{\{\phi\}} f\{\phi\}P\{\phi\} \quad (3)$$

where Z is the partition function and $P\{\phi\}$ is the conformer probability. The energy of each conformation of the chain $U\{\phi\}$ is well-defined; for alkanes with a 3-state RIS scheme, the only variable in the computation of $\langle f\{\phi\} \rangle$ is the temperature.(39)

For a chain with n-1 bonds, the end-to-end vector $\bar{r}$ connecting the first and last atoms of the chain is defined as the sum of the bond vectors ℓ_i : $\bar{r} = \sum \ell_i$. Figure 4 shows the calculated contraction of dodecane's rms dimensions $\langle \bar{r}\bar{r} \rangle$ from its maximum value in the all- trans state at low temperature; with increasing temperature as the gauche states are populated (increasing flexibility), the chain dimensions rapidly attain the high temperature limiting value.

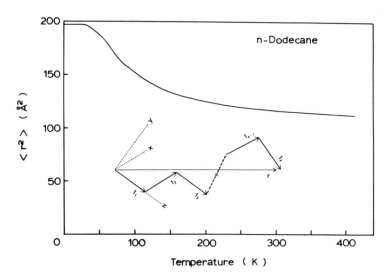

Fig. 4. The rms end-to-end vector of n-dodecane calculated as a function of the temperature using the Rotational Isomeric State conformational average.

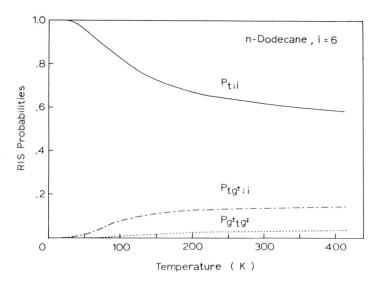

Fig. 5. The singlet trans probability, $p_{t;i}$, the doublet probability p_{tg} and, the kink probabilities $p_{g^{\pm}tg^{\mp}}$ computed for the central bond of n-dodecane are shown as a function of the absolute temperature.

The RIS probabilities also reflect the increase in flexibility with increasing temperature. The singlet probabilities $p_{t;i}$, a measure of finding dihedral angle ϕ in the trans RIS is defined by

$$p_{t;i} = \sum_{\{\phi'\}} \exp(U\{\phi\}/RT)/Z, \qquad (4)$$

where $\{\phi'\}$ signifies a sum over configurations with ϕ_i in the t state. The doublet probability, p_{tg}, the frequecy of finding the consecutive bonds i and i+1 in RIS t and g', respectively, and the "kink" probability, $p_{g't g''}$ can also be computed. These probabilities are shown for the central bond of dodecane as a function of the temperature in Figure 5.

It is clear from Figures 4 and 5 that the chain dimensions and probabilities have essentially reached their asymptotic values at modest temperatures. We should expect, therefore, that in melts of semiflexible PLCs, unless there are extreme extenuating circumstances (i. e., perfect core orientational ordering), the dimensions of a spacer chain and the conformer probability distributions within a spacer chain should be similar to those quantities in an alkane liquid. The attachment of a spacer at both ends will of course perturb the dimensions and conformer probabilities of the spacer. Clearly the magnitude of this perturbation of the spacer is the key to understanding the role of the spacer in a PLC.

Via the spacer chain, a given core may influence the orientational ordering of successive cores in the polymer. From such a perspective, a spacer chain may stabilize a linear PLC according to its inherent ability to propagate orientational correlations from one end to the other. As it is common to try to modify the physical properties of PLCs by changing the nature of the spacer chain (e.g., by introducing heteroatoms into the spacer), it is of interest to see how the intrinsic propensity for propagating orientational correlations are governed by the chemical constitution of the spacer. We examine this aspect of the spacer by computing the bond correlation function (BCF), $\langle P_2(\cos\theta_j)\rangle = \langle(3\cos^2\theta_j - 1)/2\rangle$. This conformationally averaged function of the angle between the first and jth bond of the chain is an indicator of the persistence of orientational correlations down the chain. For the all- trans conformation of an alkyl chain, for example, the BCF will be unity for j = 1,3,5,...; all odd numbered bonds are parallel to j = 1. For j even, the valence angle determines the value of the BCF. In the case of free rotation about successive dihedral angles, the BCF rapidly (j > 5) approaches zero. Positive values of the BCF indicate a preference for bond j to align on average along the direction of bond 1; negative BCF values correspond to average alignment of the bond at right angles to the first bond. The BCF computed for isolated chains may be used to get a qualitative picture of the degree of "coupling" between successive cores in the polymer in the absence of external constraints, and presumably such computations will have some bearing on the behavior of a spacer in a uniaxial condensed PLC phase.

Figure 6 shows the bond correlation function for three types of spacer chains calculated with standard RIS parameters:(39) 1) The alkyl spacer exhibits a monotonic oscillatory decay characteristic of the tetrahedral valence angles and sampling of the three RIS (trans and gauche) having the conventional statistical weighting. 2)The ethylene oxide chain loses correlations more rapidly; the low energy gauche RIS preferred at the C-C bonds is the primary cause of the faster loss of correlation and the reversal of sign for the BCF at $j = 7$.   3) The dimethylsiloxane spacer shows the most abrupt loss of orientational correlation; this is due to the presence of two un-equal valence angles in the chain.

The results in Figure 6 suggest that the cores in a PLC with a dimethylsiloxane spacer chain would behave essentially independent of one another (i.e., approach the behavior of a MLC) after a spacer length of approximately four monomer units (including the decay of the BCF from both ends).   By contrast, approximately twenty methylene units would be necessary for core independence in an alkyl spacer PLC.   Insofar as thermodynamic properties reflect the degree of core orientational correlations, experimental values of transition temperatures as a function of spacer length in linear PLCs (the magnitude of the even-odd effect) seem to bear out the trends computed in Figure 6.(40-42)   While similar kinds of considerations should obtain for sidechain PLCs, the situation is further complicated by

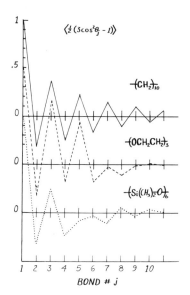

Fig. 6 The conformationally averaged bond correlation functions of three types of chains; the ordinate is shifted for each case.

the intervention of a fragment of the polymer backbone chain. In the sequence: core-spacer-backbone-spacer-core, orientational correlations will be profoundly influenced by the stereochemical idiosyncrasies of the backbone fragment. Lastly, this type of spacer characterization could be extended to PLCs with spacers having simple substituents on the spacer chain. Calculated BCFs for substituted spacers in conjunction with the corresponding synthesis and determination of transition temperatures may help define those features of the polymer which are critical to mesophase formation.

CONCLUDING REMARKS

A limited number of physico-chemical properties of polymer liquid crystals have been considered here. It appears that while there are many similarities between MLCs and PLCs, there are unique properties of the latter which may be conveniently investigated via phenomenological modeling and directed synthetic modifications. Generally speaking, PLCs may provide convenient routes to exploring properties that are not as dramatically altered in MLCs.

Outside of alluding to the importance of PLCs in the fabrication of ultra-high strength organic materials, I have not addressed the applications of PLCs, a major source of the intensive interest in PLCs. The slower time scales for mesophase reorientation will limit the use of PLCs in dynamic electro-optical devices, but the possibility of locking in liquid crystal textures into stable solid polymer films renders PLCs attractive alternatives in static optical devices, e.g., as polarizers, notch filters and non centro-symmetric matrices for non-linear optical phenomena.

REFERENCES

1. D. Chandler, J. D. Weeks and H. C. Anderson, Science 220, 787 (1983).

2. P. Flory, J. Chem. Phys. 17, 303 (1949); Principles of Polymer Chemistry Cornell Univ. Press, Ithaca, N. Y. (1953), p. 602; D. G. Ballard, J. Schelten and G. D. Wignall, Eur. Polymer J. 9, 965 (1973); J. P. Cotton et al., Macromolecules 7, 863 (1974); R. G. Kirste, V. A. Kruse and K. Ibel, Polymer 16, 120 (1975).

3. R. B. Bird and C. F. Curtiss, Physics Today 37, 6 (1984); P. G. de Gennes, Physics Today 35, 40 (1982).

4. H. Kelker and R. Hatz, Handbook of Liquid Crystals, Verlag Chemie, Weinheim (1979).

5. Liquid Crystalline Order in Polymers, A. Blumstein, ed., p. 105, Academic Press, New York (1978).

6. *Polymer Liquid Crystals*, A. Cifferri, W. R. Krigbaum and R. B. Meyer, eds., Academic Press (1982).

7. A. Blumstein and E. C. Hsu, *supra* (ref. 5), p. 105.

8. H. Finkelmann, *supra* (ref. 6), p. 35, and *J. de Chim. Phys.* 80, 163 (1983).

9. H. Ringsdorf and A. Schneller, *British Polymer J.* 13, 43 (1981).

10. E. Perplies, H. Ringsdorf and J. H. Wendorff, *J. Polym. Sci. Polym. Lett. Ed.* 13, 243 (1975).

11. A. Roviello and A. Sirigu, *J. Polym. Sci. Polym. Lett. Ed.* 13, 455 (1975).

12. W. R. Krigbaum *supra* (ref. 6), p. 175.

13. L. Liebert, L. Strzelecki, D. Van Luyen and A. M. Levelut, *Eur. Polym. J.* 17, 71 (1981).

14. G. W. Calundann and M. Jaffe, Proceedings of the Robert A. Welch Foundation Conference on Chemical Research XXVI. SYNTHETIC POLYMERS, Houston, Texas, Nov. 15-17 (1982).

15. S. Chandrasekhar, *Adv. Liq. Crys.* 5, 47 (1982).

16. J. E. Zimmer and J. L. White, *Adv. Liq. Cryst.* 5, 157 (1982).

17. H. Finkelmann, B. Luhman, G. Rehage and H. Stevens, *Liquid Cryst. Order. Fluids* 4, 715, Plenum (1984); and in *J. Colloid Sc.* 56, 260 (1982).

18. B. Gallot *supra* (ref. 5), p. 192.

19. See the following articles in *Macromolecules* 16, (1983): Funabashi et al., p. 1; Isono et al., p. 5; and Matsushita et al., p. 10.

20. J. C. Wittmann, B. Lotz, F. Candau and A. J. Kovacs, *J. Pol. Sci. Pol. Phys. Ed.* 20, 1341 (1982).

21. A. Elliott and E. J. Ambrose, *Discuss. Faraday Soc.* 9, 246 (1950).

22. J. D. Bernal and I. Fankuchen, *J. Gen. Physiol.* 25, 111 (1941).

23. C. Robinson, *Mol. Cryst. Liq. Cryst.* 1, 467 (1966).

24. E. Samulski supra (ref. 5), p. 167.

25. E. T. Samulski, Physics Today, 35, 40 (1982).

26. L. Onsager, Ann. N. Y. Acad. Sci. 51, 627 (1949).

27. P. J. Flory, Proc. Roy. Soc. (London) 234, 60 (1956).

28. A. Yu. Grosberg and A. R. Khokhlov, Adv. Polym. Sci. 41, 53 (1981).

29. C. Counsell and M. Warner, Mol. Cryst. Liq. Cryst. 100, 307 (1983).

30. G. Ronca and D. Y. Yoon, J. Chem. Phys. 76, 3295 (1982).

31. A. Ten Bosch, P. Maissa and P. Sixou, J. Chem. Phys. 79, 3462 (1983).

32. G. Sigaud, Do Y. Yoon and A. C. Griffin, Macromolecules 16, 875 (1983).

33. M. M. Gauthier, R. B. Blumstein, A. Blumstein and E. T. Samulski, Macromolecules 17, 479 (1984).

34. E. T. Samulski and R. Y. Dong, J. Chem. Phys., 77, 5090 (1982); and E. T. Samulski, Israel J. Chem. 23, 329 (1983).

35. H. Geib, B. Hisgen, U. Pschorn, H. Ringsdorf, and H. W. Speiss; J. Ame. Chem. Soc., 104, 917 (1982).

36. Ch. Boeffel, B. Hisgen, U. Pschorn, H. Ringsdorf and H. W. Spiess, Israel J. Chem. (1983).

37. A. Blumstein and O. Thomas, Macromolecules, 15, 1264 (1982) and references cited therein.

38. See also A. C. Griffin and J. S. Havens, J. Polym. Sci., Polym. Phys. Ed., 19, 951 (1981); J. I. Jin, S. Antoun, C. Ober and R. W. Lenz, Br. Polym. J., 12, 132 (1980); L. Strzelecki and D. Van Luyen, Eur. Polym. J., 16, 299 (1980); and K. Iimura, N. Koide and R. Ohta, Prog. Polym. Phys. Japan, 24, 23 (1981).

39. P. J. Flory, Statistical Mechanics of Chain Molecules, Wiley (Interscience), New York (1969).

40. A. Blumstein, J. Asrar, and R. B. Blumstein, Ordered Fluids and Liquid Crystals 4, 311 (1984).

41. L. Bosio, B. Fayolle, C. Friedrich, F. Lanpretre, P. Meurisse, C. Noel and J. Virlet, Ordered Fluids and Liquid Crystals $\underline{4}$, 401 (1984).

42. C. Agulera, J. Bartulin, B. Hisgen and H. Ringsdorf, Makromol. Chem. $\underline{184}$, 253 (1983).

CONFORMATIONAL PROPERTIES OF MACROMOLECULES OF

MESOGENIC POLYMERS IN SOLUTIONS

V.N. Tsvetkov and I.N. Shtennikova

Institute of Macromolecular Compounds
of the Academy of Sciences of the USSR
Leningrad, U.S.S.R.

It has been experimentally found by using a whole complex of physical methods that some polymers can exhibit lyotropic and thermotropic mesomorphism.

The investigations of polymer mesomorphism have been carried out for a wide variety of polymeric systems and for various types of polymers and copolymers[1].

Considerable interest in the study of this phenomenon in polymers is due to various reasons. First, the mesomorphism of a polymer solution plays a major role in the development of some technological properties of polymer materials. For example, the formation of fibers of high strength with an ultra-high modulus is mainly determined by the formation of a liquid crystalline phase in concentrated solutions. Another aspect of this phenomenon which is also of great importance is the possibility of establishing a relationship between the structure of the monomer unit and the macromolecule as a whole and the appearance of mesomorphism in solution or in polymer bulk[2].

This paper deals with the second aspect of the problem: the establishment of the structural features of the monomer unit and the conformation of the macromolecule as a whole determining the appearance of mesomorphism in polymers.

Intramolecular orientational order in polymer molecules was investigated by a comprehensive study of the conformation of these molecules in the unassociated state (i.e., in a dilute solution). These investigations were carried out for macromolecules of various chemical nature with various chemical structures, but exhibiting mesomorphism in bulk or in concentrated solutions. This enabled us to establish the general features of the molecular structure and the conformational properties leading to the mesomorphism of the polymer. Various types of polymer molecules containing mesogenic groups in the side chains and the main chains and those containing no mesogenic groups were considered.

The analysis of conformational properties of macromolecules with various chemical structures was carried out by using a combination of various physical methods of investigation of molecular structure. The methods of molecular hydrodynamics, optics and electro-optics of polymers were all used[3].

Among polymer molecules exhibiting lyotropic mesomorphism the properties of synthetic polypeptides, polyalkylisocyanates, cellulose derivatives and para-aromatic polyamides have been investigated in greatest detail.

The investigation of hydrodynamic properties, sedimentation and translational and rotational friction of dilute solutions, of these polymers using modern theoretical concepts of behavior of macromolecules in solution[4] made it possible to characterize the geometric properties of macromolecules, their size and equilibrium rigidity.

The study of the entire complex of hydrodynamic properties as a function of molecular weight showed that the molecules of mesogenic polymers exhibit considerable equilibrium rigidity characterized by the length of the Kuhn segment A of several hundred angstroms (Table 1).

To provide a more complete characterization of the order of molecular structure of mesogenic polymers, the concept of the degree of intramolecular orientational order S has been introduced[5]. It was shown that this value is a unique function of the degree of equilibrium rigidity of a chain molecule; at equal chain lengths L, the molecules of rigid chain polymers are characterized by a higher degree of orientational order than flexible chain polymers[1,5]. The function s(x), where x=2L/A, has been presented in the analytical form in ref. 1 and is shown graphically in Fig. 1 (curve 1).

In accordance with the definition of the function S(x), and in order to determine experimentally the degree

Table 1. Optical Properties of Some Mesogenic Polymers.

Polymer	$A \cdot 10^8$ (cm)	$-\frac{[n]}{[\eta]} \cdot 10^{10}$ c.g.s.	$\alpha_1 - \alpha_2 \cdot 10^{25}$ (cm^3)	$\Delta a \cdot 10^{25}$ (cm^3)
1	2	3	4	5
Poly(γ-benzyl-L-glutamate)	2000	1100	3650	34
Poly(butyl iso-cyanate)	500	300	4000	15
Poly(chlorohexyl isocyanate)	390	250	3100	17
Ladder poly(phe-nylsiloxane)	200	-160	-1800	-25
Ladder poly)clo-rophenylsiloxane)	300	-300	-4700	-40
Poly(p-phenylene terephthalamide)	280	380	4300	330
Poly(p-amide benzoxazole)	320	430	5760	300
Poly(amide ben-zimidazole)	240	320	4320	360
Poly(p-amide hydrazide)	250	280	3380	200
Polystyrene	17	-13	-145	-18

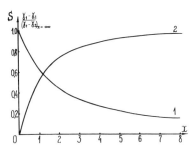

ig. 1. Degree of orientational order S (Curve 1) and relative optical anisotropy $(\gamma_1 - \gamma_2)/\gamma_1 - \gamma_2)_{x \to \infty}$ (curve 2) vs. parameter $x = 2L/A$.

of intramolecular order S, it is necessary to establish
the molecular weight M of the molecule and the degree of
its equilibrium rigidity A. However, experimental data
show[1] that the most direct and highly sensitive measure
of orientational axial and orientational polar order of
structural elements of the molecule is its optical anisotropy and dipole moment.

In this paper, the optical anisotropy and dipole
moment experimentally determined from flow birefringence
and electrical birefringence (the Kerr effect)[3] were used
for the quantitative evaluation of the degree of intramolecular order of the molecule.

The experimental value of flow birefringence (FB)
$[n]/[\eta]$ is determined[3] by the optical anisotropy of the
molecule, $\gamma_1-\gamma_2$, the value of which directly depends on
the degree of orientational axial order in the molecule
(Fig. 1, curve 2)[1] ($[\eta]$ is the intrinsic viscosity of the
solution and $[n]$ is the intrinsic value of flow birefringence).

Experiment shows that in the range of relatively
high molecular weights, M, when the shear optical coeffi-
cient, $\Delta n/\Delta\tau \equiv [n]/[\eta]$, is independent of M, lyotropic
mesomorphic polymers are characterized by a high value
of $\Delta n/\Delta\tau$ and hence by a high optical anisotropy of the
molecule, $\gamma_1-\gamma_2$[3] (Table 1).

Since in the Gaussian range of conformations the
optical anisotropy of the molecule as a whole is determined by that of the components and the degree of equilibrium rigidity of the molecule[6], the following equation
is valid:

$$\frac{[n]}{[\eta]} = \frac{\Delta n}{\Delta\tau} = B(\alpha_1-\alpha_2) = B\beta A = BS_0 \Delta a \qquad (1)$$

where $B=4\Pi(n^2+2)^2/45kT$ is the optical coefficient, n is
the refractive index of the solution, k is Boltzmann's
constant, T is the absolute temperature, $\alpha_1-\alpha_2$ is the
difference between segment polarizabilities, β is the optical anisotropy of unit length of the molecule, Δa is
the optical anisotropy of the monomer unit and S is the
number of monomer units in a statistical segment,

$$\alpha_1-\alpha_2=\Delta a S_0 \qquad (2).$$

The data given in Table 1 conclusively show that
mesogenic polymers exhibit a general property: high op-

tical anisotropy reflecting the high orientational axial order in mesomorphic polymers on a molecular level.

The study of the molecular-weight dependence of FB and orientation angles in combination with hydrodynamic methods makes it possible[7] to evaluate quantitatively the degree of this orientational axial order in the macromolecule.

Fig. 2 shows the theoretical dependence of the relative value of FB $\Delta \equiv ([n]/[\eta])/([n]/[\eta])_\infty$ on $x=2L/A$ for an assembly of rigid random coil molecules with the evaluation of the polydispersity of molecules according to conformations. The points represent experimental values of Δ at the corresponding molecular weights for one of the mesogenic polymers, polyamide benzimidazole[8], in concentrated sulfuric acid (o) and in dimethylacetamide ($\square$).

The coincidence of experimental points and the theoretical curve makes it possible to determine the value of equilibrium rigidity A and optical anisotropy of the monomer unit of the molecule Δa. The use of the values of A, the molecular weights and the dependence $S=f(x)$ allows the determination of the degree of orientational axial order in the molecule.

Table 2 gives the experimental data on the degree of intramolecular orientational axial order S for a series

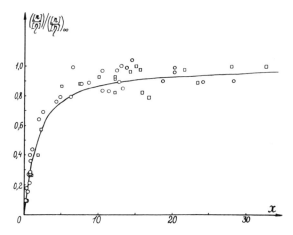

Fig. 2. Values of $([n]/[\eta])/[n]/[\eta]_\infty$ vs. $x=2L/A$ for polyamide benzimidazole in sulfuric acid (o) and in dimethylacetamide ($\square$).

Table 2. Intramolecular Orientational-Axial Order Degree S for Some Mesogenic Polymers.

Polymer	$A \cdot 10^8$ cm	$M \cdot 10^{-3}$	S
1	2	3	4
Poly(γ-benzyl-L-glutamate)	2000	16-160	0.02-0.52
Poly(butyl iso-cyanate)	500	8-80	0.75-0.19
Poly(chlorohexyl isocyanate)	390	12-120	0.63-0.15
Ladder poly(phenyl-siloxane)	200	15-150	0.53-0.08
Ladder Poly(chloro-phenylsiloxane)	300	18-180	0.62-0.15
Poly(p-phenylene terephthalamide)	280	2.5-25	0.54-0.07
Poly(p-amide benzoxazole)	320	1-10	0.65-0.19
Poly(amide benzi-midazole)	240	1-10	0.56-0.10
Poly(p-amide hydrazide)	250	1-10	0.59-0.10
Polystyrene	17		0.04-0.0004

of mesogenic polymers. Apart from the segment length A, the range of changes in molecular weights M and in S corresponding to changes in chain length from 200 to 2000 Å are listed. The data show that the molecules of poly-γ-benzyl-L-glutamate, polyisocyanates and para-aromatic polyamides are characterized by the highest degree of order. The values of S for these polymers are close to those in thermotropic mesomorphic substances[5].

This fact explains the capacity of these polymers for forming a stable lyotropic mesophase in concentrated solutions.

Another physical method, the application of which was of fundamental importance in the analysis of properties of mesogenic polymers, was the Kerr effect - electric birefringence (EB).

As a result of the chain structure of the macromolecules, they may exhibit not only uniaxial order observed in the nematic mesophase but also the orientational polar order. The physical value directly characterizing the orientational polar order in a chain molecule is its dipole moment. If the monomer unit exhibits a dipole moment μ_0 directed along the chain, then the molecule as a whole in a given conformation is characterized by the dipole moment μ_0 [3].

The theory of electric birefringence[3] makes it possible to relate the value of μ to the experimentally determined Kerr constant $K \equiv \lim(\Delta n/cE^2)$ where Δn_E is the value of birefringence in an electric field of strength E, and c is the concentration of the polymer solution.

In the Gaussian range in which $X \to \infty$, the Kerr constant is related to the square of the dipole moment of the monomer unit μ_0, and the optical anisotropy of this unit Δa by the equation

$$K_\infty = \frac{2 \Pi \, NA(n^2+2)(\varepsilon+2)^2}{1215n(kT)^2} \Delta a(\mu_0^2 \cos\theta) S_0^2/M_0 \qquad (3)$$

where ε is the dielectric permittivity of the solution, NA is Avogadro's number, M_0 is the molecular weight of the monomer unit, and θ is the angle formed by the dipole moment of the monomer unit μ_0, and the chain direction.

The study of the Kerr effect in solutions of a series of mesogenic polymers showed a high value of EB coinciding in sign with flow birefringence as follows from eq. (3). This can be seen in Table 3, which lists the Kerr constants K_∞ for some mesogenic polymers.

According to the theory[3], the high electric birefringence (high values of K_∞) corresponds to the values of dipole moments μ of the molecule higher by two or three orders of magnitude than those of non-mesogenic polymers (Table 4). This experimental fact shows that the molecules of mesogenic polymers are characterized not only by a high orientational axial order (high values of $\Delta n/\Delta\tau$ and $\alpha_1-\alpha_2$) but also by a high orientational polar order on molecular level.

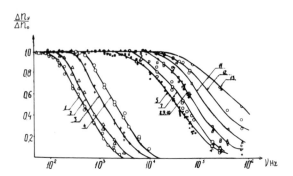

Fig. 3. $\Delta n\nu/_{\Delta n_0}$ vs. ν plot for cellulose carbanilate
fractions. Curves correspond to the following
molecular weights: $1-8.7\times10^5$, $2-6.8\times10^5$,
$3-4.4\times10^5$, $4-2.8\times10^5$, $5-0.84\times10^5$, $7-0.66\times10^5$,
$8-0.44\times10^5$, $9-0.46\times10^5$, $10-0.39\times10^5$, $11-0.34\times10^5$,
$12-0.28\times10^5$, $13-0.24\times10^5$.

Table 3. Magnitude and sign of the characteristic values
of dynamic $(\Delta n/_{\Delta\tau})$ and electric (K) birefringence
of some mesogenic polymers in solution.

Polymer	Solvent	$(\Delta n/_{\Delta\tau})\cdot10^{10}$ $(g^{-1}cm\cdot s^2)$	$K\cdot10^{10}$ $cm^5g^{-1}\left(\dfrac{v}{300}\right)^{-2}$
1	2	3	4
Poly(γ-benzyl glutamate)	Dichloro-ethane	+ 1100	+ 30,000
Poly(butyl-isocyanate)	Tetrachloro-methane	+ 300	+ 25,000
Poly(chloro-hexyl iso-cyanate)	Tetrachloro-methane	+ 250	+ 6,400
Ethyl cellu-lose	Dioxane	+ 30	+ 25
Cellulose carbanilate	Dioxane	– 144	– 110
Ladder poly (phenylsilox-ane)	Benzene	– 160	– 12.5

A characteristic feature of all mesogenic polymers

is the dispersion of electric birefringence observed at various frequencies. It should be noted that the dispersion range of these polymer molecules depends on molecular weight; the relaxation curves are displaced towards higher frequency ν with decreasing molecular weight (Fig. 3)

Fig. 3 shows the experimental dependences of the ratio of electric birefringence Δn_ν at the field frequency ν to the value of Δn_0. Physically, this phenomenon means that the orientation of molecules of mesogenic polymers is of the dipole type; the molecule is oriented in the electric field as a whole.

The investigation of the relaxation process and corresponding relaxation times suggests that the molecules of mesogenic polymers are characterized not only by high equilibrium rigidity (Table 1) but also by high kinetic rigidity of the molecule.

The quantitative measure of kinetic rigidity of the molecule is the time required for a change in its conformation. It is the kinetic rigidity that determines whether the molecule is oriented as a whole by the orientational mechanism or by the deformational mechanism. This is reflected experimentally in the value of relaxation time τ.

The experimentally determined relaxation times of dipole orientation make it possible to obtain an important hydrodynamic characteristic of the molecule: its rotational diffusion coefficient $D_r = \frac{1}{2}\tau$. For kinetically rigid molecules the following equation is obeyed[1,3]

$$D_r = \frac{FRT}{M\,[\eta]\,\eta_0} \tag{4}$$

where R is the universal gas constant, η is the viscosity of the solvent, and F is the numerical coefficient characterizing the shape of the particles. The coefficient F, calculated according to eq. (4) by using the experimental values of D_r for a series of mesogenic polymers, not only lies in the range predicted by the theory for kinetically rigid particles, but also decreases with decreasing M (Fig. 4) in accordance with the change in the shape of the molecule from a random coil to a rod. This is a criterion for the kinetic rigidity of macromolecules of mesogenic polymers.

The mesomorphism of polymers has been tentatively related to various parameters of the macromolecule. The early theories of lyotropic mesomorphism (Onsager and Flory) employed a rodlike molecular mode. However, as the data

Table 4. Dynamo- and electro-optical characteristics of cellulose derivatives. (γ-degree of substitution)

Substitution Group	γ	$\frac{[n]}{[\eta]} \cdot 10^{10}$	$(\alpha_1 - \alpha_2) \cdot 10^{25}$	$\Delta a \cdot 10^{25}$	μD
1	2	3	4	5	6
$-NO_2$	2.9	-70.6	-880	-22	1700
$-C_2H_5$	2.55	25	310	7.8	117
$-\underset{O}{\overset{\parallel}{C}}-\bigcirc$	3.0	-80	-914	-23	165
$-\underset{O}{\overset{\parallel}{C}}-CH\big\langle\bigcirc\bigcirc$	1.45	115.5	1360	34	23
$-\underset{O}{\overset{\parallel}{C}}-NH-\bigcirc$	1.9	-90	-1100	-27	114
Polystyrene for comparison		-13	-145	-18	

in Table 1 show, the molecules of the polymers presented here do not retain rodlike conformation with increasing molecular weight over a wide molecular weight range. In both their hydrodynamic and optical behavior they exhibit finite flexibility which is characterized by finite segment length A.

The change in the value of reduced FB, $\Delta = \left(\frac{[n]}{[\eta]}\right)\Big/\left(\frac{[n]}{[\eta]}\right)_\infty$, shows clearly, for example, for the molecules of mesogenic polymers of amide benzimidazole (Fig. 2) and butyl isocyanate (Fig. 5), that in the low molecular weight range the molecules behave as rigid rods (linear part $\Delta = f(x)$

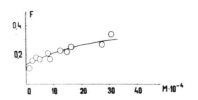

Fig. 4. F vs. M plot for poly(chlorohexyl isocyanate) fractions in tetrachloromethane.

whereas with increasing M the molecules become flexible and after a certain range of M for each polymer (depending on its degree of flexibility) is attained, the molecule adopts the conformation of a draining Gaussian coil (range in which Δ is independent of x).

It may be assumed that the decisive part in the appearance of lyotropic mesomorphism in a concentrated polymer solution is played by the parameter A which determines the degree of its intramolecular orientational axial order. Experimental data show that in contrast to low molecular weight liquid crystals, in this case, the mesomorphism is determined by the length of the part that is of rodlike shape rather than by the chain length as a whole.

The data in Table 1 suggest that for the appearance of lyotropic mesomorphism in a concentrated solution of a rigid chain polymer, the value of A should be of the order of magnitude of several hundreds of angstroms.

Hence, for the development of theoretical concepts of the mesomorphism of polymers, it is necessary to take into account quantitatively the flexibility of the chain in accordance with the data of conformational analysis in dilute solutions.

The study of optico-mechanical properties of one of ladder polyphenylsesquioxanes in bulk has revealed the complex nature of the photoelasticity effect in the films of this polymer[9]. Ladder polymers are characterized by high equilibrium rigidity of double stranded chains[10],

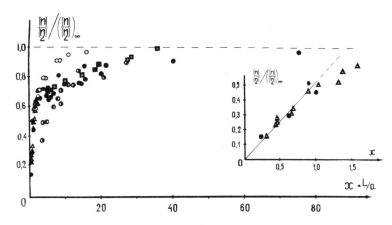

Fig. 5. Values of $([\eta]/[\eta])/([\eta]/[\eta])_\infty$ vs. x for poly(butyl isocyanate) fractions.

but no data on the appearance of lyotropic or thermotropic mesomorphism of these polymers have been available in the literature.

The study of birefringence in films of cyclolinear polyphenylsiloxane (CLPhS) gave the following results: in a polarizing microscope with crossed Nicol prisms it was distinctly seen that the CLPhS films not subjected to external mechanical influences are on the whole optically inhomogeneous. The film consists of birefringent parts, the optical axes of which are oriented randomly over the entire sample.

The anisotropic parts are not separated from each other by sharp boundaries as in polycrystalline solid bodies. In contrast, the local optical anisotropy of the film varies continuously both in value and in the direction of the axis on passing from one part to the other. Hence, the observed optical picture is similar to that for a weakly anisotropic nematic liquid crystal unoriented by the external field. This picture does not undergo any drastic changes in the temperature range from $20°$ to $200°$C.

Quantitative investigations of photoelasticity (birefringence in films due to the action of uniaxial stretching stresses) showed (Fig. 6) that birefringence in CLPhS films is of a complex type.

1. In the range of smallest stretching stresses a slight increase in load leads to a great increase in negative birefringence, Δn, and the higher the experimental temperature, the more pronounced this effect.

2. As the stress of the film is increased further, the increase in negative birefringence Δn comes slower as a result of the development of positive birefringence. The value of Δn_{eff} passes through zero (at room temperature) and goes into the positive range.

The films of common amorphous polymers in the glassy state are optically isotropic. In contrast, spontaneous birefringence observed in the unloaded CLPhS films indicates that a definite orientational order exists in the chain arrangement (in the absence of the crystalline spatial lattice) similar to the nematic order in mesomorphic liquids. This property is a direct consequence of the high rigidity and regular structure of ladder polymer molecules[10] favoring the formation of supermolecular regions with mesomorphic structure.

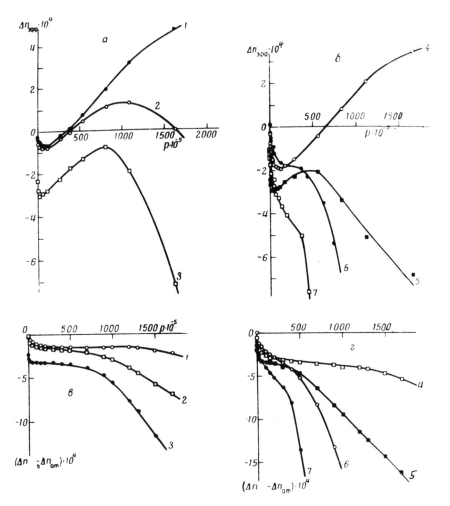

Fig. 6. Equilibrium value Δn_{eff} and $\Delta n_{eff}-\Delta n_{at}$ vs. p plot for CLPhS (p-shear stress dyne cm^{-2}).

However, the part of the polymer substance exhibiting the mesomorphic structure is not large because the maximum observed anisotropy is lower by three orders of magnitude than that of nematic liquid crystals. In other words, it should be assumed that the degree of orientational order of this "quasi-mesomorphic" structure is lower than for common low molecular weight nematics.

These experiments show that the photo-elastic effect in films of a ladder polymer is of a complex molecular and supermolecular nature. The appearance of mesomorphic properties in these films, never observed previously, is due to the rigid chain ladder structure of the molecules.

The experimental information[11] on the lyotropic meso-morphism of a step-ladder polymer, polyacenaphthylene (PAcN) may be understood from the analysis of dynamo-optical data for this polymer. The study[12] of FB of PAcN has made it possible to establish the most probable conformations of PAcN chains corresponding to negative segmental anisotropy $\alpha_1-\alpha_2=-300 \cdot 10^{-25} cm^3$ and the segment length A=(40±5) Å shown in Fig. 7 by comparing similar structures.

In the first conformation (Fig. 7) all possible axes of rotation are parallel as in the crankshaft structure characteristic of para-aromatic polyamides[2]. This molecular conformation would ensure very high rigidity of the entire molecule and high optical anisotropy. In the second conformation (Fig. 7b), the axes of rotation are place at an angle to each other which would lead to high flexibility of the molecule and a lower value of $\alpha_1-\alpha_2$.

Since the rigidity of the entire molecule is not very great - two or three times higher than for common flexible molecules - it is possible that the PAcN molecule consists

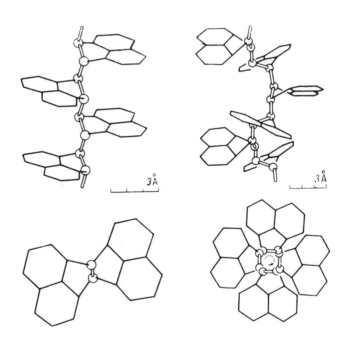

Fig. 7. Conformation of poly(acenaphthylene) chain:
a. extended; b. coiled.

of a random array of rigid and flexible parts. The true polymer in a concentrated solution is a random array of such labile structures. The array of these structures predicted from the data on birefringence and rigidity can ensure the appearance of the mesophase in a concentrated PAcN solution. This principle of molecular structure is similar to the principle of building polymers with mesogenic groups in the main chain separated by flexible portions.

Another class of mesomorphic polymers widely studied at present are comb-like polymers containing mesogenic groups in the side groups leading to thermotropic mesomorphism.

The study of optical and hydrodynamic properties of comb-like polymers was started in the series of alkylacrylates and alkylmethacrylates[1].

The theoretically and experimentally developed analysis of the dependence of optical anisotropy on the length of the side chain made it possible to distinguish the parts played by the side groups and the main chain in the generation of intramolecular order. Even the first investigations of poly(alkylmethacrylates) have led to the conclusion that their alkyl side groups exhibit high equilibrium rigidity greatly exceeding that of the main chain[1].

Important information on the generation of the orientational axial and orientational polar order in comb-like polymers on molecular level was obtained in the analysis of the properties of molecules containing chemical groups in the side chains favoring the formation of the liquid crystalline phase in low molecular weight compounds.

Table 5 gives the optical and electro-optical characteristics of a series of comb-like polymers based on the methacrylic chain. These data show that the rigidity of the main chain increases slightly with increasing length of the side chain and its increasing complexity (introduction of mesogenic groups).

Table 5 lists the number of monomer units in a segment $s = A/\lambda$, where λ is the length of projection of the monomer unit on the axis of the molecule. For polyacrylic chains, λ is 2.5 Å. In the series presented in the Table, the value of optical anisotropy increases sharply: thirty-fold. As shown previously, this reflects the generation of orientational order in the molecule. In this case, attention is drawn to the fact that the increase in

Table 5. Dynamo- and electro-optical properties of mesogenic side chain polymers.

Monomer unit		S_o	$\frac{[\eta]}{[\eta]}$ $\cdot 10^{10}$	$(\alpha_1-\alpha_2)$ $\cdot 10^{25}$	Δa $\cdot 10^{25}$	K $\cdot 10^1$
1		2	3	4	5	6
1	$n=1$	7	1.8	+ 2	+ 0.3	
2	$n=4$	6.8	- 1.0	- 12	- 2.0	
3 $-CH_2-\overset{CH_3}{\underset{O=C-O-C_nH_{2n+1}}{C}}-$	$n=6$	8.9	- 2.3	- 30	- 3.4	
4	$n=8$	8.6	- 3.0	- 40	- 4.6	
5	$n=16$	17.5	-11.5	- 160	- 9.0	
6 $-CH_2-\overset{CH_3}{\underset{O=C-O-\bigcirc-\overset{O}{C}-O-C_{16}H_{33}}{C}}-$		26	-35	- 445	-18	-0.5
7 $-CH_2-\overset{CH_3}{\underset{O=C-O-\bigcirc-\overset{O}{C}-O-\bigcirc-CN}{C}}-$			-20	- 240		
8 $-CH_2-\overset{CH_3}{\underset{O=C-O-\bigcirc-\overset{O}{C}-O-\bigcirc-O-CH_3}{C}}-$			-40	- 500		
9 $-CH_2-\overset{CH_3}{\underset{O=C-O-\bigcirc-O-\overset{O}{C}-\bigcirc-O-C_3H_7}{C}}-$			-25	- 320		
10 $-CH_2-\overset{CH_3}{\underset{O=C-O-\bigcirc-O-\overset{O}{C}-\bigcirc-O-C_6H_{13}}{C}}-$			-30	- 370		-2
11 $-CH_2-\overset{CH_3}{\underset{O=C-O-\bigcirc-O-\overset{O}{C}-\overset{NO_2}{\bigcirc}-O-C_6H_{13}}{C}}-$			-95	-1200		-7
12 $-CH_2-\overset{CH_3}{\underset{O=C-O-\bigcirc-O-\overset{O}{C}-\bigcirc-O-C_9H_{19}}{C}}-$		16	-50	- 600	-40	-6

	1	2	3	4	5	6
13	CH_3 $-CH_2-C-$ $O=C-O-\bigcirc-C-O-\bigcirc-O-C_9H_{19}$	25	-220	-2700	-110	-8
14	CH_3 $-CH_2-C-$ $O=C-O-\bigcirc-C-O-\bigcirc-O-C_{12}H_{25}$		-190	-2350		
15	CH_3 $-CH_2-C-$ $O=C-O-\bigcirc-C-O-\bigcirc-O-C_{16}H_{33}$	24	-220	-2700	-110	-4-20
16	Copolymers 7:3	24	-85	-1050	-44	-6
17	15+5 1:1	24	-55	--680	-28	-3.4
18	CH_3 $-CH_2-C-$ $O=C-O-\bigcirc-C-O-\bigcirc-O-C-$ $-O-C_{16}H_{33}$	50	-390	-4900	-100	
		20	-245	-3000	-150	

the optical anisotropy of the monomer unit, Δa, rather than to the rigidity of the main chain (S_o).

As has been shown for the series of alkyl- and methyl-acrylates and the copolymers of comb-like polymers[1], this fact, in turn, is due to the increasing interaction between the side groups. This interaction increases sharply with the introduction of mesogenic groups, i.e., moieties containing two benzene rings and an aliphatic chain ensuring the comb-like structure of the molecule (polymers 11-18). The introduction of two rings alone without increasing length of the side chain does not cause this great increase in anisotropy.

Only the introduction of mesogenic groups with an important increase in the length of the side chain is accompanied by the appearance of the orientational-polar order in the molecule (i.e., the high Kerr effect which is 10-15 times higher than for the corresponding monomer and always coincides in sign with the Maxwell effect-flow birefringence.

For polymer 6, the value of the Kerr effect K is close in order of magnitude to that for the monomer and

Table 6. Dynamo- and electro-optical properties of polymers with mesogenic side chain groups.

Polymer	S_0	$\dfrac{[\eta]}{[\eta]} \cdot 10^{10}$	$(\alpha_1-\alpha_2) \cdot 10^{25}$	$\Delta a \cdot 10^{25}$	$K \cdot 10^{10}$
1	2	3	4	5	6
1 $n=10$	20	-7.7	-95	-4.7	
2 $-CH_2-CH-$ $n=16$ $O=C-O-C_nH_{2n+1}$	22	-13.4	-164	-7.5	
3 $n=18$	28	-18.9	-232	-8.0	
4 $-CH_2-CH-$ $O=C-O-\bigcirc-\overset{O}{C}-O-\bigcirc$		-42--45	-520		
5 $-CH_2-CH-$ $O=C-O-\bigcirc-\overset{O}{C}-O-\bigcirc-O-CH_3$		-43	-520		
6 $-CH_2-CH-$ $O=C-O-\bigcirc-\overset{O}{C}-O-\bigcirc-O-C_2H_5$		-50--54	-630		
7 $-CH_2-CH-$ $O=C-O-\bigcirc-\overset{O}{C}-O-\bigcirc-O-C_3H_7$		-65--70	-780 -840		
8 $-CH_2-CH-$ $O=C-O-\bigcirc-\overset{O}{C}-O-\bigcirc-O-C_4H_9$		-60--70	-720 -840		
9 $-CH_2-CH-$ $O=C-O-\bigcirc-\overset{O}{N=N}-\bigcirc$	24	-36	-450	-19	
10 $-CH_2-CH-$ $O=C-O-\bigcirc-\overset{O}{N=N}-\bigcirc-O-C_4H_9$	24	-40--47	-510	-21	
11 $-CH_2-CH-$ $\bigcirc-NH-\overset{O}{C}-\bigcirc-O-C_9H_{19}$	25	-210	-2500	-140	-140
12 $-CH_2-CH-$ $O=C-O-$ (steroid)	20	-25	-300	-15	

also for other low molecular weight liquids with a similar
molecular structure (e.g., amorphous-liquids para-azoxy-
anisol). However, for polymer 6, K is negative (whereas

for monomers it is positive) and coincides in sign with FB. This implies that for all these macromolecules the longitudinal axes of the side groups are oriented normally to the field (and in monomers they are oriented along the field). This fact means that the orientation of the side chains is not quite free but correlates with that of the main chain.

Table 6 gives the data for a series of comb-like polymers. As before, the total length of the side chain plays an important part in the formation of intramolecular order. A characteristic feature of electro-optical properties of 11-18 polymers (i.e., the polymers in which the interaction between side groups leads to high intramolecular order) is the dependence of the Kerr constant K on molecular weight M. Fig. 8 shows typical plots for polymer 15 (Table 5) at two temperatures and polymer 12 in tetrachloromethane. This fact implies that the orientation of the polar groups of these polymers in an electric field occurs by the mechanism of large-scale intramolecular motion characteristic of rigid-chain polymers. A polymer with one benzene ring does not exhibit this dependence.

Nonequilibrium electro-optical properties of polymers with mesogenic side groups are also very distinctive. In alternating electric fields, low-frequency dispersion of the Kerr effect is observed. The range of dispersion depends on the molecular weight of the fraction, just as for lyotropic polymers. Fig. 9 shows the dependence of the relative value of the Kerr constant at the field frequency ν and at $\nu=0$ on the field frequency for solutions of

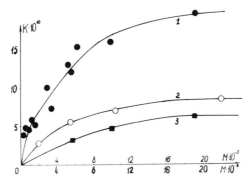

Fig. 8.   K vs. M plot for solutions of polymer 15 (Table 5) at two temperatures (curves 1,3) and 12 (curve 2) in tetrachloromethane.

polymer 13 in tetrachloromethane (Table 5).

Each curve is characterized by the average time of dipole relaxation which depends on molecular weight as follows[13]:

$$\tau = GM\,[\eta]\,\eta_o\,/RT \qquad\qquad (5)$$

This dependence of τ on M implies that the major part is played by the orientational mechanism of solution polarization. However, with increasing molecular weight the intramolecular mobility of the chain appears and the contribution of deformation effect increases. This is revealed in the decreasing dependence of τ on M and is experimentally reflected in the decrease in coefficient G in eq. (5).

Fig. 9b shows the change in the coefficients G calculated from the experimental values of τ, M and $[\eta]$ and eq. (5) for fractions of polymer 13 in tetrachloromethane. Broken straight lines in Fig. 9b represent the theoretical values of G for strong hydrodynamic interaction in the chain: 1 $G_1 = 0.422$ - the first mode - the motion of a kinetically rigid macromolecule as a whole (one kinetic segment);

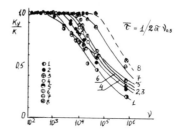

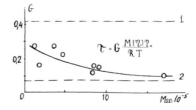

Fig. 9. a) Relative Kerr constant Kν/K vs. frequency of the electric field ν for polymer 13 fractions (Table 5) in tetrachloromethane; b) Coefficient G (eq. 5) vs. M plot of polymer 13 fractions.

2 G_2=0.071 - third mode - rotation of parts of a three-kinetic segment chain.

This means that with increasing M, smaller scale motion than that of the whole molecule appears (it might be assumed that in the investigated molecular weight range, the molecules of polymer 13 contain up to three kinetic segments). Probably, these kinetic segments including large portions of the main chain are formed as a result of the interaction between the side groups and mesogenic parts. With increasing M the properties of the molecules of this polymer vary from those of a kinetically rigid one to those of a a kinetically flexible coil.

Experimental data show that the ordering of the macromolecule is affected not only by the structure of the mesogenic group but also by the type of its insertion into the side chain (Table 7).

In polymers 1 and 6, the mesogenic group is separated from the main chain by a flexible polymethylene spacer. For these polymers, the optical anisotropy and the Kerr constant are lower by one order of magnitude than for polymers with a similar structure of the mesogenic side chain but without a flexible unit.

In polymer 1, the Kerr effect coincides in sign with the Maxwell effect, which means that the orientation of the side chains in an electric field correlates with the direction of the main chain. However, both the small values of K and $[n] / [\eta]$ and the absence of dispersion of the Kerr effect show that the chains of polymers 1 and 6 exhibit no high orientational axial and orientational polar orders characteristic of polymers with mesogenic side groups. In these polymers (1 and 6), the Kerr effect is due to small scale chain motion; the kinetic elements are much smaller in size and their mobility is much higher than, for example, for polymer 7.

This means that the separation of the mesogenic part of the side groups from the main chain by a spacer (e.g., by a polymethylene part) leads to a sharp decrease of the correlation in orientation of the side group and the main chain and, as a result, to a decrease in intramolecular order. Moreover, the weakening of the orientational correlation between the main chain and mesogenic side groups of the comb-like molecule in the bulk of the polymer facilitates the generation of orientational order between the side groups of different molecules and thus favors the formation of the mesomorphic structure in polymer bulk.

Table 7. Influence of spacer in side chain group on optical properties of mesogenic polymers.

Polymer	$\frac{[n]}{[\eta]}\cdot 10^{10}$	$K\cdot 10^{10}$
1 $\begin{array}{c} CH_3 \\ -CH_2-C- \\ O=C-O-(CH_2)_{10}-\overset{O}{\overset{\|}{C}}-O-\langle O\rangle-\overset{O}{\overset{\|}{C}}-O-\langle O\rangle-O-C_4H_9 \end{array}$	-6.5	-0.8
2 $\begin{array}{c} -CH_2-CH- \\ O=C-O-\langle O\rangle-\overset{O}{\overset{\|}{C}}-O-\langle O\rangle-O-C_4H_9 \end{array}$	$-60-70$	
3 $\begin{array}{c} -CH_2-CH- \\ O=C-O-\langle O\rangle-N\overset{O}{\overset{\uparrow}{=}}N-\langle O\rangle-O-C_4H_9 \end{array}$	$-40-47$	
4 $\begin{array}{c} CH_3 \\ -CH_2-C- \\ O=C-O-\langle O\rangle-O-\overset{O}{\overset{\|}{C}}-\langle O\rangle-O-C_3H_7 \end{array}$	-25	
5 $\begin{array}{c} CH_3 \\ -CH_2-C- \\ O=C-O-\langle O\rangle-\overset{O}{\overset{\|}{C}}-O-\langle O\rangle-O-CH_3 \end{array}$	-40	
6 $\begin{array}{c} CH_3 \\ -CH_2-C- \\ O=C-O-(CH_2)_2-O-\langle O\rangle-\overset{O}{\overset{\|}{C}}-O-\langle O\rangle-O-CH_3 \end{array}$	-10	
7 $\begin{array}{c} CH_3 \\ -CH_2-C- \\ O=C-O-\langle O\rangle-\overset{O}{\overset{\|}{C}}-O-\langle O\rangle-O-C_9H_{19} \end{array}$	-220	-8

Hence, the analysis of conformational properties of an extensive class of comb-like polymers with mesogenic side groups showed that the appearance of thermotropic mesomorphic structures in the melts of these polymers is determined by the spontaneous formation of liquid crystalline fragments on a molecular level as a result of the interaction between the side groups.

At present, polymers with mesogenic groups in the main chain are being built according to the same principles as comb-like polymers with mesogenic groups in the side chain. Their important components are aromatic rings and ester groups. The knowledge of the degree of rigidity of the ester group was considered to be important for the analysis of the conformation of these polymers.

It might be assumed, on the basis of indirect evidence, that the equilibrium rigidity of the chains of aromatic polyesters is relatively high because the compounds consisting of a combination of ester groups and para-aromatic rings tend to form the liquid crystalline phase. However, no information on the equilibrium rigidity and molecular structure of these polymers has been available because they are insoluble in both organic solvents and strong acids.

These difficulties could be avoided by using copolymers containing polyesters as rigid fragments "diluted" with flexible chain fragments. The optical properties (flow birefringence, $[n]/[\eta]$) have been investigated for three component copolymers containing the following components:

N   HO—◯—◯—OH

T   ClOC—◯—CO—Cl

D   HO—◯—C(CH₃)₂—◯—OH

The flexibility of the copolymers is determined by the presence of dimethylene groups $-C(CH_3)_2$. The equilibrium rigidity of their molecules and the dependence of $[n]/[\eta]$ on the composition of the copolymer have been determined[14] for a series of these copolymers containing components N, T, D. The chain rigidity of a para-aromatic polyester has been determined by extrapolation of the value of $1/[n]/[\eta]$ (inversely proportional to the degree of equilibrium rigidity A) to the 100% content of the rigid component[14]. The segment length A=500-800 Å found from the intercept with the ordinate (Fig. 10) corresponds to the rigidity of para-aromatic polyamides and is determined by the rigidity and complanarity of the ester group, just as in the case of the planar trans-structure of the amide group[2].

The conformational properties of polymer molecules with mesogenic groups in the main chain were studied in the series of alkylene-aromatic thermotropic polyesters in which the length of flexible methylene parts was varied: polydeca-, penta- and tetramethylene-tere-phthaloyl-di-para oxybenzoate (P-IOMTB, P-5MTB, P-4MTB).

Table 8 gives the values of flow birefringence,

Table 8. Optical characteristics of samples of polyethy-
leneterephtalate and poly-deca- (penta- and
tetra-) methylene-terephtaloyl-para-oxybenzoate
in dichloroacetic acid.

Monomer Unit	$\dfrac{[\eta]}{[\eta]}$ $\cdot 10^{10}$	$\alpha_1-\alpha_2$ $\cdot 10^{25}$	A Å	A Å$_F$
1	2	3	4	5
1 $-CH_2-CH_2-O-\overset{O}{\underset{\parallel}{C}}-\langle O\rangle-\overset{O}{\underset{\parallel}{C}}-O-$	+3.9	48.7	15	
2 n=10 $-(CH_2)_n-O-\overset{O}{\underset{\parallel}{C}}-\langle O\rangle-O-\overset{O}{\underset{\parallel}{C}}-\big(\langle O\rangle-\overset{O}{\underset{\parallel}{C}}-O-\big)_2$	+16.4 +16.0 +14.7	200	50	60
3 n=5	+20.8	250		
4 n=4	+22.3	280		

[η] / [η] , obtained for these polymers in dichloracetic
acid and their structural formula. This Table also lists
the corresponding values of [η] / [η] for polyethylene terephthalate (PETPh). The structural formulae indicate
that PETPh ahd P-IOMTB have approximately the same percentage of rigid (phenyl ring and ester group) and flexible (methylene groups) parts or the same percentage of
para-aromatic rings in the repeating unit. This fact
suggests that the anisotropy per unit length is equal for
these two polymers.

In accordance with eq. (1), the ratios of the [η] / [η]
values for P-IOMTB and PETPh show that for p-10 MTB the
length of the statistical Kuhn segment A is four times
greater than for PETPh and is equal to 60 Å. This value
is in good agreement with the results of the evaluation
of A from hydrodynamic investigations[14].

This increase in the equilibrium rigidity of the investigated polymers with mesogenic groups in the main
chain is probably due to the fact that at an approximately
equal percentage of para-aromatic rings in the repeating
unit in P-IOMBT the rigid parts alternating with flexible
parts are much longer than in PETPh. It is these long
rigid parts that ensure the mesomorphism of this class of
polymers. As can be seen from the FB data, a decrease in
the length of flexible methylene parts does not lead to
an important change in the conformation of the molecule
as compared to P-IOMBT.

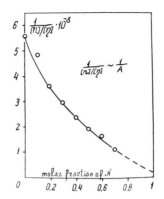

Fig. 10. Dependence of reduced birefringence ($[n]$ / $[\eta]$)$^{-1}$ on the molar fraction of the component N in the copolymer.

P-IOMBT in dichloroacetic acid was investigated in an alternating electric field. Experimental data show that the molecules of this polymer, just as those of other mesomorphic polymers, are characterized by a relatively high Kerr constant and the corresponding high dipole moment and orientational polar order.

The relationship established in the analysis of conformational properties of mesogenic polymers on molecular level made it possible to formulate the principle of synthesis of polymers capable of forming a liquid crystalline phase. The main principle of the synthesis is the introduction into the macromolecule of "rigid" moieties favoring the formation of liquid crystalline structures alternating with flexible fragments. This principle is valid in the introduction of mesogenic groups into the main chain and in that of "rigid" mesogenic groups into the side chains of comb-like polymers. Molecular analogies of low molecular weight liquid crystalline compounds may be used as "rigid" fragments.

This principle provides ample possibilities for obtaining the desired conformations and degrees of order of the molecule and hence provides an approach to the formation of new polymeric liquid crystals with variable properties.

REFERENCES

1. V.N. Tsvetkov, E.I. Rjumtsev, I.N. Shtennikova, Liquid Crystalline Order in Polymers, A. Blumstein, Ed., Academic Press, N.Y. (1978).

2.   V.N. Tsvetkov, I.N. Shtennikova, Macromolecules, <u>11</u>,
     N 2, 306 (1978).
3.   V.N. Tsvetkov, L.N. Andreeva, Advances in Polymer
     Science, <u>39</u>, 95 (1981).
4.   V.N. Tsvetkov, Vysokomol. Soed., <u>A25</u>, N 8, 1571 (1983)
5.   V.N. Tsvetkov, Vysokomol. Soed., <u>A11</u>, N 1, 132 (1969).
6.   W. Kuhn, H. Kuhn, Helvetica Chim. Acta, <u>26</u>, 1394
     (1943).
7.   V.N. Tsvetkov, Dokl. Ak. Nauk, <u>266</u>, 670 (1982).
8.   I.N. Shtennikova, T.B. Peker, V.N. Tsvetkov, et al.,
     Vysokomol. Soed., <u>A23</u>, N 11, 2510 (1981); <u>A25</u>,
     N 8, 1643 (1983).
9.   V.N. Tsvetkov, K.A. Andrianov, N.G. Vitovskaja et al.,
     Vysokomol. Soed., <u>A14</u>, N 12, 2603 (1972).
10. V.N. Tsvetkov, Makromol. Chemie, <u>160</u>, 1 (1972).
11. Sh.A.A. Aharoni, J. Macromol. Sci. Phys., <u>21(1)</u>,
     105 (1982).
12. V.N. Tsvetkov, M.G. Vitovskaja et al., Vysokomol.
     Soed., <u>A24</u>, N 11, 2275 (1982).
13. N.V. Pogodina, V.N. Tsvetkov, Vysokomol. Soed., <u>A24</u>,
     N 11, 2275 (1982).
14. V.N. Tsvetkov, L.N. Andreeva et al., Europ. Polm. J.
     (in press).
15. L.N. Andreeva, E.V. Beljaeva et al., Vysokomol. Soed.,
     (in press).

MOLECULAR THEORY OF MESOMORPHIC POLYMERS

A. Ten Bosch, P. Maissa, and P. Sixou

Laboratoire de Physique de la Matière Condensée
Parc Valrose
06034 Nice Cedex

INTRODUCTION

Molecular theories of mesomorphic polymers may serve to give a better insight into the mechanisms involved in the formation of ordered phases in macromolecular liquids. The importance of parameters such as chain length, chain flexibility... can be explored and the results used in synthesis of new systems with well-defined characteristics. Studies of the influence of solvents on the mesomorphic phase transition and the possibility of ordering by external fields (electric, magnetic, flow) are problems of technological interest in the spinning of fibers and in the formation of new polymer blends.

In the present paper, we first review briefly the rigid rod models for liquid crystalline phase transitions. In these models, emphasis is placed on the anisotropic form and on the orientation dependent intermolecular interactions between rigid particles. Conformational studies on isolated chains have shown that liquid crystalline polymers are rather semi-rigid in character although only a narrow range of deformations is possible due to intrachain interactions. The effect of chain flexibility on the formation of liquid crystalline phases has been pointed out both experimentally and theoretically[1].

The chain flexibility can be a function of temperature or solvent concentration and possible conformational changes in mesomorphic polymers have been both predicted[2] by theoretical calculations and observed in experiment. The elastic chain model[3] takes intrachain interactions into account in a very simple way but neglects fine details of macromolecular structure. We present some results within this model for liquid crystalline behaviour and finally discuss the effects of external ordering fields.

RIGID ROD MODELS

A transition to an ordered state can occur in a fluid of small rigid rods of sufficiently high concentration due to steric repulsion alone, as was shown by Onsager[4]. In the lattice model of Flory[5], strong steric interaction is imposed by the requirement that the monomers occupy only lattice sites. Repulsive hard rod interaction between spherocylinders has also been investigated in continuum models[6,7] which are not restricted to small sets of discrete orientations and improve the quantitative agreement with experiment[8]. Maïer and Saupe[9] first pointed out the importance of attractive orientational interactions such as the induced dipolar forces. Later work included both the hard core repulsion and Van der Waals attraction in a mean field approximation. The total effective pseudopotiential is found by expansion of the intermolecular interaction in spherical harmonics for cylindrical rods. The anisotropic part therefore includes all interactions with the symmetry of the quadrupolar Legendre polynomial $P_2(\cos\theta)$.

$$V(\theta) = V_o - V_1 \, S \, P_2 \, (\cos\theta) \quad + \ldots$$

These theories all predict a first order nematic-isotropic phase transition, and a weakly temperature dependent order parameter. In rigid rod Maïer-Saupe theory, the order parameter is given by the angle of the rod to the direction of prefered orientation (fig. 1a) :

$$S = \left\langle \frac{3}{2} \, \cos^2\theta - \frac{1}{2} \right\rangle \text{ where } \langle \, \rangle = \text{thermodynamic}$$

average.

At the transition S_c is a universal constant (0.43) and the transition temperature, T_c is given simply by the parameter V_1 with

$$\frac{V_1}{KT_c} = 4.55$$

The theory extended to solutions[10] leads to phase diagrams quasi-linear in the liquid crystal concentration. This has indeed been found in many compatible systems.

In a first step towards a comparable theory of liquid crystalline polymers, we study the effect of inherent chain flexibility with a Maïer-Saupe theory.

SEMI-FLEXIBLE L.C.

a) Flexibility

In the elastic chain model of polymers, the polymer is described by a continuous elastic line or space curve r (n) where n denotes the distance to the origin along the backbone (fig. 1b).

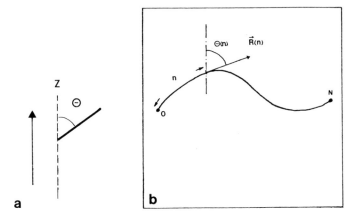

Fig. 1 : description of the configuration of a rigid rod a)
and of a worm-like chain b)

All distances will be given in monomer units. The degree of polymerization is denoted by N . Three elastic constants can be used to
measure the elastic energy of deformation of the chain : stretch,
bend and twist[11]. In the following, we concentrate on molecules
for which the bend elasticity (constant B) is dominant. This leads
to the wormlike chain model[3], with an end-to-end distance
$<R^2>$ = qN ($q/N \rightarrow 0$) and q = B/kT, is here the persistence length .
In this model, the chain elasticity is automatically given by the
value of the persistence length of the isolated chain and is inversely proportional to the temperature.

The worm like chain model has been' applied to interpret data on
intrinsic viscosity[12], light scattering[13]... in dilute solutions.
These measurements can be used to fix approximately the value of q
for a given polymer system, although the solvent often affects the
numerical value due to excluded volume interactions. A connection
to the flexibility introduced in the lattice model has been given[14].

b) Intermolecular interactions

The introduction of chain flexibility into the mean field continuum model is straight forward[15]. The order parameter is now defined along the chain curve (fig. 1 b).
Locally $S(n) = \frac{1}{2}$ $<3 \cos^2 \theta (n) - 1>$ where θ (n) is the tangent angle   at point n. As an average over the entire
chain, per monomer $S = 1/N \int S(n)dn$. The potential at n is then
$V(r(n)) = V_0(n) - V_1(n)S(3/2\cos^2 \theta(n)-1/2)$
If only orientational dependence is considered, all thermodynamic
properties of interest can be calculated from the weight function

G $(RR'N)^{16}$ where

$$G \ (RR'N) \ = \ \int \mathbf{D}r(n) \ P \ (r) \quad \delta r(0) \ - \ R) \quad \delta \ (rN) \ - \ R')$$

and $\quad P(r) \quad = \quad \exp \left(- \dfrac{1}{KT} \right) \quad \int dn \left[\left\{ \dfrac{\mathbf{d \ r}}{\mathbf{d \ n}} \right\}^2 \dfrac{B}{2} \ + \ V \ (n) \right]$

is the Boltzmann probability which includes bending elastic energy,
and the orientation dependent chain attraction. This function can
be shown to fulfill Schrödinger equation :

$$\left[\dfrac{\delta}{\delta n} - \dfrac{1}{B \ KT2} \ \Delta_R \ + \ V(R) \right] G \ (RR' \ n) \ = \quad \delta(R \ - \ R') \quad \delta(n)$$

A solution is found in an expansion in eigenfunctions[17]. The spherical symmetry suggest use of spherical harmonics. This leads finally to a matrix equation which can be solved numerically. In another
approach[18] which has proven successful in calculation of phase diagrams , a Landau de Gennes expansion for semi-rigid polymers has
been derived. The partition function $Z = \int \mathbf{D}r \ P \ (r)$ and free energy $F = - KT \ln \ Z$ are expanded in the order parameter and

$$F \ = \ F_0 \ + \ a_2 \ S^2 \ + \ a_3 \ S^3 \ + \ a_4 S^4$$

where the expansion coefficients are given by expressions in the
parameters q, V_1, N of the model. The order parameter is found from
$\delta F / \delta S = 0$. The transition (or in mixtures, pseudo-transition)
temperature T_c follows from the additional condition : $F = F_0$
The rigid rod case $(B \rightarrow \infty \ )$ corresponds to the original Maier-
Saupe model[19].

c) Results

Pure L.C. polymer

The effect of the flexibility has been studied for a fixed
chain length[20]. The order parameters and transition temperatures
are found to decrease with increasing flexibility (fig. 2). The
temperature dependence of the peristence length can be of importance. From the wormlike chain q $\sim$ 1/T , although excluded volume
effects could modify this simple behaviour.

From a discussion of the limiting cases q/N $\rightarrow$ o, N/q $\rightarrow$ 0 the
chain length dependence of the critical properties (order parame-
ter, transition temperature) for a constant value of q can be discussed, for example initial increase and then flattening at high
degree of polymerization of the transition temperature T_c was shown
to occur. This has indeed been found in many liquid crystalline
polymers[21].

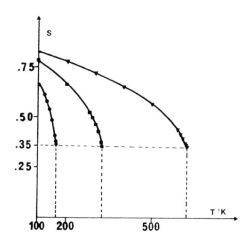

Fig. 2. Order parametr S of the polymer as a function of
temperature for different values of the polymer chain
flexibility. The degree of polymerization N = 100 and
the interchain interaction energy is V_1 = 0.4 kcal/mole
The chain flexibility is given by the worm-like persistence length q (● q = 5, ■ q = 10, ▼ q = 20).

Solutions

Phase diagrams in non-interacting solvents can also be studied[22]. Here due to the temperature dependence of the persistence
length $T_c \sim \sqrt{x} \, T_c^\circ$. This can also be given as a critical
concentration.

$$x_c \sim \left(\frac{T_c}{T_c^\circ} \right)^2$$

where T_c° is the pure transition temperature (fig.3)

In a solution with a (Maïer-Saupe) rigid rod liquid crystal[17]
the phase diagrams are non-linear with a plateau or dip towards
the rigid rod component (fig. 4). The Landau-de Gennes calculation
is more rapid and gives 3% agreement. The order parameter per monomer of the semi-rigid polymer is smaller than that of the rigid
rod. The local order parameter S (n) illustrates ordering along the
chain and shows a large plateau in the middle of the chain with a
sharp decay at the free ends. We could also demonstrate induced rigidity in the liquid crystalline phase with a large increase in
the end-to-end distance at T_c.

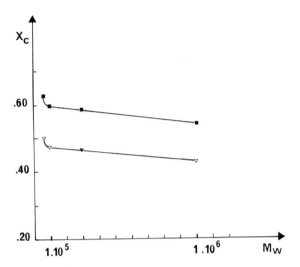

Fig. 3 . Critical concentration as a function of molecular
weight for hydroxypropylcellulose in solution in DMAC
at different temperatures   T = 100°C (■), 60°C (▼)

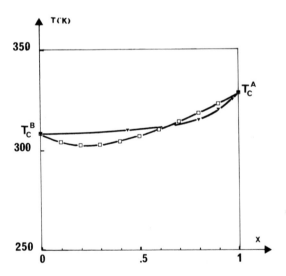

Fig. 4 . Phase diagram for mesomorphic phase transition for a
liquid crystal polymer A (degree of polymerization
N = 100, persistence length q = 10) in solution in a liquid
crystal solvent B (▼)   exact (□) Landau de Gennes
approximation.

External fields

Finally quadrupolar external fields can be included in the calculation[23]. Examples are elongational flow, magnetic or electric
fields. An additional external potential is then present

$$V_e (n) = -V_e \quad (\frac{3}{2} \quad Cos^2 \; \theta \; (n) - \frac{1}{2} \;)$$

where V_e = fg elongational flow

$$= \frac{1}{3} \frac{\Delta \varepsilon}{4\pi} E^2 \qquad \text{electric field E (non polar, zero conductivity)}//z$$

$$= \frac{1}{3} \Delta X \; H^2 \qquad \text{magnetic field H}//z$$

Here f is the coefficient of friction, g the velocity gradient
$\Delta \varepsilon = \varepsilon // - \varepsilon \perp$ and $\Delta X = X // - X \perp$ anisotropy of dielectric constant or susceptibility.
We find an increase in the transition temperature linear in Ve, and
a continuous "second order" transition above a certain critical external field. The external field can cause ordering in the isotropic phase (fig. 5). As an application, the magnetic birefringence in a isotropic liquid crystalline polymer melt has been calculated and good agreement with measurements of Maret[24] is found.

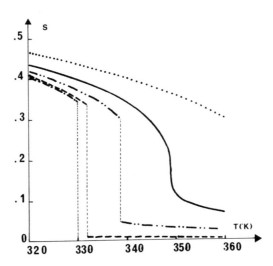

Fig. 5 Order parameter S as a function of temperature on variation
of the ratio Ve/v_1 (-.-.-) 0, (-----) 10^{-3}
(-..-..-) $45 \cdot 10^{-3}$,(——) Vec/v_1= 10^{-2}, (......)
$2.4 \cdot 10^{-2}$. The chain parameters are persistence length
q = 10, degree of polymerization, N = 20,
T_c = (Ve = 0) = 330 K.

CONCLUSIONS

Many qualitative effects found experimentally in semi-rigid polymers have been reproduced in this simple model which presents certain advantages over the lattice model. Chain flexibility is easily included and can be varied continously between the rigid rod and flexible polymer limits. More complex conformations such as helical chains could also be investigated and any changes in conformation such as induced rigidity can be shown to occur. Local properties of experimental interest can be calculated ; for example, the local order parameter along the chain backbone. No restrictions on the orientation or the order parameter are needed and a simple relation to the earlier rigid rod Maïer-Saupe theory can be given. Non-uniform phases and cholesterics can be treated within the same model.   The problems which remain to be investigated are :

- inclusion of steric excluded volume effects. Consequences would be more information on biphasic regions, incompatibility, screening, (chain length dependence of the potential) interfaces..

- Side chain polymers : differences and similarities.

- Dynamic properties (viscosity) should be explored within this model, and compared with results on rigid rod models[25].

REFERENCES

1. P.J. Flory, Proc. Roy . Soc. A 234, 60 (1956) .
2. P. Pincus, P.G. de Gennes, J.Polym. Sci. 65, 85, (1978)
3. N. Saito, K. Takahasi, Y. Yunoki, J. Phys. Soc. Jap., 22, 219, (1967).
4. L. Onsager, Ann NY Acad. Sci., 51, 627, (1949).
5. P.J. Flory, Proc. Roy. Soc. A234 , 73, (1956).
6. M.A. Cotter, Mol. Cryst. Liq. Cryst. 97, 29, (1983).
7. J.G. Ypma, G. Vertogen, Phys. Rev. A 17, 1490, (1978).
8. M. Nakagawa, T. Akahane, J. Phys. Soc. Jap., 52, 399 (1983).
9. W. Maïer, A. Saupe, Z. Naturforschung 14a, 882 (1959).
10. R.L. Humphries, P.G. James, G.R. Luckhurst Symp. Faraday Soc. 5, 107 (1971).
11. A. Miyake, Y. Hoshino, J. Phys. Soc. Jap. 52, 399 (1983).
12. S. Dayan, P. Maïssa, M.J. Vellutini, P. Sixou, Polymer 23, 800, (1982).
13. F. Fried, G. Searby, M.J. Seurin, S. Dayan, P. Sixou, Polymer, 23, 1755, (1982).
14. S. Dayan, P. Maïssa, M.J. Vellutini, P. Sixou, J. Polym. Sci., 20, 33 (1982).
15. F. Jähnig, J. Chem. Phys. 70, 3279 (1979).
16. K.F. Freed, Adv. Chem. Phys. 22, 1 (1972).
17. A. Ten Bosch, P. Maïssa, P. Sixou, J. Chem. Phys., 79, 3462 (1983).
18. A. Ten Bosch, P. Maïssa, P. Sixou, J. de Phys. (France) 44, 105, (1983).

19. P. Maïssa, A. Ten Bosch, P. Sixou, J. Chem. Phys. (submitted).
20. A. Ten Bosch, P. Maïssa, P. Sixou, Phys. Letters 94A, 298 (1983).
21. M.J. Seurin, A. Ten Bosch, P. Sixou, Polymer Bulletin 10, 438, (1983).
22. M.J. Seurin, A. Ten Bosch, P. Sixou, Polymer Bulletin 9, 450, (1983).
23. P. Maïssa, A. Ten Bosch, P. Sixou, J. Polym. Sci., Polymer Letters Ed. 21, 757, (1983).
24. G. Maret, Liquid Crystalline Polymers, Washington, 1983.
25. G. Marrucci, Mol. Cryst. Liq. Cryst. 72, 153, (1982).

CONCEPTS IN RHEOLOGICAL STUDIES OF

POLYMERIC LIQUID CRYSTALS

Donald G. Baird

Department of Chemical Engineering and
Polymer Materials and Interfaces Laboratory
Virginia Polytechnic Institute and State University
Blacksburg, VA 24061-6496

INTRODUCTION

The intent of this chapter is not to merely review the
rheology of liquid crystalline polymers (LCP) as this has been done
elsewhere[1], but rather it is to compare their behavior with
flexible chain polymer systems and to describe various rheological
tests which can be used to characterize LCP. We will assume that
the reader is somewhat familiar with the concepts of polymer
rheology such as fluid elasticity, memory, and normal stresses.
With this background the uniqueness of the rheology of polymeric
liquid crystalline systems, which at first sight is not
significantly different from most polymer systems, will be more
fully appreciated. An excellent review of both the theory of
polymeric liquids and their flow characteristics is given in the
book by Bird, Armstrong and Hassager[2]. Although this chapter
will not be masked in the mathematics surrounding non-Newtonian
fluids, it is certainly helpful if the reader is familiar with
concepts which are associated with the rheology of flexible chain
polymer systems. Finally, it is not our intention to use the
results of others in general, so that much of the illustrative data
will come from our laboratory.

One of the most unique aspects of liquid crystalline fluids is
the fact that most of the orientation and texture generated during
flow is maintained in the solid state[3,4]. Hence, the flow
characteristics (i.e. the rheology) are directly related to the
properties of injection-molded specimens or fibers. We shall,
therefore, review some of what has been reported on the rheology of
liquid crystalline fluids in an effort to identify any behavior

119

which indicates that these fluids become oriented during flow and
do not loose this orientation in the time scale of flexible chain
polymers. We then discuss several transient shear flow experiments
which may help to identify the orientation during flow and the lack
of significant loss of this orientation on cessation of flow. In
essence, we are proposing various rheological tests which could be
useful in identifying liquid crystalline polymer systems with
desirable processing characteristics.

Two Basic Rheological Flows

In order to characterize polymeric fluids and to test
rheological equations of state it is customary to use simple, well
defined flows. The two main flows are simple shear and simple
elongational. These are shown schematically in Figure 1. In shear
flow, material planes (see Figure 1) move relative to each other
without being stretched, whereas in extensional flow the material
elements are stretched. These two different flow histories
generate different responses in not only flexible chain polymers
but in liquid crystalline polymers. When these flows are carried

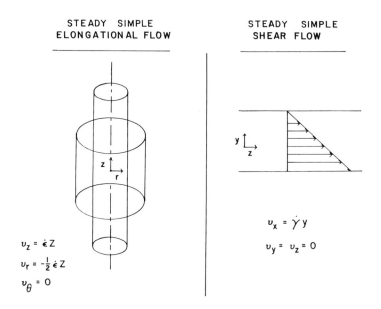

**STEADY SIMPLE
ELONGATIONAL FLOW**

**STEADY SIMPLE
SHEAR FLOW**

$$v_z = \dot{\epsilon} z$$
$$v_r = -\tfrac{1}{2}\dot{\epsilon} z$$
$$v_\theta = 0$$

$$v_x = \dot{\gamma} y$$
$$v_y = v_z = 0$$

Fig. 1.   Schematic drawing of two flows commonly used in rheological studies.

120

out in a manner in which the rheological properties no longer change with time, the flows in Figure 1 are termed steady simple shear flow or simple elongational flow.

In shear flow it can be shown using certain symmetry arguments that there are three independent quantities of stress which depend only on the properites of the fluid. These are the primary normal stress difference (N_1) defined as

$$N_1 = \sigma_{11} - \sigma_{22} \tag{1}$$

the secondary normal stress difference (N_2)

$$N_2 = \sigma_{22} - \sigma_{33} \tag{2}$$

and the shear stress

$$\sigma_{12} = \sigma_{21}. \tag{3}$$

Here σ_{ij} are components of the extra stress tensor referred to the cartesian axes in Figure 1 and $\sigma_{ii} > 0$ for tensile stresses. Three material functions can also be defined for shear flow:

$$N_1 = \phi_1(\dot{\gamma})\,\dot{\gamma}^2 \tag{4}$$

$$N_2 = \phi_2(\dot{\gamma})\,\dot{\gamma}^2 \tag{5}$$

$$\sigma_{12} = \eta(\dot{\gamma})\,\dot{\gamma} \tag{6}$$

where ϕ_1 is the primary normal stress difference coefficient, ϕ_2 is the secondary normal stress coefficient, η is the viscosity, and $\dot{\gamma}$ is the shear rate.

In simple elongational flow we define only the extensional viscosity ($\bar{\eta}$) as follows

$$\sigma_{33} - \sigma_{11} = \bar{\eta}(\dot{\varepsilon})\dot{\varepsilon} \tag{7}$$

where $\dot{\varepsilon}$ is the extensional rate.

Typically, simple shear flow is generated in cone-and-plate, plate-plate, or couette devices. Simple extensional flow is generated by devices in which a cylindrically shaped specimen is deformed in such a manner that the length increases exponentially with time. The topic of rheometry is reviewed and discussed in several places[2,5,6] and will not be discussed in further detail here.

Besides the experiments which are carried out under steady state conditions, there are also a number of transient flow experiments which are used to characterize complex fluids. These flow experiments and the corresponding material functions are summarized

in Figure 2. Several of the flows are particularly suited for
studying the relaxation processes which take place in liquid
crystalline fluids. In particular, we note such experiments as
stress growth on the inception of shear flow, stress relaxation on
cessation of flow, and suddenly imposed strain. Again, these are
all described in Figure 2 and further details are available
elsewhere[2].

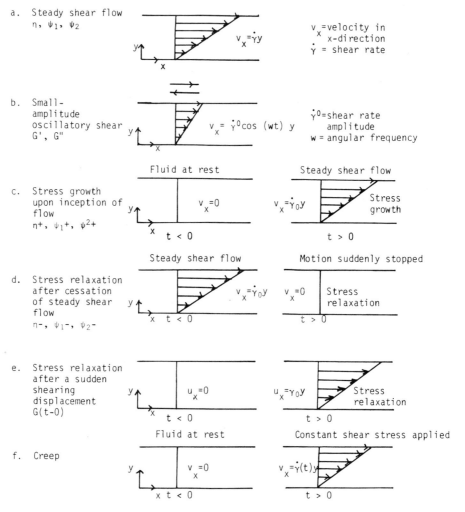

Fig. 2. Shear flow experiments and corresponding material functions.

There are several other transient flow experiments which can be instructive in identifying relaxation of orientation and structure. These experiments are summarized in Figure 3. In the interrupted shear experiment the fluid is initially sheared and then allowed to rest for a time (τ_R). The amount of recovery of the stress overshoot at the start up of flow is then used as an indication of the time for recovery of orientation and structure. There are two types of flow reversal experiments. In the first one the direction of shear is instantaneously reversed from steady flow conditions. In the other experiment, the flow is stopped and the stresses are allowed to relax before the flow is reversed. Again the stress growth behavior at the start up of flow is then monitored.

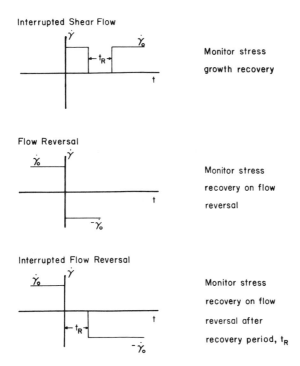

Fig. 3. Additional transient shear flow experiments for studying relaxation of structure and orientation.

Orientation, Texture and Flow

Probably the most important aspect about polymeric liquid crystals is their ability to orient during processing and maintain this orientation during the solidification process. Flexible chain polymers also orient during the flow process, but relaxation of this orientation occurs so fast that it is difficult to capture all the orientation. The structure of flexible chain polymers is governed typically by an entanglement network. The concept has been envoked to explain many of the important rheological aspects of flexible chain polymers such as shear thinning viscosity, normal stresses, elastic recoil, fading memory, and long relaxation times. On the other hand, the question arises as to what the structure of liquid crystalline polymers is and how this structure accounts for rheological properties.

On a microscopic scale just as for low molecular weight liquid crystals, nematic and cholesteric mesophases have been identified. Whether the smectic structure actually exists for polymeric systems is still open to question. However, it seems to be the macroscopic structure which accounts for the rheological properties of LCP. The common view of the structure of liquid crystalline polymers at rest is that of randomly oriented domains of highly ordered regions embedded a matrix of lesser order. We have shown this idea in the schematic drawing in Figure 4. The domain concept is substantiated

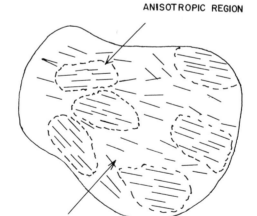

ANISOTROPIC REGION

ISOTROPIC REGION

Fig. 4. Schematic of domain structure thought to exist for polymeric liquid crystals of the nematic type.

by the opaqueness of the melt in visible light and as discussed later by the rheological properties and by the appearance of the melt in a polarizing microscope[7].

The orientation of each domain is described by a vector ($\underset{\sim}{n}$) called the director. Before the start of flow these domains are randomly oriented if no previous flow history effects are present. However, once flow starts, these domains orient in the flow direction and depending on the type of flow,they deform and eventually the domain structure could be destroyed. In which case the microscopic order could then be more significant.

The structure and orientation developed during flow is certainly one of the significant aspects surrounding liquid crystalline polymers. The type of flow (i.e. shear flow versus extensional flow) not only influences the orientation generated but the texture and morphology. To illustrate this point we consider the wide angle X-ray (WAXS) diffraction patterns shown in Figure 5 obtained from an injection molded liquid crystalline copolyester. The outer layers of an injection molded plaque are primarily subjected to extensional flow at the advancing front. The WAXS patterns show the material is highly oriented. At intermediate layers between the outer edge and the center, the flow is shear flow. The WAXS patterns also show orientation but the molecules are not as highly oriented along the flow direction as they are in the case for extensional flow. At the center of the plaque very little orientation is observed. In Figure 6 are presented electron micrographs of etched layers of the same system subjected to WAXS analysis. Here we can see that the highly oriented layers have a fibrous texture which is considerably different from the texture of the layers subjected to shear flow. Hence, even though orientation is observed in samples subjected to both shear and extensional flow, the textures are apparently different.

Several researchers[8,9,10] have attempted to study the orientation and texture developed during flow using rheo-optical techniques. Zachariades[9] used glass patterns and polarized light to view the texture of sheared copolyester melts. However, only the qualitative appearance of the melt was recorded. Onogi and coworkers[8] also used rheo-optical techniques to record the intensity of transmitted polarized light during flow and on cessation of flow of lyotropic systems. Besides observing an increase in intensity on the start up of flow, Onogi and coworkers[8] found that on cessation of flow the intensity decreased much more slowly than the stress. This suggested that there are two separate relaxation processes. One is the relaxation of orientation and the other is the time for stress relaxation. The time for relaxation of orientation can be of the order of minutes to hours while the stresses relax in a matter of seconds. Further studies of this nature are certainly of interest but they

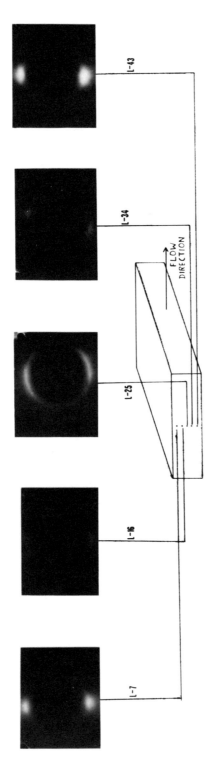

Fig. 5. Wide angle X-ray diffraction patterns of layers of an injection molded plaque (From reference 7).

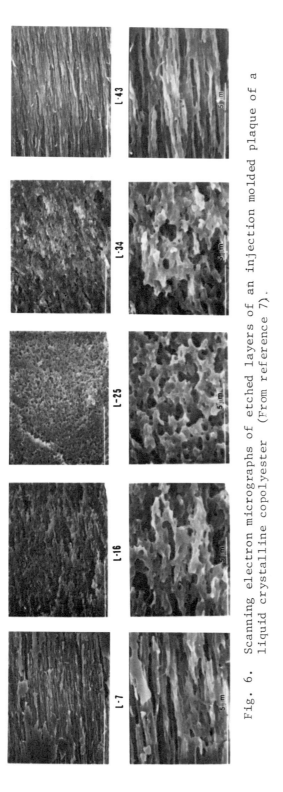

Fig. 6. Scanning electron micrographs of etched layers of an injection molded plaque of a liquid crystalline copolyester (From reference 7).

are difficult to carry out because the samples are typically so
turbid that extremely thin samples must be used. For thermotropic
melts, it may be possible to use quenched samples and analyze the
samples using solid state techniques[10].

RHEOLOGICAL PROPERTIES OF LIQUID CRYSTALLINE POLYMERS

Steady Shear Flow

Our intent, as mentioned earlier, is not to review all the
studies concerned with liquid crystalline fluids but to compare
their properties with flexible chain polymers, interpret their
properties in terms of the domain structure, and look for
correlations between flow characteristics and processing
conditions. We first examine the behavior of liquid crystalline
copolyesters in steady shear flow and in small strain dynamic
oscillatory flow.

Typically the viscosity of liquid crystalline fluids is highly
dependent on shear rate over many orders of magnitude of $\dot{\gamma}$. This
shear dependence of viscosity is illustrated by the data in Figure
7 for a 60 mole % PHB/PET copolyester. In this figure, data

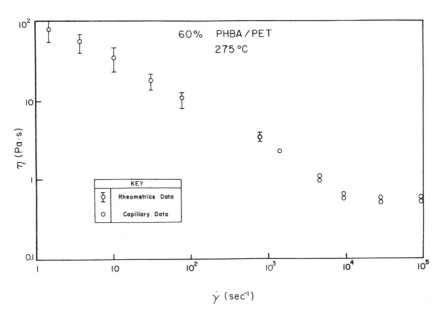

Fig. 7.   Viscosity versus shear rate for 60 mole % PHB/PET obtained
from cone-and-plate and capillary rheometers.

obtained by means of both rotational rheometers (cone-and-plate and plate-plate) and capillary rheometers are included. The shear rate spans nearly 6 orders of magnitude. At low shear rates (e.g. $\dot{\gamma}$ ~ 0.05 sec^{-1}) the viscosity continues to rise rather than reaching a low shear rate limit, which is referred to as the zero shear viscosity (η_0). At high shear rates ($\dot{\gamma} \sim 10^5$ sec^{-1}) η seems to reach a limiting high shear rate value (η_∞) of about 0.6 Pa.s. There seems to be no dependence of η on the dimensions of the rheometer or type of rheometer at this temperature.

In comparison to the behavior of flexible chain polymers we note that the viscosity dependence on $\dot{\gamma}$ is similar. However, for flexible chain polymers of similar molecular weight, η is not highly $\dot{\gamma}$ dependent. Hence, at processing conditions, η of LCP can be two to three orders of magnitude lower than that of flexible chain isotropic systems. Furthermore, isotropic polymer systems which exhibit pseudoplastic behavior at such low shear rates usually have values of η_0 two to three orders of magnitude higher than the values of η measured at low $\dot{\gamma}$ for LCP.

Some researchers[1,11] consider the viscosity versus shear rate curve to consist of three regions. At low $\dot{\gamma}$, η depends on $\dot{\gamma}^{-1}$ because of the presence of yield stresses. In region II there is a plateau where η is nearly independent of $\dot{\gamma}$. Finally, in region III, η depends on $\dot{\gamma}$ to some power (n) between -1.0 and 0 which is similar to flexible chain isotropic systems. It is our contention that yield stresses (σ_0) are not common features of pure liquid crystalline fluids. For several different lyotropic systems, values of η_0 are observed at low $\dot{\gamma}$[12,13]. It is possible that only when a mixture of liquid crystalline and solid regions exists, are yield stresses observed. We have replotted the viscosity data versus shear stress (σ) in Figure 8 for the copolyester system described previously in Figure 7. In addition, we have added data from measurements at two other temperatures. Here we see that η does not rise to infinity at a limiting shear stress as would be the case when a yield stress is present. Other experiments which are described later also confirm the lack of a yield stress in the copolyester system.

The shape of the flow curve (i.e. η vs. $\dot{\gamma}$) is important for several reasons. First, if during processing flows, orientation of the director is shear rate dependent, then the flow curve will determine the distribution of orientation. For example, if the slope (n) of ln η versus ln ($\dot{\gamma}$) approaches -1.0, then the fluid would pass through a die as a plug leading to no shear induced orientation. On the other hand, if the LCP system were highly oriented in the entry region of a die then having a value of n approaching -1.0 might be advantageous since the orientation generated in the entry region would not be disturbed during flow in the capillary. The pressure drop in a flow region is also dependent

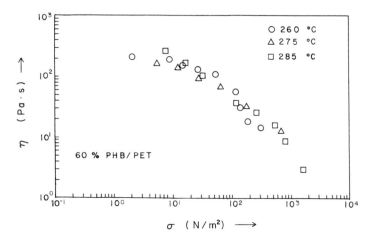

Fig. 8. Viscosity versus shear stress for 60 mole % PHB/PET at
various temperatures.

on the shape of the flow curve. Finally, the magnitude of die swell
at the exit of a die is related to the stress distribution in the
cross section of a die which is also a function of the distribution
of $\dot{\gamma}$.

Normal stress differences generated in shear flow are common
features of isotropic macromolecular fluids. They are associated
with fluid elasticity and are related to a number of phenomena such
as elastic recoil, die swell, orientation, and melt fracture. In
fact, the ratio, $N_1/2\sigma$, is related to the ultimate recoverable
shear strain which can be measured in a cone-and-plate device on
cessation of flow. Liquid crystalline fluids also exhibit positive
values of N_1 (see Figure 9) and in some cases negative values of
N_1[1,14]. Values of N_1 are presented for a 60 mole % PHB/PET
system in Figure 9. This data by itself may not be very
meaningful and so we ask: how elastic are LCP relative to
isotropic systems? The instantaneous elastic recovery is related
to the compliance (J_e) which is given by N_1/σ^2. In Figure 10
we compare J_e for 60 mole % PHB/PET with values for PET, and we
find that J_e for the LCP is nearly two orders of magnitude higher
than it is for PET. This same result has been reported for a
lyotropic system[12]. Hence, at first appearance we might think
that LCP systems are highly elastic.

On the other hand, LCP systems have been reported to exhibit
no die swell (i.e. the diameter of the jet (D_j) and the diameter

130

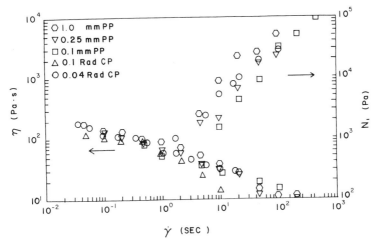

Fig. 9. Viscosity and normal stress data obtained from a parallel plate rheometer using various gaps and a cone-and-plate rheometer for 60 mole % PHB/PET at 275°C.

of the capillary (D) are identical) or in some cases the extrude even contracts[12]. Some representative data for a 60 mole % PHB/PET copolyester are presented in Figure 11. Data is presented for several temperatures and we see that die swell increases with increasing temperature which is contrary to flexible chain isotropic polymers. Values of die swell based on a theory proposed by Tanner (15) and calculated from measured rheological properties have been compared with measured values of D_j/D_o[14]. The theory, which treats die swell as a phenomenon of free recovery, predicts values of D_j/D_o which range from 1.1 to 1.5 even when no swell occurs. The increase of die swell at elevated temperatures may be due to the increase in isotropic regions. Hence, although LCP systems exhibit relatively high normal stresses, they apparently exhibit negligible elasticity over some temperature range.

Dynamic Oscillatory Shear Flow

One of the most common transient shear flows used to analyze the rheological characteristics of macromolecular fluids is that of small strain dynamic oscillatory flow. Measurements are usually carried out at strain levels where the stresses are directly proportional to strain. In this experiment, components of stress in phase (G", which is called the loss modulus) with strain rate and those out of phase (G', which is called the storage modulus) are measured as a function of angular frequency (w). G' represents

131

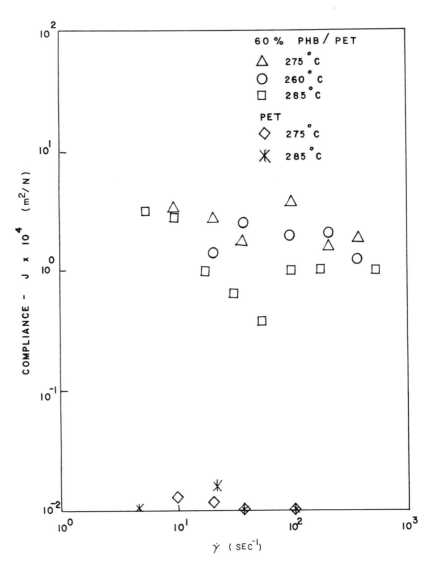

Fig. 10.   Compliance versus shear rate for 60 mole % PHB/PET and
          PET.

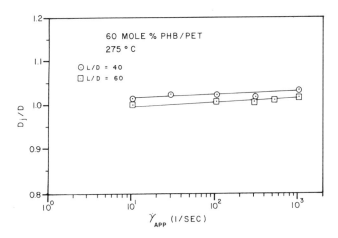

Fig. 11. Die swell versus shear rate for 60 mole % PHB/PET.

the amount of energy stored per cycle of deformation
while G" represents the energy loss per cycle of deformation. One
can also define a complex viscosity (η^*)

$$\eta^* = \eta'' - i\,\eta'$$

where η' and η'' can be shown to be related to G' and G". The
interpretation of these functions is described in detail
elsewhere[16].

Liquid crystalline fluids exhibit both values of G' and G" as
illustrated in Figures 12 and 13 for a 60 mole % PHB/PET
copolyester. The curves are observed to be frequency dependent
over the whole range. This is indicative of their fluid-like
behavior and lack of yield stresses. In the case of flexible chain
polymers G' and G" crossover (with G' > G") at a frequency which is
similar to $\dot{\gamma}$ where pseudoplastic behavior occurs. We observe for
the LCP in the temperature range of 250°C to 285°C that G' is
always less than G" and that any tendency for G' to crossover the
G" curves occurs at values of w at least four orders of magnitude
higher than the onset of pseudoplastic response. This suggests
that the mechanism responsible for shear thinning behavior is
independent of the relaxation processes which account for the
linear viscoelastic properties. Finally, we note the lack of
temperature dependence of G' and G" in the temperature range of
250°C to 285°C. This is probably indicative of the onset of the
formation of isotropic regions.

133

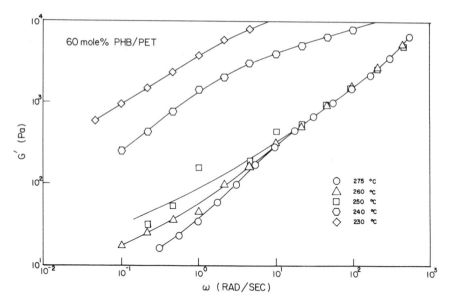

Fig. 12. Storage modulus versus angular frequency for 60 mole %
PHB/PET at various temperatures.

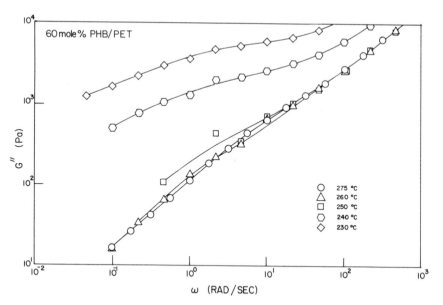

Fig. 13. Loss modulus versus angular frequency for 60 mole % PHB/PET

In the limit as w or $\dot{\gamma}$ approach zero $|\eta*|$ and η and 2G' and N_1 are predicted to coincide. Experimentally this has been found to be true for many flexible chain systems. In some cases $|\eta*|$ and η agree over a large range of $\dot{\gamma}$ and w. For liquid crystalline fluids both agreement and disagreement between the steady shear and dynamic material functions has been found[1]. In Figures 14 and 15 we compare η and $|\eta*|$ and N_1 and 2G', respectively, for the copolyester system of 60 mole % PHB/PET. Here we observe remarkable agreement between the two sets of material functions. Although this is not generally the case for LCP, we find the agreement rather interesting since it may say something about whether the domains actually orient in shear flow or just tumble.

Transient Shear Flow Experiments
_____

The transient shear flow experiments described in Figure 2 may provide the most insight into development of orientation and structure in LCP. We first look at stress growth at the start up of shear flow. In this experiment the stress build up at the start up of flow is monitored as a function of time. Some representative data for a 60 mole % PHB/PET copolyester are presented in Figure 16. At this particular temperature we observe two stress peaks. The first occurs almost instantaneously (less than 2 strain units) on the start up of flow while the second peak occurs at much higher strains (40 to 50 units). The double peak behavior is not necessarily

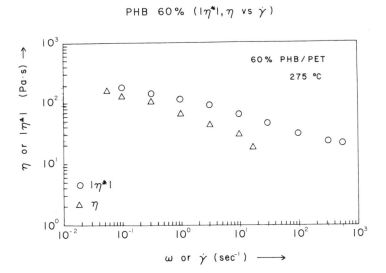

PHB 60% ($|\eta*|, \eta$ vs $\dot{\gamma}$)

60% PHB/PET
275 °C

O $|\eta*|$
△ η

η or $|\eta*|$ (Pa·s) $\longrightarrow$

ω or $\dot{\gamma}$ (sec^{-1}) $\longrightarrow$

Fig. 14. Comparison of the steady shear viscosity and the dynamic viscosity versus angular frequency or shear rate.

$(N_I \text{ vs } \dot{\gamma})$

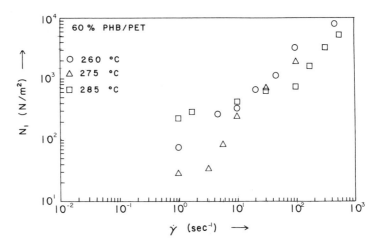

Fig. 15.   Primary normal stress versus shear rate for 60 mole %
PHB/PET.

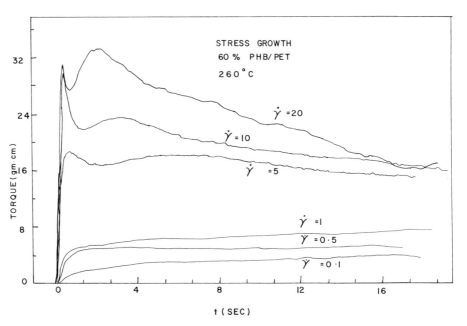

Fig. 16.   Stress growth (shear stress) at the start up of shear flow
for 60 mole % PHB/PET.

a common feature of all liquid crystalline fluids but may be related to the existence of a two phase system which arises for these copolyesters. In flexible polymers, overshoot is associated with the entanglement network which restricts the motion of chains. In the LCP systems the first peak seems to be associated with the development of orientation of the liquid crystalline domains and texture. Evidence for orientation comes from wide angle X-ray diffraction patterns obtained by quenching sheared discs[10].

The interrupted shear test described in Figure 3 may be useful in identifying the time for the orientation to relax. In this test a LCP is sheared until steady state stresses are obtained. The flow is then stopped and the fluid allowed to rest for various lengths of time before flow is started up again. The stress growth behavior after various periods of relaxation is then compared with the initial response. Some representative data for the 60 mole % PHB/PET system is presented in Figure 17. After three minutes of rest time in the melt we see that the first peak is not recovered. However, the second peak is nearly fully recovered. In fact we have observed that the second peak recovers in about 6 seconds. This test may then represent a method for determining the time for loss of orientation which is difficult to measure using rheo-optical techniques.

Whereas orientation and texture generated in flow take long periods of time (minutes) to relax, the stresses are found to relax in a matter of several seconds. In Figure 18 we have plotted shear stress versus time on cessation of flow. Here we see that the shear stress (it is plotted in Figure 18 as the reduced time dependent viscosity, η^-/η_s) relaxes faster with increasing $\dot{\gamma}$. However, by $\dot{\gamma} = 10.0$, it relaxes to zero in a time of less than 2 seconds. The normal stresses, which are plotted in figure 19 relax slower than the shear stresses which is what is observed for flexible chain polymers. The key point is that the time for the stresses to relax is much faster than would be expected based on a relaxation time determined from the flow curve or the time for relaxation of orientation.

Suddenly Imposed Shear Strain

The final experiment that may be of value in characterizing the flow characteristics of LCP is that of stress relaxation following a suddenly imposed shear strain. In this experiment various strain amplitudes are imposed very rapidly. In this way no flow occurs. At low strain levels intermolecular forces which can result in yield stresses may be more readily identified. From this type of experiment we determine the strain dependent relaxation modulus defined as follows:

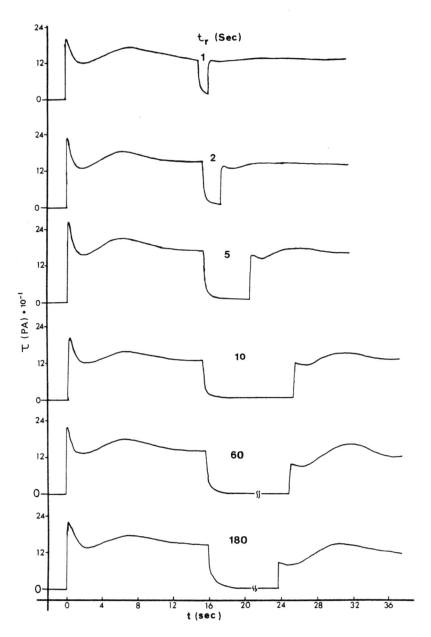

Fig. 17.   The use of the interrupted shear test to illustrate the
recovery of stress growth for 60 mole % PHB/PET at 275°C.

138

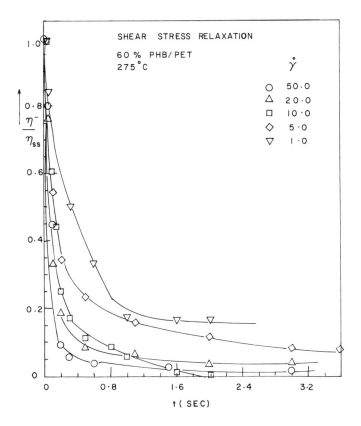

Fig. 18.   Shear stress relaxation versus time for various shear
           rates for 60 mole % PHB/PET.

$$G(t - t', \gamma_0) = \frac{\tau(t - t')}{\gamma_0} \tag{8}$$

Data for a 60 mole % PHB/PET system are shown in Figure 20. We
observe at all strain levels that G continues to relax to zero
rather than approach a plateau as would be the case when a yield
stress exists. The relaxation modulus also seems to be highly
stain dependent which is in contrast to the fact that dynamic and
steady shear material functions agree so well. For flexible chain
polymers the strain dependence of G is associated with the rate of
loss of entanglements. For LCP it is not clear as to the
significance of the strain dependence of G.

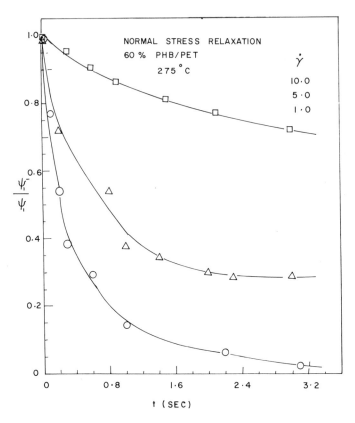

Fig. 19. Normal stress relaxation following steady shear flow for 60 mole % PHB/PET.

CONCLUDING REMARKS

There are now numerous compositions of liquid crystalline polymers under consideration as fiber spinning and injection molding materials. However, the problems involved in processing these systems are similar. In particular, how can one process these polymers to yield desirable isotropic properties or at least have biaxial orientation; how can one achieve the optimum properties from a given composition; and how does the chemical composition and structure affect the properties? In flexible chain systems one must quench in orientation in a time scale which is faster than the relaxation process of the molecules. Typically there is a distribution of relaxation times in which the longest relaxation time is a matter of a few seconds. This longest relaxation time also governs a number of other flow characteristics.

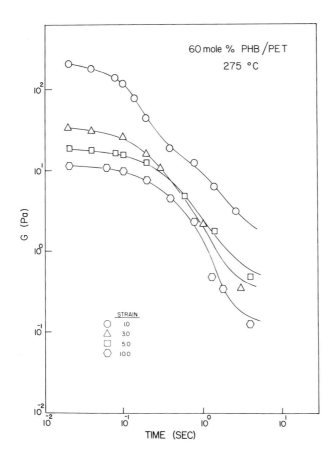

Fig. 20.   Shear modulus versus time for various strain levels for
60 mole % PHB/PET.

For example, if the time for the process to occur is faster than
the longest relaxation time, then the fluid behaves more like an
elastic solid.   For the liquid crystalline systems there seems to
be two relaxation times which are important.   One is the time for
relaxation of orientation and the other is the time for relaxation
of stress.   Whereas these phenomena are connected for flexible
chain polymers, they seem to separate for liquid crystalline
polymers.   In other words, there are several stress free states for
LCP.   Some of the behavior observed may be partly due to the
copolymer nature of thermotropic systems.   However, the lyotropic
systems based on the polyamide structure also exhibit similar
behavior.

What we have attempted to do here is to present rheological tests for identifying the development and relaxation of orientation and structure in liquid crystalline polymers. Because these fluids are typically quite turbid, it is difficult to use rheo-optical techniques. The interpretation of the rheological tests must then come partly from studies on quenched solid specimens. In summary, it is believed that a detailed set of rheological tests based on the transient response of LCP can be used to evaluate various liquid crystalline polymers and identify processing conditions which will lead to the optimum physical properties.

This chapter should not be concluded without a few comments about the comparion of the rheological properties of LCP with those of low molecular weight liquid crystals (LC). In general the viscosity of LC is independent of $\dot{\gamma}$ and at low $\dot{\gamma}$ can be 4 to 5 decades lower in magnitude than that of LCP. At high $\dot{\gamma}$, the difference in η is only 1 to 2 orders of magnitude. LC typically are not viscoelastic although they have on some occasions been reported to exhibit normal stresses. They tend not to exhibit any significant transient response as observed for LCP. However, the transient response of LC has not been studied in any detail. This is primarily due to the difficulty in measuring the response of fluids with such low viscosity. The major similarity between LCP and LC comes in their ability to orient in flow and maintain orientation on cessation of flow. Polymer systems tend to maintain this orientation longer because Brownian motion has less of an effect on randomizing the large domains.

Finally, it should be pointed out again that the results presented here, which are based on measurements of 60 mole % PHB/PET, are not meant to be generally representative of liquid crystalline polymers. The results are intended only to illustrate the response of a given LCP to various types of flow experiments and to point out the value in making these measurements. For most LCP systems only steady shear flow measurements have been carried out. Hence what we are discussing in this chapter is somewhat futuristic. That is, we are pointing out the need to carry out a detailed set of experiments as described here and the value of these experiments in assessing the processing characteristics of a given liquid crystalline polymer. At the same time some of the measurements described here are discussed in several of the following papers. Hence, it is hoped that this chapter will serve not only to make the following papers more meaningful but to stimulate additional rheological studies on liquid crystalline polymers.

ACKNOWLEDGEMENTS

Support for this research by the Army Research Office (Grant No. DAA29-80-K-0093) is gratefully acknowledged.

142

REFERENCES

1. K. F. Wissbrun, *Journal of Rheology,* 25(6), 619 (1981).
2. R. B. Bird, R. C. Armstrong, and O. Hassager, "Dynamics of Polymeric Liquids: Fluid Mechanics," Wiley, New York (1977).
3. W. J. Jackson, Jr. and H. F. Kuhfuss, *J. Polym. Sci.,* Polym. Chem. Ed., 14:2043 (1976).
4. F. E. McFarlane, V. A. Nicely and T. G. Davis, "Contemporary Topics in Polymer Science," E. M. Pearce and J. R. Schaefgen, Eds., Vol. 2, Plenum Press, New York (1977).
5. K. E. Walters," Rheometry," Chapman and Hall, London (1975).
6. J. M. Dealy, "Rheometers for Molten Plastics," Van Nostrand Reinhold Company, New York (1972).
7. E. G. Joseph, G. L. Wilkes, and D. G. Baird, Submitted to *Polymer.*
8. Y. Onogi, J. L. White, and J. F. Fellers, *J. Non-Newt. Fluid Mech.,* 7, 121 (1980).
9. A. E. Zachariades, *Polym. Eng. and Sci.,* 23(15), 797 (1983).
10. D. G. Baird, E. Joseph, R. Pisipati, G. Viola, and G. L. Wilkes, *Annual Tech. Con., Soc. Plas. Eng., Preprints,* New Orleans (1984).
11. S. Onogi, and T. Asada, "Rheology," G. Astarita, G. Marucci, and L. Nicolais Eds., Vol. 1, Plenum Press, New York (1980).
12. D. G. Baird, *J. Rheology,* 24(4), 465 (1980).
13. G. Kiss, and R. S. Porter, *J. Polym. Sci., Polym. Symp.,* 65, 193 (1978).
14. A. Gotsis, M.S. Thesis, Virginia Tech (1984).
15. R. I. Tanner, *J. Polym. Sci.,* A-2, 8:2067 (1970).
16. J. D. Ferry, "Viscoelastic Properties of Polymers," Wiley, New York (1961).
17. R. E. Jerman, and D. G. Baird, *J. Rheology,* 25(2), 275 (1981).

DO CONDIS CRYSTALS EXIST?

Bernhard Wunderlich and Janusz Grebowicz

Department of Chemistry
Rensselaer Polytechnic Institute
Troy, New York 12181

Definition of Condis Crystals

In a condis crystal cooperative motion between various conformational isomers is permitted. In the CD-glass this motion is
frozen, but the conformationally disordered structure remains.
For a condis crystal it is not necessarily expected that all possible conformations can be reached, but all conformations of the
same type are involved in the condis crystal motion. If conformational isomers of low energy exist which leave the macromolecules
largely in a parallel, extended, low energy conformation, conditions for the formation of condis crystals are given. The conformational changes involve more or less hindered rotations about
backbone bonds or side chain bonds and are thus to some degree related to the orientational motion in plastic crystals.

General Statements About Condis Crystals

In order to find condis crystals, one looks for molecules
with high temperature polymorphs that have high conformational
entropies. For small molecules, there may be difficulties to distinguish between condis and plastic crystals. For macromolecules,
there may be condis crystals of relatively rigid backbone or sidechain polymers that have, at higher temperature (or with the proper
solvent), a liquid crystal polymorph. Furthermore, one must distinguish crystals which develop practically independent motion of
isolated side-groups from condis crystals. Such independent motion
is excited over a wide temperature range and does not give rise to
a mesophase transition. A typical example is the rotation of
methyl groups in macromolecules. Such rotation starts gradually
without a thermally noticeable transition.

No general rules about the entropy of transitions, as were found for liquid and plastic crystal transitions, can be set up for condis crystals. Two typical examples may illustrate this point. Polytetrafluoroethylene has a relatively small room-temperature transition entropy on its change to the condis state and a larger transition entropy for final melting. Polyethylene has, in contrast, a higher condis crystal transition entropy than melting entropy.

Theories about the condis state can be based on the kink-model of the solid state[1]. Rotational isomers which do not disturb the parallel molecular arrangement too severely are introduced into the crystal. Pechhold and Blasenbrey[2] suggested, for example, for the condis crystals of trans-1,4-polybutadiene a cooperative, statistical treatment of pairs of straight and singly kinked chains. This model leads to large kink-blocks without torsion above a first order transition. On final melting these break into smaller kink blocks with torsional defects, jogs and folds[3]. Most likely the number of computed conformational defects is too small for the condis phase and a change of parameters may be needed. Baur[4] permits such a mixture of chains with larger numbers of defects (rotational isomers) which comes closer to the condis crystal situation. The key of the theory is the evaluation of the number of different arrangements of the conformationally disordered chains at constant energy. This value is needed for the calculation of the partition function. The solution is closely related to the two dimensional Ising model with external field. In the condis crystal, the equivalent effect of the external field on the dipoles of the Ising model is the excess energy for the new rotational isomers. Since this model cannot be solved exactly, Baur used the Bragg-Williams and the quasi chemical methods for analysis.

Examples of Condis Crystals

In the following, a series of macromolecules and small molecules will be discussed which were reported to have some type of mesophase behavior. We find that they fit well into the general class of condis crystals or CD-glasses. It is of special interest that the condis crystal state permits increased crystal perfection due to its higher mobility.

The special properties of a condis state of polyethylene were observed in our laboratory in 1964 when crystallizations were attempted at hydrostatic pressures above 300 MPa[5]. Extented chain crystals of close to 100% crystallinity resulted under these, otherwise more restrictive, crystallization conditions of elevated pressure. It was suggested from morphological evidence that during the crystallization high mobility exists in the crystal phase which leads to the chain extension. Later, a special, highly symmetric (hexagonal) high pressure phase was discovered which was linked with the extended chain crystallization[6].

The enthalpy of the reasonably reversible, first order transition from the orthorhombic to the hexagonal condis phase of polyethylene is 3.71 kJ/mol at about 500 MPa pressure which is about 80% of the total heat of fusion. The entropy of disordering is 7.2 J/(K mol). The volume change at the transition is 0.075 cm^3/g, which is about 8% of the crystalline volume. The suggestion that the high pressure phase of polyethylene is a mesophase (assumed at that time to be a nematic liquid crystal) was first made by Yasuniwa and Takemura[7] after optical microscopy at elevated pressure. In the meantime, it is proven that the disorder in polyethylene consists of statistically disordered conformations. By Raman spectroscopy[8] evidence of many gauche conformations in the chain was brought. Similarly, ultrasonic studies[9] showed a decrease of the shear modulus to the liquid level at the transition to the condis phase. All is in accord with the X-ray data which show an expansion normal to the chain axis and a contraction in the chain axis direction.

Polytetrafluoroethylene condis crystals exist at atmospheric pressure between about 303 K and the melting temperature (600 K). Starkweather[10] proposed that this phase I may be a smectic mesophase. This was concluded on the basis of the rheological behavior. Below about 300 K no flow could be detected within the instrumental limitations. As soon as the condis crystal was formed, the apparent shear viscosity decreased rapidly with increasing temperature, to increase again on melting. The shear-rate dependence was found to be that observed for small molecule smectics. As polyethylene, polytetrafluoroethylene has no mesogenic groups and is flexible enough to show normal melting at 600 K. The melt crystallizes in a similar fashion as polyethylene at elevated pressure; i.e. the condis crystal, once grown in a chain folded macroconformation, can thicken to extended chain crystals of high crystallinity. The condis crystals show sharp X-ray diffraction spots only on the equator. The combined volume changes of both room temperature transitions is 1.3%, the transition entropy is 3 J/(K mol), of which more than 80% are gained at 295 K. The transition entropy to the isotropic melt at 600 K is 5.7 J/(K mol).

Isotactic polypropylene is known[11] to crystallize when cooled quickly from the melt into a crystal form which was called a smectic, mesomorphic form. The X-ray diffraction pattern resembled clearly that of a smectic material. Later, this structure was also called para-crystalline to indicate the poor crystalline order. It was found that this structure is metastable below 335 K. At room temperature it has been reported to persist for over 18 months[12], so that it should be called a CD-glass. The proposed transition mechanism to the stable crystal form involves intramolecular helix perfection (removal of helix reversals) and intermolecular alignment. The transition is exothermic with about -0.7 kJ/mol enthalpy of transition. The heat of fusion of the stable crystal is, in contrast, about 6.9 kJ/mol. The question of interest remains whether there is at higher temperature and pressure a region of stability for the CD-glass. It is of interest to note that helix reversals can thus be removed in the condis state, but errors in inclination (isoclined or anticlined) cannot.[13] In order to change inclinations, the whole crystallized chain must be reversed, a process only possible on fusion of the chain and new crystallization.[14] One can thus remove the energetically more difficult helix reversals without recrystallization, but not the energetically less significant inclination disorder. As a result, many vinyl polymers have disordered crystal structures relative to inclinations.[13] The role of the condis state as a metastable intermediate in crystallization has not been much investigated. It is possible that in many crystallizations involving helices, the four possible ways of arranging isotactic helices are chosen randomly. The helix reversals can then be corrected as long as the condis state persists.

Trans-1,4-polybutadiene was one of the first macromolecules for which a reversible phase transition to a state of high conformational mobility was proposed[15] and proven by NMR experiments. The condis crystal (crystal form II) is found between the two transitions. The heat of transition from the rigid crystal (form I) to the condis crystal is about twice as big as the transition of the condis crystal to the isotropic melt. On transition to the pseudo-hexagonal condis crystals at about 350 K, the chains remain in their relative positions, but increase their chain separation by 7.5% and shorten the chain length by 4% and have an overall volume increase of 9%. Estimations of the number of rotational isomers accessible in the condis crystal range from 9 per repeating units to 4 per two repeating units which would lead to conformational entropies of 18.3 to 5.8 J/(K mol), respectively. The total number of possible isomers is 27. The entropy change due to expansion has been estimated based on expansivity and compressibility data to be 3.2 J/(K mol). Transition temperatures of crystals I to condis crystals and then to the melt are 356 and 437 K. The corresponding entropies are 22 and 8.4 J.(K mol), respectively. The total entropy of fusion of 30.4 J/(K mol) corresponds to 3 moles of bonds gaining conformational freedom as expected empirically. It agrees also with the entropy of fusion of the cis-1,4-polybutadiene [32 J/(K mol)] which does not form stable condis crystals.

Poly(diethyl siloxane) was suggested by Beatty et al.[16] based on DSC, dielectric, NMR, and X-ray measurements to possess liquid crystalline type order between about 270 and 300 K. The macro-molecule shows two large lower temperature first order transitions, one at about 200 K, the other at about 270 K. The transition of the possible mesophase to the isotropic liquid at 300 K is quite small and irreproducible, so that variable, partial crystallinity was proposed [measured heat of transition about 150 J/mole]. Very little can be said about this state which may even consist of residual crystals. It is of interest, however, to further analyze the high temperature crystal phase between 200 and 270 K. It is produced from the, most likely, fully ordered crystal with an estimated heat and entropy of transition of 5.62 kJ/mol and 28 J/(K mol), respectively [calculated from calorimetric data as-suming 60% crystallinity as determined by NMR]. The transition to the melt (or possible mesophase) at 270 K has a similarly estimated heat and entropy change of 1.70 kJ/mol and 6.3 J/(K mol), respectively. The entropy gain at 200 K is so large, that a condis crystal seems likely. Some indication of higher mobility is also given in the NMR data.

Alkoxy and aryloxy polyphosphazenes present a group of macro-molecules which frequently have a wide temperature range of a stable mesophase. Schneider et al.[17] list 16 polymers of this type. In the light of the present discussion they should be clas-sified as condis crystals. The polymers are flexible with low intrinsic barriers to rotation, as is indicated by their low glass transitions, which range from about room temperature down to 180 K, depending on substitution. The thermal behavior is dominated by the two first order transitions limiting the condis crystal state.

Small molecules may also form condis crystals, provided they possess suitable conformational isomers. It is of interest to note that several of the organic molecules normally identified as plastic crystals are probably better described as condis crystals. Their motion is not the complete reorientation of the presumed rigid molecule, but rather an exchange between a limited number of conformational isomers. Examples are 2,3-dimethylbutane, cyclohexanol and cyclohexane.

A series of other condis crystals are the larger cycloalkanes, analyzed by Grossmann[18]. The melting transition of cyclotetracosane has only about 1/4 the heat of transition at the disordering to the condis phase. The condis phase has a much higher symmetry and fewer X-ray diffraction lines. Infrared and Raman spectroscopy indicate that practically no additional conformational isomers are introduced on final melting. In rings with larger numbers of carbon atoms (cyclohexanonacontane, for example) the melting tran-sition becomes the dominant transition.

Conclusions about Condis Crystals

The examples document that condis crystals as defined above do, indeed exist. A comparison between the various condis crystals shows that large variations in the amount of conformational disorder and motion is possible even in similar molecules. The tritriacontane in the condis state possesses about 3 gauche conformations per 100 carbon atoms. For cyclodocosane which is in its transition behavior similar to tetracosane, one estimates about 16 gauche conformations per 100 carbon atoms. The concentration of gauche conformations in cyclodocosane and polyethylene condis crystals are close to the equilibrium concentration in the melt, while the linear short chain paraffin condis crystals are still far from the conformational equilibrium of the melt.

The other thermodynamic mesophases are well enough understood to propose with the addition of the condis state a subdivision into six types. Depending on the type of disorder, they are called liquid crystals, plastic crystals or condis crystals (positional and if applicable conformational disorder, orientational disorder, and conformational disorder, respectively). For the corresponding glasses, which represent the frozen-in mesophases, the names LC-, PC-, and CD-glasses are proposed. For macromolecules not only equilibrium states, but multiphase, metastable equilibrium states must be considered.

Condis crystals and glasses of macromolecules are thus a newly recognized type of mesophase. The mobility in this mesophase may lead to chain extension and crystal perfection and as a corollary, it may be possible that mechanical deformation can cause the stabilization of the condis state. Several examples of stable condis crystals are documented, but there seem to be also examples of metastable condis crystals which are produced as intermediates to crystallization. The size of the condis crystal transitions vary depending on the number of conformational isomers involved in the cooperative transitions.

Acknowledgments. This work was supported by the Polymers Program of the National Science Foundation (Grant Number DMR 78-15279). A full report on the mesophase transition has been given in the Adv. Polymer Sci.[19]

References

1. B. Wunderlich, "Macromolecular Physics" Vol. 2, Academic Press, New York, 1976.
2. W. Pechhold and S. Blasenbrey, Angew. Makromol. Chem., 22, 3 (1972).

3. W. Pechhold and S. Blasenbrey, Rheol. Acta, 6, 174 (1967).
4. H. Baur, Colloid and Polymer Sci., 252, 641 (1974).
5. B. Wunderlich and T. Arakawa, J. Polymer Sci., A2, 3697 (1964).
6. For example, D. C. Bassett, Dev. Cryst. Polymers, Vol. 1, 115 (1982).
7. M. Yasuniwa and T. Takemura, Polymer, 15, 661 (1974).
8. H. Tanaka and T. Takemura, Polymer J., 12, 355 (1980).
9. K. Nagata et al., Jap. J. Appl. Phys., 19, 985 (1981).
10. H. W. Starkweather, J. Poly. Sci., Polymer Phys. Ed. 17, 73 (1979).
11. G. Natta et al., Rend Accad. Naz. Lincei, 24, 14 (1959).
12. R. L. Miller, Polymer, 1, 135 (1960).
13. See for example, B. Wunderlich, "Macromolecular Physics" Vol. 1, Academic Press, New York, 1973.
14. P. Corradini et al. Europ. Polym. J., 19, 299 (1983).
15. G. Natta and P. Corradini, J. Poly. Sci. 39, 29 (1959).
16. C. L. Beatty et al., Macromolecules, 8, 547 (1975).
17. W. S. Schneider et al., in A. Blumstein ed. "Liquid Crystalline Order in Polymers." Academic Press, New York, 1978.
18. H.-P. Grossmann, Polymer Bulletin, 5, 137 (1981).
19. B. Wunderlich and J. Grebowicz, Adv. Polymer Sci., 60/61, 1 (1983).

X-RAY STUDIES OF THE STRUCTURE OF HBA/HNA COPOLYESTERS

R.A. Chivers[*], J. Blackwell[*], G.A. Gutierrez[*],
J.B. Stamatoff[**] and H. Yoon[**]

[*]Department of Macromolecular Science
Case Western Reserve University, Cleveland, Ohio 44106
[**]Celanese Research Company
Summit, New Jersey 06987

ABSTRACT

X-ray diffraction patterns of fibers spun from liquid crystal-line melts of p-hydroxybenzoic acid (HBA) and 2-hydroxy-6-naphthoic acid (HNA) show a high degree of axial orientation. Several meridional maxima are detected which are aperiodic and also change in position and number with the monomer composition. The positions of these maxima can be predicted by calculating the theoretical scattering of random copolymer chains, in which the residues are represented by points separated by the monomer lengths. Both peak positions and intensities are reproduced when intraresidue interferences are allowed for in an atomic model for the random chains. This procedure also allows determination of the stiff-chain persistence (or correlation) length from the breadth of the maximum at $d = 2.1\text{Å}$ which increases from 9 to 13 residues as the HBA content is increased from 25 to 75%.

INTRODUCTION

Copolymers of p-hydroxybenzoic acid (HBA) and 2-hydroxy-6-naphthoic acid (HNA) are converted to highly anisotropic liquid crystalline melts in the temperature range 250 – 400°C. [1] These materials can thus be processed from the melt, e.g. as high strength fibers and self-reinforced injection molded plastics in contrast to the homopolymers, poly(HBA) and poly(HNA), which are infusible, largely intractable materials. These wholly aromatic copolymers have stiff, extended conformations due to the 1,4 and 2,6 ester linkages, and this is thought to be responsible for the highly

153

ordered structures in the melt. The kinks in the chain produced
by the non-linear 2,6-naphthyl unit lead to the disruption of the
crystalline structure in homopoly(HBA), thereby lowering the melting
point. Similar effects have been seen in copolyesters of HBA with
a variety of other monomers. [2,3]

The aim of this work is to investigate the structure of these
copolyesters at the molecular level both in terms of the sequence
distribution and conformation of the individual molecules and how
these pack in three dimensions. This paper describes X-ray diffraction work on these copolymers, and primarily addresses the problem
of the chain structure. X-ray diffraction patterns of melt-spun
fibers of HBA/HNA copolyesters (fig. 1) show a high degree of
molecular orientation parallel to the fiber axis. Meridional maxima
can be seen which are aperiodic, i.e. they are not orders of a
simple fiber repeat; these maxima also change in number and position
with the monomer ratio. This is seen clearly in fig. 2 which shows
$\theta/2\theta$ meridional X-ray diffractometer scans for five HBA/HNA compositions. We have calculated the theoretical diffraction patterns of
random chain copolymers of HBA and HNA, and details of these calculations are given below.

DIFFRACTION BY APERIODIC COPOLYMER CHAINS

The stereochemistry of the aromatic units and ester links leads
to a stiff chain in which there are only minor deviations from
linearity, and this will probably be enhanced by the development
of liquid crystallinity and the effect of shear during processing.
Figure 3a shows a short chain segment generated using standard bond
lengths and angles. [4] The only conformational variables are the
phenyl-carboxyl inclination angles, which were set at $\pm30°$. The
meridional intensity is derived from the projections of chains of
this type on the fiber axis (z), and this is very nearly independent
of the conformation. Thus, we can consider the molecule as linear
and, as a reasonable approximation, we can assign a single set of
atomic coordinates for the HBA and HNA residues, wherever they occur
in the chain.

The scattered intensity from a chain of N atoms each of
scattering factor f_j and located at z_j is given by:

$$I(Z) = |F_c(Z)|^2 \qquad (1)$$

where

$$F_c(Z) = \sum_{j=1}^{N} f_j \exp(2\pi i Z z_j) \qquad (2)$$

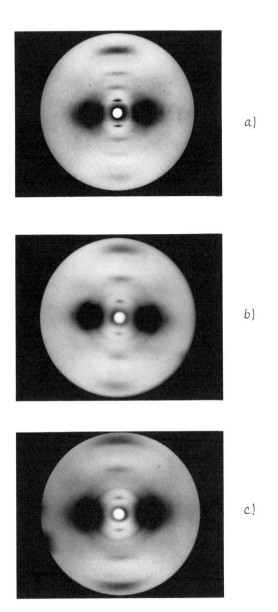

a)

b)

c)

Figure 1

X-ray fiber diagrams of HBA/HNA copolyesters in the following
monomer ratios: (a) 30/70   (b) 58/42   (c) 75/25.
(Reproduced from reference 5.)

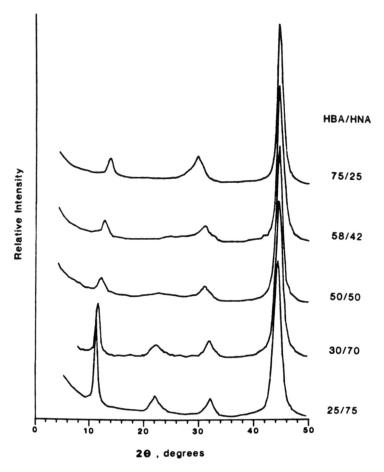

Figure 2

Meridional θ/2θ diffractometer scans of fibers of five HBA/HNA monomer ratios.

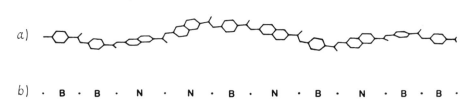

a)

b) · B · B · N · N · B · N · B · N · B · B ·

Figure 3

(a) Model of a typical random sequence of HBA/HNA copolymer.
(b) Point model representation of the sequence in (a).
(Reproduced from reference 7).

156

In our first analyses, we simplified the calculations by approximating the residues to single points, located for convenience at the ester oxygen and with a scattering factor of unity. [5] Each chain now contains only M points, separated from their neighbors by the appropriate monomer lengths. The point model for the sequence in fig. 3a is shown in fig. 3b. Using this type of model we can look for agreement between the observed and calculated peak <u>positions</u> along the meridian. Comparison of observed and calculated <u>intensities</u> requires consideration of the intra- as well as inter-residue interferences, using an atomic model for the chains. However, it is shown below that the atomic model may be generated readily using the point model as a precursor.

If we wish to consider an assembly of random chains of M units, we must average over all the R possible sequences:

$$I = \sum_{c=1}^{R} p_c |F_c(Z)|^2 \tag{3}$$

where p_c is the probability of the cth sequence. This assumes there is random staggering of the chains along the fiber axis. For a reasonable value of M, R becomes very large and calculation of I(Z) via equation 3 is a lengthy procedure. Initially, we considered only <u>some</u> of the R possible sequences, chosen using a random number generator (e.g. in ref. 5 we used 200 chains of 15 residues). However, the terms in equation 3 can be grouped together by performing the calculations via the autocorrelation function of the chains Q(z). This is the neighbor probability distribution and is shown in fig. 4 for the monomer ratio 58/42.

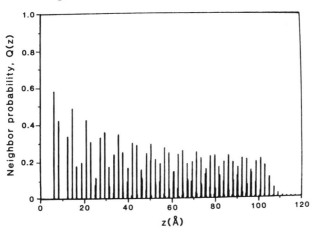

Figure 4

Neighbor probability distribution for a random infinite chain of HBA/HNA: 58/42 out to the 15th neighbors. (Reproduced from ref. 7.)

157

The scattered intensity is now given by:

$$I(Z) = \sum_{\ell} Q(z_{\ell}) \exp(2\pi i Z z_{\ell}) \qquad (4)$$

Not only does this method greatly reduce computation time, but it also produces a result which is not subject to the statistical variations of the random number generator approach.

An atomic model can be generated from the point chain simply by convolving the appropriate residue with the point representing it. Equation 2 becomes:

$$F_c(Z) = \sum_A \sum_j F_A(Z) \exp(2\pi i Z z_{j,A}) \qquad (5)$$

where A represents the type of residue and $z_{j,A}$ is the location of the jth residue of type A in the chain. F_A is the transform of the atoms in residue A located at the chain origin:

$$F_A(Z) = \sum_{k=1}^{N_A} f_k \exp(2\pi i Z z_k)$$

If the calculation for the atomic model is to be performed via equation 4, however, each term in $Q(z)$ represents a pair of neighbors. Thus, the atomic model is obtained by multiplication of the $Q(z)$ terms in equation 4 by the Fourier transforms of the cross convolutions of the appropriate pairs of neighboring monomers A and B. In the case of copoly(HBA/HNA), $Q(z)$ is effectively divided into four components corresponding to HBA-HBA, HBA-HNA, HNA-HBA and HNA-HNA neighbor relationships:

$$Q(z) = \sum_A \sum_B Q_{AB}(z) \qquad (7)$$

The scattered intensity is then given by:

$$I(Z) = \sum_{\ell} \sum_A \sum_B Q_{AB}(z_{\ell}) F_{AB}(Z) \exp(2\pi i Z z_{\ell}) \qquad (8)$$

where:

$$F_{AB}(Z) = \sum_a \sum_b f_{A,a} f_{B,b} \exp[2\pi i Z(z_{B,b} - z_{A,a})] \qquad (9)$$

The subscript pairs A,a and B,b designate the ath atom in residue A and the bth atom in residue B, respectively, when both residues are located at the origin.

EXPERIMENTAL

Specimens of copoly(HBA/HNA) were prepared according to the method described by Calundann [1] with the following HBA/HNA mole ratios: 25/75, 30/70, 50/50, 58/42, and 75/25. Fibers were drawn from the molten chips and X-ray diffractions were recorded on films using a Searle toroidal focusing camera. In addition, $\theta/2\theta$ diffractometer scans were recorded for large bundles of parallel fibers along the direction corresponding to the meridian of the fiber diagram, using a Philips D-76 diffractometer.

I(Z) was calculated via the autocorrelation functions for random copolymer chains using the equations described above and the results were compared with the observed meridional data in terms of the peak positions, intensities, and profiles. The monomer lengths were taken to be 6.35Å for HBA and 8.37Å for HNA and the atomic coordinates used are given in ref. 5, and are based on standard bond lengths and angles, consistent with the crystal structures of model aromatic esters. [4]

Calculations for models with different chain lengths showed that chains of about ten residues gave reasonable agreement with the experimental data. In order to make the model more realistic, calculations were performed for a normal distribution of chain lengths of mean ten residues, and standard deviation one residue. The intensities calculated for the atomic models were corrected by Lorentz and polarization (Lp) factors to allow comparison with the observed $\theta/2\theta$ diffractometer scans.

RESULTS AND DISCUSSION

Fiber diagrams for three copolymer compositions are shown in fig. 1, and meridional $\theta/2\theta$ diffractometer scans for all five are shown in fig. 2. The data indicate that there is a high degree of axial orientation of the molecules in the fibers. The fiber diagrams show intense equatorial scatter at $d \approx 5.0 - 2.5$Å which contains some very sharp maxima: equatorials at $d = 4.6$ and 2.6Å and an off-equatorial at $d = 3.3$Å. These data indicate that there is some three-dimensional order, but the diffuse equatorial scatter shows that much of the lateral packing is irregular. The question of the chain packing can only be addressed when we have a model for the chains themselves, which will be derived from the meridional data (fig. 2).

For all five samples, the most intense maximum is at $2\theta \approx 43°$ $(d \approx 2.1$Å$)$. This changes in position only very slightly with composition although it can be seen to become sharper with increasing HBA content. In addition, all samples show maxima in the region of $2\theta \approx 12°$ $(d \approx 7$Å$)$ and $2\theta \approx 30°$ $(d \approx 2.9$Å$)$. These can be seen to move steadily closer together on increasing the content of HBA. When there is 50% or more HNA present, a fourth maximum can be seen at $2\theta \simeq 22°$ ($d \simeq 4.1$Å) and this becomes more intense with increasing HNA content. The peak positions are given in table 1.

Table 1

Composition HBA/HNA	Experimental d-Spacings* (Å)		Calculated d-Spacings[†] (Å)	
			Atomic Convolution	Point Convolution
	Film	Diffractometer		
25/75	8.09 ± 0.07	8.11 ± 0.07	8.09	7.98
	4.08 ± 0.04	4.15 ± 0.02	4.20	4.17
	2.77 ± 0.03	2.85 ± 0.01	2.85	2.84
	2.05 ± 0.03	2.09 ± 0.01	2.10	2.10
30/70	7.95	7.89	8.01	7.88
	4.11	4.09	4.21	4.17
	2.83	2.87	2.86	2.85
	2.06	2.09	2.10	2.10
50/50	7.43	7.49	7.61	7.41
			4.31	4.11
	2.84	2.95	2.95	2.93
	2.02	2.09	2.10	2.10
58/42	7.35	7.19	7.45	7.19
			4.40	4.01
	2.98	2.96	2.99	2.98
	2.05	2.08	2.11	2.10
75/25	6.78	6.70	7.04	6.75
	3.03	3.09	3.09	3.09
	2.03	2.09	2.11	2.11

*Experimental errors are given for the 25/75 preparation and are similar for the other compositions.

[†]The positions of peak maxima have been interpolated from the discrete calculated points.

Figure 5 shows the calculated intensity distributions for point models of all five compositions. The peak positions are given in table 1, where they can be seen to be in good agreement with those observed. These data are part of a larger series of intensity curves calculated over the entire range of composition from HBA/HNA 0/100 to 100/0. [6]   These data show a steady trend in the peak position as the monomer ratio is varied.   Three peaks are predicted for homo-poly(HBA) (100%) in the range considered here: at $d = 6.35$, 3.10, and 2.12Å.   As the HNA content is increased the first two of these move apart and, at 50/50, an additional peak is generated between them at $d = 4.19$Å.   Homopoly(HNA) (0/100) gives four peaks at 8.37, 4.19, 2.89, and 2.09Å.   The peak at $d \simeq 2.1$Å is approximately indepen-dent of the monomer ratio across the entire range because of the approximately 3:4 ratio of the monomer lengths.

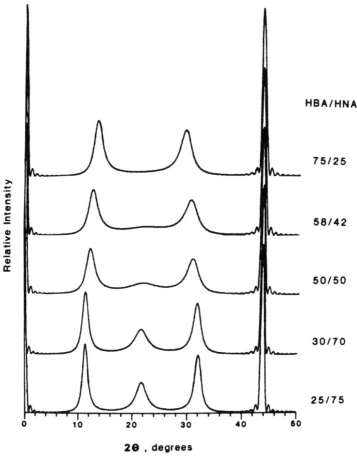

Figure 5

Meridional intensity I(Z) plotted against 2θ for point models of five compositions of HBA/HNA.

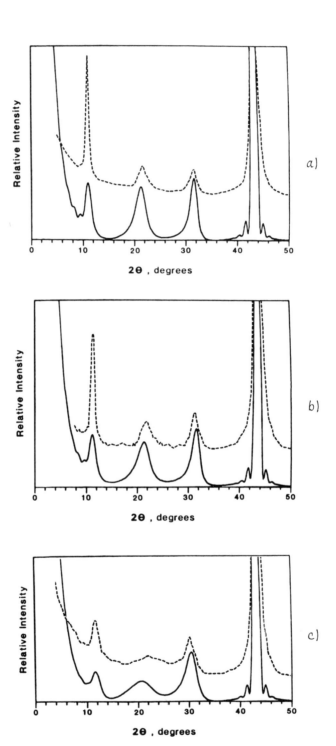

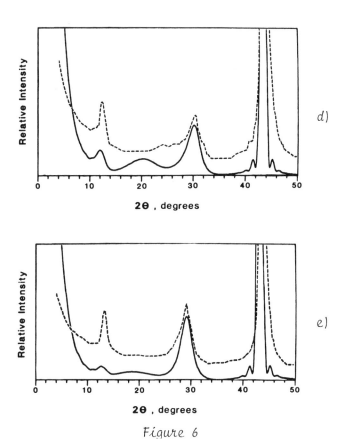

Figure 6

Meridional intensity distribution I(Z) plotted against 2θ as calculated from the atomic model (solid line) and experimentally obtained from a diffractometer (dashed line). The HBA/HNA monomer ratios are: (a) 25/75 (b) 30/70 (c) 50/50 (d) 58/42 (e) 75/25. (Modified from reference 7.)

The intensity distributions calculated for the atomic models (using equation 8) are shown in fig. 6, together with the diffractometer traces overlaid for direct comparison. The calculated data are for a distribution of chain lengths centered on $M = 10$, and have been corrected for the Lorentz and polarization effects. The intensity agreement is very good, especially for the peaks at $d \sim 4\text{Å}$, $\sim 3\text{Å}$, and $\sim 2\text{Å}$. The maximum at $d \sim 7\text{Å}$ is observed to have a higher relative intensity than that calculated. This may be due to the assumption that all the residues have their ester oxygen--ester oxygen vectors parallel to the chain axis. Examination of fig. 3c shows this is only approximately the case and that in fact there is a distribution of residue orientation, such that the "average" residue will be tilted away from the parallel position. Gutierrez et al. [8] have shown that the intensity agreement for the analogous copolymers prepared from

HBA, 2,6-dihydroxynaphthalene and terephthalic acid can be improved by refining the average residue orientations with respect to the fiber axis, and it is likely that this procedure will also improve the agreement for copoly(HBA/HNA). In addition, the meridional intensities may be affected by interchain interferences as a result of a preferred axial stagger in parts of the specimens, which is indicated by the presence of sharp equatorial and off-equatorial Bragg reflections.

As mentioned above, a special feature of the HBA/HNA system is that the monomer lengths are in the approximate ratio 3:4 and, as a result of this, the peak at $d \simeq 2.1\text{Å}$ scarcely moves with polymer composition. This peak may therefore be considered as a Bragg maximum: the point model can be thought of as a lattice repeating every 2.1Å but with only 25 – 33.3% occupancy. As such, the peak at $d \simeq 2.1\text{Å}$ is affected by the dimensions of the lattice (in this case the length of the chain). Figure 7 shows calculated intensity distributions for atomic models for chains of different length from $M = 6$ to 15 for HBA/HNA = 58/42. Also shown is the diffractometer scan for the same composition. The curves have all been normalized to give the same intensity at $Z = 0$. The data show that whereas peaks at higher d-spacings are almost independent of chain length, $d = 2.1\text{Å}$ changes in width, from 2.0° (2θ) for six residues to 0.8° for fifteen residues. The observed half width for this composition is 1.1°, (after correction for instrumental broadening), and matches the theoretical half width for a chain of approximately eleven residues. Application of the Debye-Scherer equation to this peak gives a lattice length of 77Å compared to 79.2Å for the length of an average chain of eleven residues at this composition. Extending these calculations to other compositions, the observed line width (from fig. 2) changes across the composition range: the corrected half widths for 25/75, 30/70, 50/50, 58/42, and 75/25 are 1.4°, 1.3°, 1.1°, 1.1°, and 1.0°, respectively. These correspond to chain lengths ranging from ∿9 to ∿13 as the HBA content increases across the series. The degree of polymerization for these copolymers is reported [1] to be approximately 150 (based on a molecular weight of 25000) and thus our calculations relate to a <u>persistence</u> length, defined as the length over which the stiff chain retains the correlation of monomer lengths. Eventually, because of inevitable variations in residue orientations and torsion angles, our approximation that residue lengths are constant will break down. These calculations show that this breakdown occurs after about eleven residues in HBA/HNA 58/42. The rest of the pattern (at higher d-spacings) is already reproducible for chains of six residues. It is not surprising that the persistence length should decrease with increasing HNA content when we consider that the 2,6 linkage in HNA is not collinear, and will introduce disruptions into the otherwise linear poly(HBA) chain.

The above calculations show that the X-ray data are consistent with a copolymer structure with completely random monomer sequence.

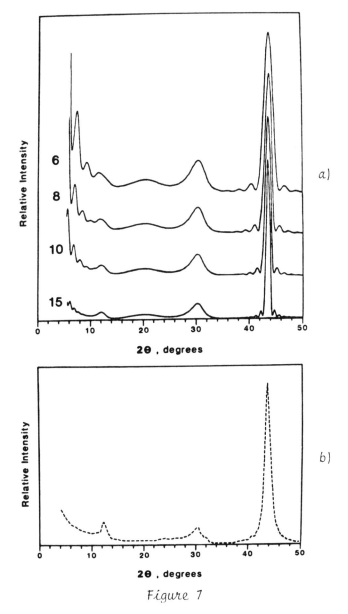

Figure 7

(a) Calculated meridional intensity distribution I(Z) plotted against 2θ for HBA/HNA : 58/42 chains of different numbers of residues.

(b) Experimental diffractometer trace for the same composition.

(Reproduced from reference 7.)

In separate work we have tested the sensitivity of the calculation to deviations from randomness by varying the reactivity ratios of the monomers in our calculations. This work will be the subject of a later paper, but it is important to note that all but small deviations from randomness lead to unacceptable changes in the position of the peaks at $\sim 7\text{Å}$ and $\sim 3\text{Å}$ and point to the random sequence as the acceptable model for the copolymer chain.

ACKNOWLEDGMENTS

The work at Case Western Reserve University has been supported by grants from the National Science Foundation (ISI81-16103) and from Celanese Research Company.

REFERENCES

1. G.W. Calundann (Celanese), U.S. Patent 4,161,470 (1979).
2. J.-I. Jin, S. Antoun, C. Ober, and R.W. Lenz, British Polymer Journal 12 132 (1980).
3. J. Preston, Ang. Makromol. Chem. 109/110 1-19 (1982).
4. B.J. Adams and S.E. Morsi, Acta Cryst. B32 1345 (1976).
5. G.A. Gutierrez, R.A. Chivers, J. Blackwell, J.B. Stamatoff, and H. Yoon, Polymer 24 937 (1983).
6. J. Blackwell, G.A. Gutierrez, and R.A. Chivers, Macromolecules, in press.
7. R.A. Chivers, J. Blackwell, and G.A. Gutierrez, Polymer, in press.
8. G.A. Gutierrez, J. Blackwell, and R.A. Chivers, Polymer, in press.

X-RAY STUDIES OF THERMOTROPIC AROMATIC COPOLYESTERS

John Blackwell, Genaro A. Gutierrez, and Robin A. Chivers

Department of Macromolecular Science
Case Western Reserve University
Cleveland, Ohio 44106

ABSTRACT

X-ray diffraction has been used to investigate the solid state structure of aromatic copolyesters that can be spun as high strength fibers from thermotropic melts. Copolymers of 4-hydroxybenzoic acid (HBA), 2,6-dihydroxynaphthalene and terephthalic acid give fiber diagrams that indicate a high degree of axial orientation. Much of the structure has disordered lateral packing but there is some three-dimensional order. The meridional maxima are aperiodic and shift in position with changes in the monomer composition. The positions of these maxima can be reproduced by a model for chains of random monomer sequence, in which the residues are represented by points separated from their neighbors by the appropriate monomer lengths. Extension of these calculations to an atomic model for the chains allows for comparison of the observed and calculated X-ray intensities. We find that refinement of the model by tilting the monomer residues with respect to the chain axis leads to good agreement between the observed intensities and those calculated for a completely random monomer sequence. An indication of the chain packing in the ordered regions comes from study of a second set of copolymers prepared by modification of poly(ethylene terephthalate) by incorporation of HBA. The latter copolymers are found to contain ordered regions with a structure similar to that for the high temperature hexagonal form of homopoly(HBA).

INTRODUCTION

We are using X-ray methods to investigate the solid state structure of a group of thermotropic aromatic copolyesters. These copolymers have stiff chain conformations and form mesomorphic

167

melts, from whence they can be processed as high strength fibers and novel molded plastics. The patent literature (see refs. 1 and 2 for reviews) describes a family of liquid crystalline copolyesters, consisting primarily of aromatic moieties such as 1,4-phenylene and 2,6-naphthylene, with the result that the monomer linkages lead to extended chain structures. The homopolymer formed by condensation of 4-hydroxybenzoic acid, poly(HBA), is an infusible crystalline polymer with a structure (chain conformation and packing) that is comparable to Kevlar, i.e. poly(p-phenylene terephthalamide) However, copolymers of HBA with 2-hydroxy-6-naphthoic acid (HNA) are melt processable, and melt spun fibers have properties that approach those of Kevlar 49. A large number of HBA copolymers have been patented and different degrees of success are reported (in terms of the properties obtained), depending on the comonomer selected. It appears that copolymerization introduces defects that break the strong intermolecular interactions of the homopolymer and lead to a melt processable polymer. In addition the defects must allow for liquid crystallinity in the melt, and probably also for development of some three-dimensional order in the solid state, as will be discussed below.

This paper will describe our analyses of the structures of copolyesters, concentrating on the system studied in most detail: the terpolymer prepared from 4-hydroxybenzoic acid, 2,6-dihydroxy-naphthalene (DHN), and terephthalic acid (TPA). We will also discuss analyses of the structure of HBA modified poly(ethylene terephthalate), in which the presence of $-CH_2-CH_2$ groups in the backbone leads to a more flexible chain. Other work on the copolymers of HBA and HNA is described in the following paper by Chivers et al. [3]

Figure 1 shows X-ray diffraction patterns for melt spun fibers of copoly(HBA/HNA/TPA) for three monomer mole ratios: 60/20/20, 50/25/25, and 40/30/30. Schematics of these X-ray patterns are shown in fig. 2, in view of the difficulties in reproduction of the original data. The patterns contain a number of sharp arcs that indicate a high degree of axial orientation of the chains. There is also intense diffuse scattering on the equator in the 6 - 3Å region that shows that the lateral packing of the chains is largely irregular. However, the presence of some sharp equatorial and off-equatorial Bragg reflections points to the presence of three-dimensional order in part of the specimens: the fact that there are off-equatorial reflections means that there must be preferred axial register for some chains or copolymer sequences.

The most striking feature of the X-ray data is that there are several meridional maxima, and that these are not orders of a single repeat distance but are aperiodic and also vary in position depending on the monomer ratio. At present there is no information on the monomer sequence distribution from NMR analysis. (The polymer has

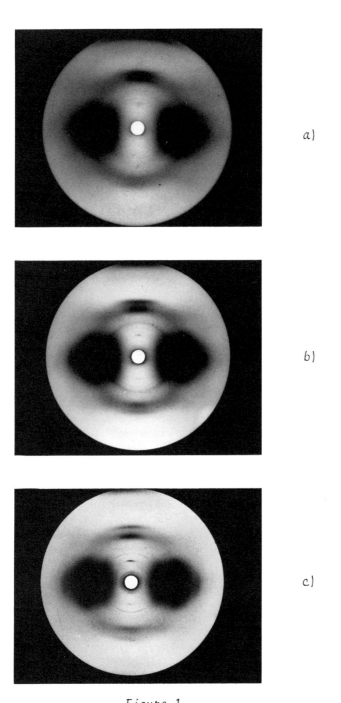

a)

b)

c)

Figure 1

X-ray fiber diagrams of copoly(HBA/DHN/TPA) in the following
monomer ratios: a) 60/20/20   b) 50/25/25   c) 40/30/30 [2]

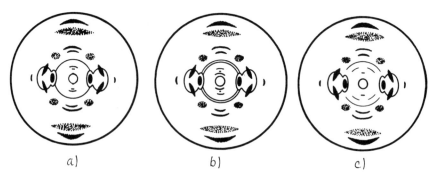

a)   b)   c)

Figure 2

Schematics of the X-ray fiber diragrams of copoly(HBA/DHN/TPA)
shown in fig. 1.   Monomer mole ratios:
a) 60/20/20   b) 50/25/25   c) 40/30/30   [2]

very low solubility in the only known solvent, pentafluorophenol,
and the monomers are very similar to each other.)   However, these
X-ray data rule out extensive block copolymer character. If this
were the case we should detect the diffraction characteristics of
each block at constant d-spacings in the three X-ray patterns.
Rather the data suggest a more random sequence of monomers.   In
the work described below we have predicted the diffraction patterns
of stiff aperiodic polymer chains and have shown that the data for
copoly(HBA/DHN/TPA) point to a completely random monomer sequence.
For HBA-modified PET the analysis for stiff aperiodic chains cannot
be applied because of the flexibility of the $-CH_2-CH_2-$ units, but
it will be seen that the data for these copolymers are also consistent with a completely random monomer sequence.   X-ray patterns
for HBA-modified PET containing 60 and 80 mole % HBA are shown
in fig. 3, and schematics of these are shown in fig. 4.   Further
details of the analyses summarized below are given in refs. 4-8.

EXPERIMENTAL

Melt spun fibers of copoly(HBA/DHN/TPA) were obtained from
Celanese Research Company for three monomer mole ratios: 60/20/20,
50/25/25, and 40/30/30.   These copolymers had been prepared by
melt copolymerization of the three monomers (using the acetoxy
derivatives of HBA and DHN) as described by Calundann [9].   Melt
spun fibers of HBA-modified PET were supplied by Tennessee Eastman
Company, and contained 60 and 80 mole % HBA.   These had been synthesized from PET and 4-acetoxy-benzoic acid by transesterification
in the melt, as described by Jackson and Kuhfuss [10].

X-ray fiber diagrams were recorded on Kodak No Screen film
using a Searle toroidal camera and Ni-filtered Cu Kα radiation.

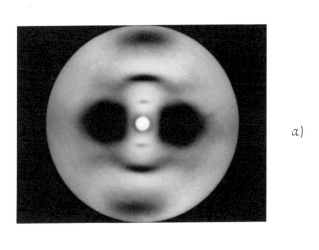

a)

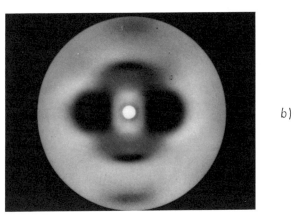

b)

Figure 3

X-ray fiber diagrams of HBA-modified PET.   Mole percentages:
a) 80% HBA   b) 60% HBA   [3]

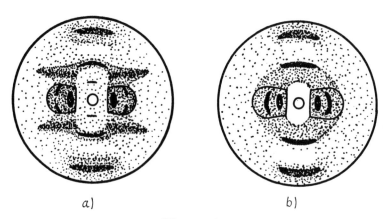

Figure 4

Schematics of the X-ray fiber diagrams of HBA-modified PET shown
in fig. 3. Mole percentages:
a) 80% HBA   b) 60% HBA   [3]

DIFFRACTION BY APERIODIC CHAINS

A model for a typical random sequence of copoly(HBA/DHN/TPA)
is shown in fig. 5a. The meridional scattering intensity is given
by the projection onto the fiber axis of this and all other possible
sequences. The chain is highly extended as a result of the geometry
of the 1,4 and 2,6 aromatic linkages and the planarity of the car-
boxyl groups. The only conformational freedom arises from rotation
about the aromatic-carboxyl linkages, and since these are approximately parallel to the chain axis, the length of the chain and hence
its projection on the fiber axis will be approximately independent
of the conformation. As a starting point we represented the residues
by points, placed for convenience at the ester oxygens, and separated
by the lengths of the residues. The lengths used for the residues
were 6.35Å for HBA, 7.15Å for TPA, and 7.83Å for DHN, based on the
structures of low molecular weight aromatic esters (see ref. 8).
Figure 5b shows the point model for the sequence in fig. 5a.

The scattering intensity I(Z) for a linear array of points is
given by

$$I(Z) = |F(Z)|^2 \qquad (1$$

where

$$F_c(Z) = \sum_1^n \exp 2\pi i Z z_j \qquad (2$$

172

Figure 5

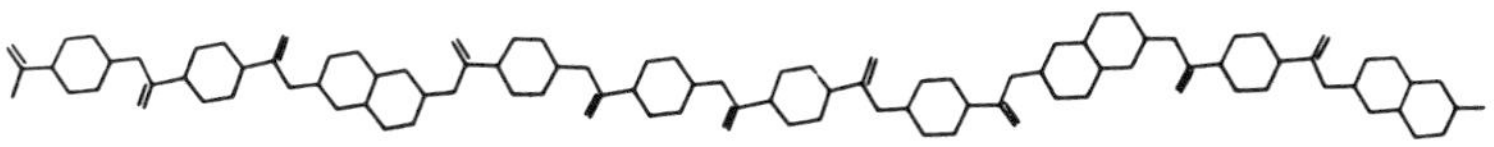

a) Model of a typical random sequence of copoly(HBA/DHN/TPA).
b) Point model for the sequence of copoly(HBA/DHN/TPA) in a). [5]

$F_C(Z)$ is the Fourier transform of a chain of n points along the axis in reciprocal space corresponding to the meridian of the X-ray fiber diagram, and z_j is the coordinate in real space of the jth point in the chain. In our first calculations we used a random number generator to set up monomer sequences and averaged I(Z) over, e.g. 200 chains of 40 monomers. The length of the chain here corresponds to a correlation of persistence length for the stiff conformation rather than the degree of polymerization, which is reported [9] to be ~ 150 for copoly(HBA/DHN/TPA).

A more systematic approach is to average $F_C(Z)$ over all N possible sequences:

$$I(Z) = \sum_{1}^{N} p_C |F_C(Z)|^2 \qquad (3)$$

where p_C is the probability of the cth chain sequence. This would be a lengthy calculation for long chains, but can be simplified by grouping the terms when the intensity is evaluated as the Fourier transform of the autocorrelation function for an average chain, Q(z):

$$I(Z) = \sum_{1}^{m} Q(z_j) \exp 2\pi i Z z_j \qquad (4)$$

where $Q(z_j)$ is the probability of point residues separated by z_j, and the summation is over all m values of z where Q(z) is non zero. Q(z) can be evaluated for copolymer chains using simple probability equations.

As will be seen, calculations based on a point model for copoly (HBA/DHN/TPA) successfully predict the positions of the observed meridional intensity maxima. However it is necessary to expand these calculations to determine whether this is also the case for an atomic model for the chains, which is a more realistic representation of the actual structure. Use of an atomic model is also essential for comparison of the observed and calculated intensities, so that we can allow for intraresidue as well as interresidue

interference effects. Conversion to an atomic model is effected
by separation of Q(z) into its components for the different residue
pairs:

$$Q(z) = \sum_A \sum_B Q_{AB}(z) \tag{5}$$

where $Q_{AB}(z)$ is the probability of a residue of type B at a separation +z from a residue of type A. The intensity is given by:

$$I(Z) = \sum_A \sum_B \sum_j Q_{AB}(z_j) F_{AB}(Z) \exp 2\pi i Z z_j$$

where $F_{AB}(Z)$ is the Fourier transform of the cross convolution of
residue B with residue A:

$$F_{AB}(Z) = \sum_a \sum_b f_{A,a} f_{B,b} \exp 2\pi i (z_{B,b} - z_{A,a}) Z \tag{6}$$

In equation (6), f is the atomic scattering factor, and the subscripts A,a and B,b designate the ath atom of residue A and the bth
atom of residue B, respectively. Atomic coordinates for the HBA,
DHN, and TPA residues are given in ref. 8, and were derived using
standard bond lengths and angles, consistent with the structures of
aromatic ester model compounds.

RESULTS

Copoly(HBA/DHN/TPA)

The calculated meridional intensity distributions for point and
atomic models for the three monomer ratios are shown in figs. 6 and
7. These data are for chains of 10 monomers with completely random
monomer sequence. We have found that at chain lengths greater than
6 - 7, the only effect on the calculated curves is in the widths of
some of the peaks, and the data for chains of 10 residues approximate well to the widths of the peaks in optical densitometer scans
of the X-ray patterns. For the analogous copoly(HBA/HNA) system,
Chivers et al. [3,11] have shown that chain lengths of 9 - 13 residues give good agreement for the peak widths, depending on the HBA
content.

The positions of the X-ray maxima are shown by arrows in fig.
6. The data (fig. 1) are characterized by three strong maxima: a
doublet in the region of 3Å that becomes more separated as the proportion of HBA is decreased, and a single maximum at $d \cong 2.0$Å for all
three compositions. The 60/20/20 and 50/25/25 patterns contain weak
doublets in the 6 - 7Å region: only a single reflection is seen in
that region of the 40/30/30 pattern. There are also a few much weaker
maxima that will not be considered further at the present time.

174

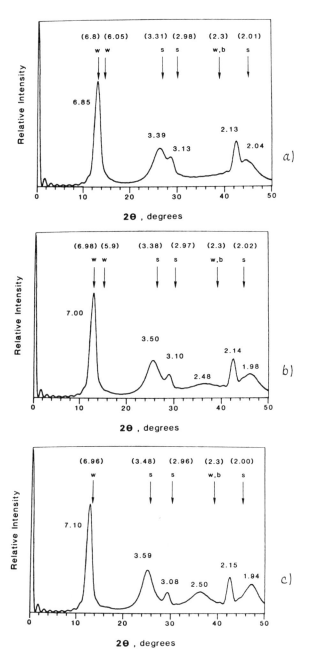

Figure 6

Meridional intensity I(Z) plotted against 2θ for point models of copoly(HBA/DHN/TPA). Monomer mole ratios:

a) 60/20/20   b) 50/25/25   c) 40/30/30   [6]

175

Examination of the intensity curve calculated for the 60/20/20 point model (fig. 6a) shows a maximum at (d=) 6.8Å, a doublet at 3.38 and 3.10Å, and a further maximum at 2.1Å. The subsidiary maxima around the origin and the peak at 2.1Å arise because the calculation are for a limited lattice of 10 points, albeit a distorted lattice. The agreement for the observed strong intensity maxima is very good, within 0.1Å, and the peak at 6.8Å is broad enough to encompass both of the weak maxima observed in the 6 – 7Å region. It must be remembered that with a point model we can only hope to match the peak positions: an atomic model is necessary before we can compare the intensities.

The intensity distributions for the 50/25/25 and 40/30/30 point models (figs. 6b and 6c) are very similar to that for 60/20/20, except that the doublet in the 3Å region becomes more separated: the separation increases from 0.3 to 0.4 to 0.5Å as the HBA content falls from 60 to 50 to 40 mole percent. For both ratios, the observed positions of the strong maxima are reproduced to within 0.1Å. The agreement between the observed and calculated peak positions could in fact be improved by refinement of the residue lengths, but this is best left until after an atomic model has been considered. It can be concluded that the X-ray data are consistent with a completely random monomer sequence, at least for the point model.

In fig. 7a, curve A shows the calculated meridional intensity for an atomic model of the 60/20/20 copolymer. The residues have the same lengths as in the point model (fig. 6), and the residues are assumed to lie with their ester oxygen--ester oxygen vectors parallel to the chain axis. The calculated intensities have been corrected for Lorentz and polarization effects, assuming the geometr of a $\theta/2\theta$ diffractometer scan. The subsidiary maxima seen for the monodisperse point model have been eliminated by averaging over a distribution of chain lengths centered on 10 residues. It can be seen that the major effect of intraresidue interferences is to reduc the intensity in the 6 – 7Å region, where a weak doublet at 3Å and the maximum at 2.1Å are still predicted, in approximately the same positions as before. There has been no prediction of new peaks nor unacceptable elimination of old peaks on conversion from the point to the atomic model. Examination of curves A in figs. 7b and 7c shows that this is also true for the 50/25/25 and 40/30/30 atomic

Figure 7 ⟶

Meridional intensity I(Z) plotted against 2θ for atomic models of copoly(HBA/DHN/TPA). Monomer mole ratios:
  a) 60/20/20   b) 50/25/25   c) 40/30/30
In each case curve A is for a model with the monomers parallel to the chain axis, and curve B is for the following monomer tilt angles: HBA 5°   DHN 25°   TPA 10°   [6]

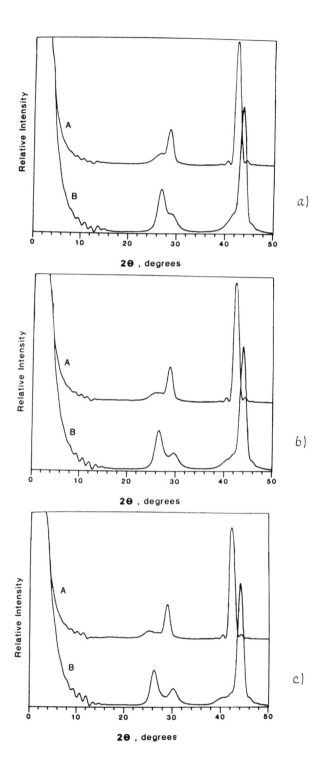

a)

b)

c)

models. The results justify our original view that, to a first
approximation, the intensity distribution can be derived from the
product of the interference function for the point model and the
transform of an "average" residue. The Fourier transforms of the
individual residues have minima at $d \cong \ell$ (the residue length) and
then oscillate slowly up to a maximum at $d \cong 2\text{Å}$. This accounts for
the low intensity calculated in the 7Å region and the strength of
the maximum at 2.1Å.

Closer examination of curves A in figs. 7a – c shows that the
intensity agreement for the 3Å doublet is imperfect in that the
inner maximum is weaker than the outer maximum, whereas the reverse
is observed in the fiber diagrams in fig. 1. Several explanations
are possible for this, but the most likely is that the model is
defective in that we have assumed that the monomers are all aligned
parallel to the chain axis. This can be seen to be only approximately correct in fig. 5a, which is the projection of a stereochemi-
cally acceptable model in which all the atomic-carboxyl inclinations
are set at 30°. It is clear that there will be a distribution of
the ester oxygen--ester oxygen vectors about the fiber axis, and
this can be modeled by tilting each residue. If each residue is
tilted by the same angle then this has no effect on the intensity
curve other than a proportional shift of the maxima to lower d-
spacings. However, examination of models shows that the DHN residues are more likely to be tilted as a result of the off-set
2,6-linkages; the HBA residues are least likely to be tilted,
with TPA intermediate between the other two. We have considered
many different tilt combinations, and the results presented in
fig. 7 as curve B for each monomer ratio are for models in which
HBA is tilted by 5°, TPA by 10°, and DHN by 25°. It can be seen
that this leads to a reversal of the intensities of the 3Å doublet,
for which good agreement is now obtained. This agreement is demonstrated in fig. 8 which shows the 3Å region of the calculated
curves compared with the optical densitometer scans of the observed
doublet. The latter were derived from fiber diagrams of specimens
tilted so as to record the meridional intensity in the region of
$d = 3.2\text{Å}$ as accurately as possible.

The above results demonstrate that we can reproduce both the
positions and the intensities of the meridional maxima using a model
for a completely random copolymer sequence. In separate work [12]
we have checked the sensitivity of the calculations to non randomness, i.e. blockiness, by calculation of Q(z) for unequal monomer
reactivities, and the results show that all but minimal blockiness
can be ruled out. Tilting the residues is only one of several
possible ways to refine the model so as to improve the intensity
match. We are presently investigating a distribution of tilt angles
(rather than a single tilt for each residue type) and details of
this work will be presented at a later date. A further factor that
could affect the meridional intensity is the presence of three-

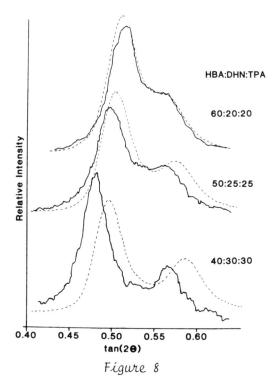

Figure 8

Comparison of the observed and calculated intensity in the 3Å meridional region for copoly(HBA/DHN/TPA). Monomer mole ratios:
    a) 60/20/20   b) 50/25/25   c) 40/30/30
Solid line: calculated intensities for tilted-atomic model.
Broken line: optical densitometer scan of meridional maxima recorded with fiber axis inclined at 76° to the beam direction.   [6]

dimensional order: this dictates a preferred axial stagger for some of the chains, rather than the random stagger assumed here. We are currently working on a three-dimensional structure, and an important clue to this is given by the data for HBA-modified PET, as described below.

HBA-Modified PET

The fiber diagrams of HBA-modified PET (fig. 3) are superficially very similar to those for the copoly(HBA/DHN/TPA) system: there are sharp meridional maxima, intense equatorial diffuse scattering, and some sharp equatorial and off-equatorial reflections [5]. These last Bragg reflections have d-spacings almost identical to those for copoly(HBA/DHN/TPA), indicating that there are features common to the packing for the ordered regions of both copolymers. However, a most interesting difference is that the meridional maxima are periodic. The 80% HBA copolymer shows meridionals at 6.3Å (weak), 3.15Å (medium), and 2.1Å (strong). In the case of the 60% HBA copolymer we only observe the 3.15 and 2.1Å reflections, and this, together with the general appearance of the X-ray pattern, suggests that this copolymer contains ordered regions that are less well developed than in the 80% HBA copolymer, but with the same basic structure.

Interpretation of these data for the copolymers is made possibl by examination of the homopolymer, poly(HBA), which is best studied by electron diffraction. Lieser [13] has shown that the poly(HBA) can exist in two different orthorhombic forms at room temperature. In addition, at high temperature, the polymer adopts a third, more open hexagonal structure [5], and the fiber diagram of this form shows all the reflections given by the copolymer. We can conclude that the fibers of HBA-modified PET (at high HBA content) contain ordered regions that have the same basic structure as the high temperature form of homopoly(HBA).

The diffraction analysis for aperiodic chains used for poly (HBA/DHN/TPA) cannot be applied to HBA-modified PET, because the flexibility of the (CH_2-CH_2-) units will make it impossible to assume constant axial lengths for the monomers. As a result we cannot say anything concerning the randomness of this set of copolymers. Economy et al. [14] and Lenz et al. [15] have argued in favor of HBA blocks in the 80% copolymer, although solution NMR work on the 60% copolymer suggests a random sequence. Nevertheless, although a blocky structure cannot be ruled out for the 80% copolymer, it is not necessary to explain the higher order because a number of ethylene terephthalate residues can probably be tolerated as defects within the open high temperature hexagonal lattice of poly(HBA). Such a model would simply require the segregation of HBA rich sequences, which will be fairly abundant in the random 80% copolymer, and less so in the 60% copolymer. For copoly(HBA/DHN/TPA the monomers are very similar to one another, and thus it is reasonable that chain packing can occur in part of the structure in a manner similar to that in the open form of poly(HBA).

ACKNOWLEDGMENTS

We thank Celanese Research Company, Summit, New Jersey, and Tennessee Eastman Corporation, Kingsport, Tennessee, for supplying the specimens used in this work, which was supported by NSF grants 81-07130 from the Polymer Program and 81-19425 (Materials Research Laboratory).

180

REFERENCES

1. J.-I. Jin, S. Antoun, C. Ober, and R.W. Lenz, Br. Polym. J. 12 132 (1980).
2. J. Preston, Ang. Makromol. Chem. 109/110 1 (1982).
3. R.A. Chivers, J. Blackwell, G.A. Gutierrez, J.B. Stamatoff, and H. Yoon, this volume.
4. J. Blackwell and G.A. Gutierrez, Polymer 23 671 (1982).
5. J. Blackwell, G. Lieser, and G.A. Gutierrez, Macromolecules 16 1418 (1983).
6. G.A. Gutierrez, R.A. Chivers, J. Blackwell, J.B. Stamatoff, and H. Yoon, Polymer 24 937 (1983).
7. J. Blackwell, G.A. Gutierrez, and R.A. Chivers, Macromolecules, in press.
8. G.A. Gutierrez, J. Blackwell, and R.A. Chivers, Polymer, in press.
9. G.W. Calundann (Celanese), U.S. Patent 4,184,996 (1980).
10. W.J. Jackson and H.F. Kuhfuss, J. Polym. Sci.-Polym. Chem. Ed. 14 2043 (1976).
11. R.A. Chivers, J. Blackwell, and G.A. Gutierrez, Polymer, in press.
12. G.A. Gutierrez and J. Blackwell, submitted to Macromolecules.
13. G. Lieser, J. Polym. Sci.-Polym. Phys. Ed. 21 1611 (1983).
14. A.E. Zachariades, J. Economy, and J.A. Logan, J. Appl. Polym. Sci. 27 2009 (1982).
15. R.W. Lenz and K.A. Feichtinger, Polym. Prepr. Am. Chem. Soc., Div. Polym. Chem. 20 114 (1979).

TRANSIENT SHEAR FLOW BEHAVIOR OF THERMOTROPIC

LIQUID CRYSTALLINE COPOLYESTERS

D. G. Baird, A. Gotsis and G. Viola

Department of Chemical Engineering and
Polymer Materials and Interfaces Laboratory
Virginia Polytechnic Institute and State University
Blacksburg, VA 24061-6496

INTRODUCTION

The most unique aspect about the processing of liquid
crystalline polymers is that the exceptional physical properties are
found in the as-spun fibers or as-injection molded parts[1,2].
Although heat treatment leads to improved properties, no further
drawing steps are required to generate highly oriented systems.
Hence, the orientation which is found in fiber and injection molded
specimens must arise during flow and be maintained during the
solidification process. The final physical properties must
therefore be directly related to the rheological properties of the
liquid crystalline fluids.

There have, however, been very few studies reported in the
literature concerned with the rheology of liquid crystalline
polymers. The most recent review concerned with this topic is given
by Wissbrun[3]. Most studies reported to date have been concerned
with the steady, unidirectional shear flow properties. Liquid
crystalline fluids apparently exhibit extremely long relaxation
times (the relaxation times may be two orders of magnitude higher
than those for very high molecular weight flexible chain polymers)
but no elasticity (i.e. no die swell is observed over certain
ranges of temperature and shear rate). In some cases these fluids
may exhibit yield stresses (τ_0) but this may not be a general
characteristic of the liquid crystalline state. It seems more
likely that the fluids are just pseudoplastic over a wide range of
shear rates (perhaps for shear rates as low as 10^{-4} sec^{-1}).
These fluids have also been observed to exhibit negative normal
stresses[3].

In steady shear flow experiments, the flow of liquid
crystalline polymeric fluids in most instances resembles that of
isotropic polymers[3]. However, there are some anomalous phenomena
such as extrudate contraction and negative primary normal stress
differences[3,6,8] which set these materials apart from isotropic
polymeric fluids. The most significant difference, however, is the
ability of liquid crystalline polymers to orient while flowing and
to maintain this orientation during the cooling process. The
textures and orientation generated during processing flows are most
likely responsible for the exceptional physical properties found in
these systems. It is believed that transient flow experiments may
reflect the development of structure and orientation and the loss of
this structure on cessation of flow. In the following paragraphs we
explore the response of two copolyesters of p-hydroxy-benzoic acid
(PHB) and polyethyleneterephthalate (PET) under various transient fl
conditions. We also look qualitatively at Ericksen's transversely
isotropic fluid theory[5] to see if it can predict some of the
observed response of these fluids.

EXPERIMENTAL

Two liquid crystalline copolyesters of 60 mole % PHB/PET and 80
mole % PHB/PET were used in this study. These samples were supplied
by Tennessee Eastman and the properties are given elsewhere[2].
Rheological properties were measured under various conditions of
shear and thermal history using a 0.1 radian cone and plate
attachment of a Rheometrics Mechanical Spectrometer (RMS). Various
types of flow histories were used in this study and these are listed
below:

1) Stress growth on the inception of shear flow

$$\dot{\gamma}_{yx} = 0 \qquad\qquad -\infty < t' \leq 0$$

$$\dot{\gamma}_{yx} = \dot{\gamma} \qquad\qquad 0 \leq t' \leq t$$

where $\dot{\gamma}$ is the shear rate which is constant, t' is the past time,
and t is the present time;

2) stress relaxation following shear flow

$$\dot{\gamma}_{yx} = \dot{\gamma} \qquad\qquad -\infty < t' < 0$$

$$\dot{\gamma}_{yx} = 0 \qquad\qquad 0 \leq t' \leq t$$

3) Reversal of shear flow

$$\dot{\gamma}_{yx} = \dot{\gamma} \qquad\qquad -\infty < t' < 0$$

$$\dot{\gamma}_{yx} = -\dot{\gamma} \qquad\qquad 0 \le t' \le t$$

RESULTS AND DISCUSSION

The first set of experiments are concerned with the response of PHB/PET copolyesters on the start up of flow. Stress growth at the start up of shear flow is presented for 60 mole % PHB/PET in Figures 1 and 2. At 240°C and 250°C the stress growth curves exhibit a single overshoot peak similar to high molecular weight flexible chain polymers. However, the maximum stress is nearly 300% higher than the equilibruim value. At temperatures of 260°C, 275°C and 285°C two peaks are observed. The first peak occurs at low strains, less than 2 strain units, while the second one occurs at 40 to 60 strain units depending on the temperature. The cause and significance of these double peaks are not clear at this point. In terms of the double peak behavior, several points should be brought out. At temperatures of 260°C and above, the texture of the melt is known to change[9]. As reported by Joseph and coworkers[9], at temperatures of 260°C and above, the melt appears under the polarizing microscope to consist of two phases both of which are

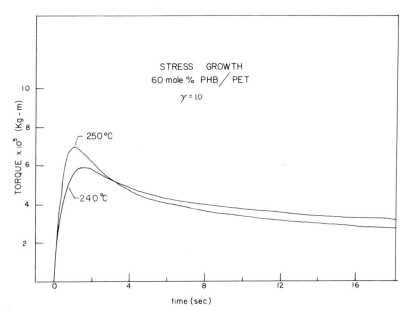

Fig. 1. Shear stress growth on the start up of shear flow for 60 mole % PHB/PET at 240°C and 250°C.

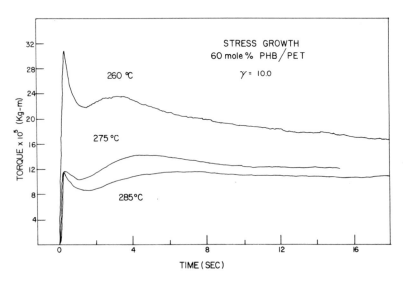

Fig. 2.   Shear stress growth on the start up of shear flow for 60
mole % PHB/PET at three temperatures.

birefringent. It may be that the first peak is associated with the
orientation of the domains which appear to be embedded in a matrix
of less ordered material and the second peak is due to the
deformation of the matrix material.

In Figure 3 we have presented the stress growth curves at 275°C
obtained at several different shear rates. The appearance of the
first peak occurs at $\dot{\gamma}$ values of about 1.0 sec.$^{-1}$ while the
second peak appears at values of $\dot{\gamma}$ of about 5.0 sec.$^{-1}$. In
flexible chain systems stress overshoot occurs at values of $\dot{\gamma}$
similar to the reciprocal of the longest relaxation time (τ). Based
on the shear dependence of viscosity[10], τ should be at least 100
sec. and hence for shear rates of the order of 0.01 sec.$^{-1}$,
overshoot should be observed. However, we observe that values of
$\dot{\gamma}$ must be of the order of 1.0 sec.$^{-1}$ before overshoot is
observed. Hence, we cannot associate the overshoot with the
relaxation processes which occur in flexible chain polymers.

In Figure 4 we have plotted values of the primary normal stress
difference (N_1) versus time. Here we observe that a single
overshoot peak is observed in N_1. The peak stress occurs at
strains of the order of 40 to 60 strain units which is similar to
the range observed for the appearance of the second peak in the
shear stress. Whereas the shear stress rises rapidly at the start
up of flow, the normal stresses rise gradually. It should be
pointed out that this is exactly the behavior predicted by the
corotational Jeffrey's model[11].

186

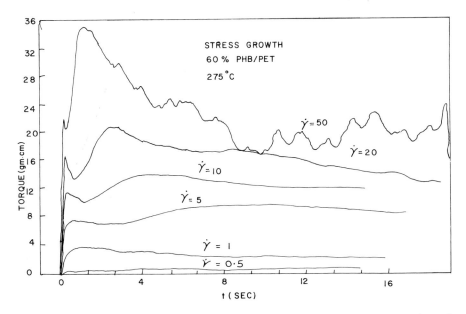

Fig. 3.   Shear stress growth at 275°C at different shear rates for
60 mole % PHB/PET.

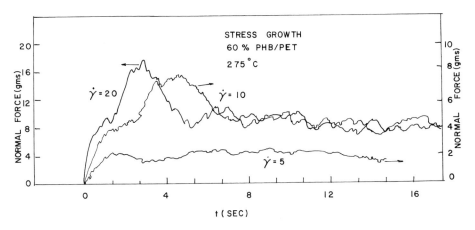

Fig. 4.   Normal stress growth on the inception of shear flow for 60
mole % PHB/PET at 275°C.

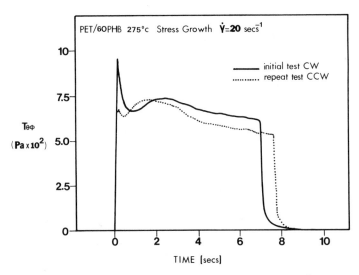

Fig. 5. Stress growth and relaxation behavior of 60 mole % PHB/PET at 275°C for an unsheared sample (——) and after reversing the flow direction (••••).

In Figure 5 are presented curves which show both the start up of flow behavior and the stress relaxation behavior of the shear stress. In the initial test (solid line) we observe that the shear stress ($\tau_{\theta\varphi}$) relaxes to zero in about one second, which is considerably faster than would be expected for a flexible polymer system with a similar value of τ. This result seems to suggest that there is no significant yield stress at least at 275°C. Once the stress relaxed to zero, the sample was subjected to a similar start-up experiment but with the direction of shear reversed. Here we note that the first peak is not reproduced but only the second peak. It is reasonable to expect the loss of the first peak, if we associate it with orientation changes during the start up of flow. Since the direction of the director ($\underset{\sim}{n}$), which describes the orientation of the liquid crystalline domains, does not change on reversing the flow direction, then it is reasonable to expect that no overshoot would occur. The appearance of the second peak suggests that it must be associated with a phenomenon which recovers within one or two seconds. It is possible that changes in texture could recover during the relaxation period of one to two seconds.

The shear reversal experiment was repeated but in this case reversal was carried out directly from steady shear flow rather than after allowing the stresses to relax. Results from this experiment are presented in Figure 6. The stress growth curve is significantly

188

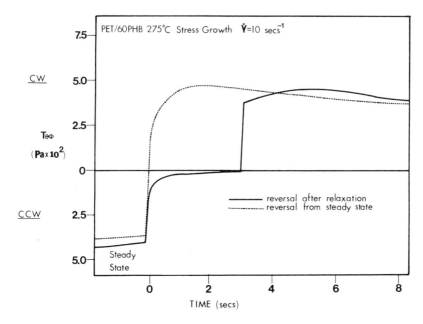

Fig. 6. Stress growth behavior of 60 mole % PHB/PET following reversal of flow from steady state (••••) and after stress relaxation (——).

altered in this case (dotted line) as the shear stress rises more gradually before passing through a maximum. The fact that no time is allowed for the texture to relax before the flow direction is reversed is reflected somewhat by the shape of this curve.

The 80 mole % PHB/PET system exhibits similar behavior for the shear stress as the 60 mole % PHB/PET system. This is illustrated in Figure 7 where two overshoot peaks are observed on the start up of flow. On reversing the flow direction, the stress almost rises instantaneously to its equilibrium value. On stopping the flow, the shear stress relaxes rapidly to zero stress.

However, the normal stress behavior at the start up of flow is significantly different as observed in Figure 8. Values of N_1 jump up to a positive value, then to a negative value, and then to a positive overshoot value. Even in steady flow conditions, N_1 values seem to oscillate about a mean value. On stopping the flow, N_1 relaxes rapidly (about 2 seconds) to a negative N_1 value. If the stress is monitored over a period of time, the residual negative values of N_1 seem to relax to zero. Hence, the 80 mole % PHB/PET system exhibits different behavior at the start up of flow than the 60 mole % PHB/PET system.

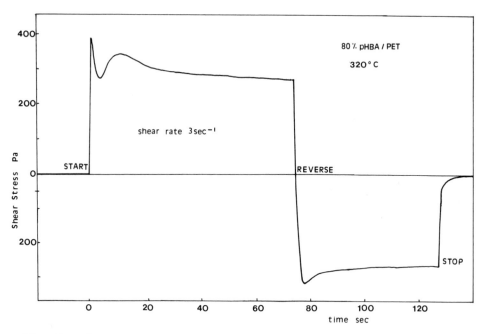

Fig. 7.   Stress growth behavior at the start up of flow and on reversal of flow of 80 mole % PHB/PET at 320°C.

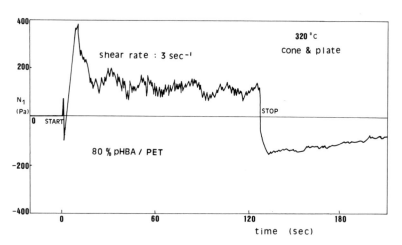

Fig. 8.   Stress growth and stress relaxation behavior of 80 mole % PHB/PET at 320°C.

Since negative values of N_1 have been observed for a cholesteric system[12], these values were investigated to a further degree. At low shear rates (for values less than 1.0 sec^{-1}), N_1 reached steady state values which were negative (see Figure 9). However, at values of N_1 greater than 1.0 sec.$^{-1}$, N_1 became positive. On cessation of flow, N_1 always relaxed to a negative value which was like a yield normal stress. The fact that only the 80 mole% system exhibited negative values of N_1 and that as the melt temperature was increased above 330°C, only positive values of N_1 were observed[6], suggested that the negative values may be associated with shear induced crystallization of PHB rich regions. Negative normal stresses have been observed for several other systems[3], but usually under slightly different circumstances suggesting the possibility of different mechanisms.

Comparison with Ericksen's Transversely Isotropic Fluid Theory
_____

Ericksen[5] proposed a theory for fluids such as nematic liquid crystals which could become anisotropic during flow. By assuming symmetry around the director, the expression for the stress tensor was somewhat simplified. We compare here briefly Ericksen's transversely isotropic fluid theory with the transient behavior observed for thermotropic copolyesters of PHB/PET.

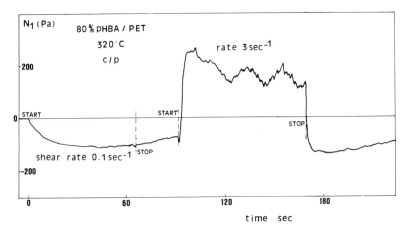

Fig. 9.   Primary normal stress difference versus time for two shear rates for 80 mole % PHB/PET at 320°C.

The extra stress ($\underset{\sim}{g}$) is given by the following expression:

$$\underset{\sim}{g} = \alpha_1\ \underset{\sim}{nn} + \alpha_2\ \underset{\sim}{\dot{\chi}} + \alpha_3\ \underset{\sim}{\dot{\chi}} \cdot \underset{\sim}{\dot{\chi}} + \alpha_4\ (\underset{\sim}{nn} \cdot \underset{\sim}{\dot{\chi}} + \underset{\sim}{\dot{\gamma}} \cdot \underset{\sim}{nn}) +$$
$$\alpha_5\ (\underset{\sim}{nn} \cdot \underset{\sim}{\dot{\chi}} \cdot \underset{\sim}{\dot{\chi}} + \underset{\sim}{\dot{\chi}} \cdot \underset{\sim}{\dot{\chi}} \cdot \underset{\sim}{nn}) \tag{1}$$

An expression for relating the orientation of the director to the flow was also derived:

$$\frac{\mathcal{D}n}{\mathcal{D}t} = \lambda\ \frac{1}{2}\ (\underset{\sim}{\dot{\chi}} \cdot \underset{\sim}{n} - \underset{\sim}{n}\ (\underset{\sim}{n} \cdot \underset{\sim}{\dot{\chi}} \cdot \underset{\sim}{n})) \tag{2}$$

where $\mathcal{D}/\mathcal{D}t$ is the Jauman derivative and λ is a material parameter which is related to the orientation of the domains. Here we have used the non-linear expression for stress and the linear equation for determining the director. $\underset{\sim}{\dot{\chi}}$ is the rate of deformation tensor and the α's are functions of the shear rate. For simple shear flow in which the velocity field is given by:

$$v_1 = \dot{\gamma}x_1 \qquad v_2 = v_3 = 0 \tag{3}$$

the various quantities of stress are

$$\sigma_{12} = \sigma_{21} = \alpha_1 n_1 n_2 + \frac{\alpha_2}{2}\ \dot{\gamma} + \frac{\alpha_4}{2}\ (n_1^2 + n_2^2)\ \dot{\gamma} + \frac{\alpha_5}{2}\ n_1 n_2\ \dot{\gamma}^2 \tag{4}$$

$$N_1 = \sigma_{11} - \sigma_{22} = \alpha_1\ (n_1^2 - n_2^2) + \frac{\alpha_5}{2}\ (n_1^2 - n_2^2)\ \dot{\gamma}^2 \tag{5}$$

$$N_2 = \sigma_{22} - \sigma_{33} = \alpha_1\ (n_2^2 - n_3^2) + \frac{\alpha_3\ \dot{\gamma}2}{4} + \alpha_4 n_1 n_2 \dot{\gamma} + \alpha_5 n_1 n_2 \dot{\gamma} \tag{6}$$

The director is obtained from the set of non-linear ordinary differential equations:

$$\frac{d\,n_1}{d\,t} = \frac{1}{4}\ \dot{\gamma}\ n_2\ (\lambda + 1 - 2\lambda\ n_1^2) \tag{7}$$

$$\frac{d\,n_2}{d\,t} = \frac{1}{4}\ \dot{\gamma}\ n_1\ (\lambda - 1 - 2\lambda\ n_2^2) \tag{8}$$

$$n_3^2 = 1 - n_1^2 - n_2^2 \tag{9}$$

The last equation is due to the constraint that $\underset{\sim}{n} \cdot \underset{\sim}{n} = 1$. These equations contain eight unknowns (5 α_i's and 3 n_i's). To solve them one needs to have both rheological data and information pertaining to the orientation of the director in flow.

Although it may be possible to measure this orientation using wide angle X-ray diffraction techniques[13], we present here only a qualitative comparison between the theory and experimental results. In Figure 10 we present the theoretical predictions of the shear stress at the start up of flow. We have assumed values of unity for

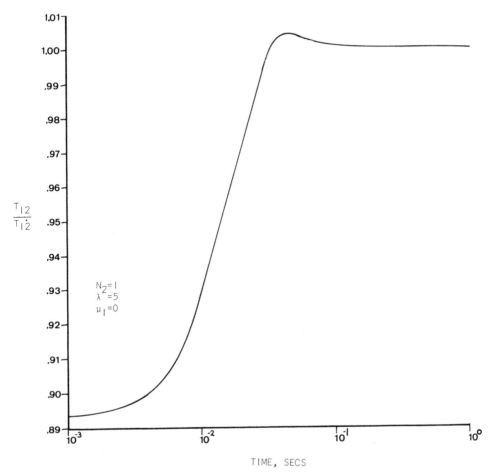

Fig. 10. Prediction of the shear stress using Ericksen's transversely isotropic fluid theory. The domains were assumed to be initially oriented along 2-direction (i.e. perpendicular to the shear surface).

the α's in eqns. (4-6) and have plotted the shear stress in reduced form. To solve the equations (7-9) one must assume an initial value for $\underset{\sim}{n}$. We have arbitrarily assumed that initially

$$\underset{\sim}{n} = n_2 \underset{\sim}{\delta}_2 \qquad (10)$$

(i.e. the molecules are oriented perpendicular to the shear planes). The shear stress is observed to rise instantaneously to an initial value of shear stress and then rise more slowly to the equilibrium

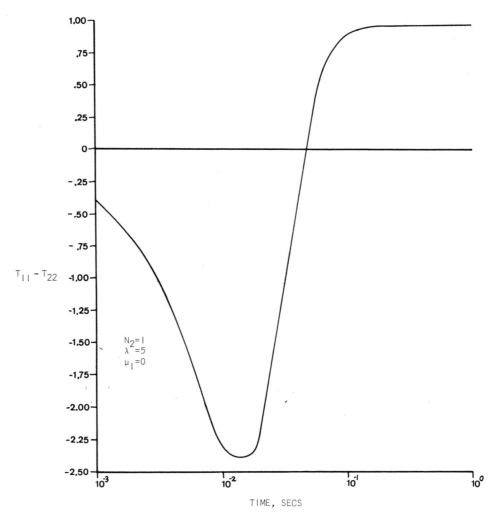

Fig. 11. Primary normal stress difference at the start up of flow
as predicted by Ericksen's theory. Values of N_1 are in
dimensionless form.

value. The primary normal stress difference is plotted versus time
in Figure 11. Here we observe that $\sigma_{11} - \sigma_{22}$ actually overshoots in
the negative direction and then rises to a positive equilibrium
value. This is in qualitative agreement with the observed behavior
of the 80 mole % PHB/PET system. (Note: values of $\sigma_{11} - \sigma_{22}$ were
estimated using arbitrarily assumed values for the α's.)

194

The Ericksen model also qualitatively predicts other observed behavior. It can predict yield stresses (see eq. (4)). It predicts that the stresses relax instantaneously to zero while the director remains oriented which seems to be what is observed. The model also predicts that the fluid orients in extensional flow in the flow direction even though it may not orient in shear flow. This is also what is observed experimentally.

CONCLUSIONS

The shear stress growth on the inception of shear flow may reflect the orientation of the liquid crystalline domains. Orientation seems to occur within less than 2 strain units in shear flow. This primary normal stress difference can exhibit different phenomena from the shear stress response. In particular for the 60 mole % PHB/PET system, values of N_1 are positive and rise gradually to the equilibrium values whereas the 80 mole % PHB/PET system can exhibit negative values of N_1. Ericksen's transversely isotropic fluid theory can qualitatively handle some of the observed phenomena. Further studies which couple the transient flow behavior to the orientation and morphology need to be carried out.

ACKNOWLEDGEMENTS

Support for this research from the Army Research Office (Grant No. DAA29-80-K-0093) is sincerely appreciated.

REFERENCES

1. J. R. Schaefgen, U.S. Patent 4,118,372 (1978).
2. W. J. Jackson, Jr. and H. Kuhfuss, J. Polym. Sci., Polym. Chem. Ed., 14:2043 (1976).
3. K. F. Wissbrun, J. Rheol., 25:6, 619 (1981).
4. Y. Onogi, J. L. White and J. F. Fellers, J. Non-Newtonian Fluid Mech., 7:121 (1980).
5. J. L. Ericksen, Arch. Ration. Mech. Analysis, 4:231 (1960).
6. A. Gotsis, M.S. Thesis, Virginia Polytechnic Institute and State University, Blacksburg, VA (1984).
7. D. G. Baird, J. Rheol., 24:465 (1980).
8. R. E. Jerman and D. G. Baird, J. Rheol., 25:275 (1981).
9. E. Joseph, G. L. Wilkes and D. G. Baird, Polymer, In Press.
10. A. Gotsis and D. G. Baird, J. Rheol., In Press.
11. R. B. Bird, R. C. Armstrong and O. Hassager, "Dynamics of Polymeric Liquids: Fluid Mechanics", John Wiley, 1977, New York.
12. G. Kiss and R. S. Porter, Mol. Cryst. Liq. Cryst., 60:267 (1980).
13. D. G. Baird, G. Viola, E. Joseph and R. Pisipati, Society of Plastics Engineers Technical Papers, New Orleans, (1984).

PRELIMINARY THERMAL AND STRUCTURAL STUDIES OF BLENDS
BASED ON A THERMOTROPIC LIQUID CRYSTALLINE COPOLYESTER
AND POLY(ETHYLENE) TEREPHTHALATE

Eugene G. Joseph, Garth L. Wilkes, and Donald G. Baird

Department of Chemical Engineering and
Polymer Materials and Interfaces Laboratory
Virginia Polytechnic Institute and State University
Blacksburg, VA 24061-6496

INTRODUCTION

Due to economic, technological, and regulatory pressures, there
has been a gradual narrowing of the chemical variety of polymers
being produced[1]. In order to obtain new materials with unique
properties, one of the approaches taken by polymer scientists is to
use polymer blends. Some of the reasons as to why these materials
are attractive are (i) the ability to obtain higher performance
materials economically, (ii) modification of performance as a market
develops, (iii) extend the performance of an expensive resin and
(iv) re-use of plastics scrap through blending and (v) generation of
a unique material in terms of processability or performance[1]. The
two principal means by which polymer blends can be prepared are
mechanical mixing or casting from solution.

Presently, there is also a significant amount of interest in
polymeric systems that exhibit liquid crystalline behavior since
they often can develop high strength - high modulus properties. For
a polymer to form a liquid crystalline phase, the molecule or a
major portion of it should be stiff and rodlike or platelike in
nature. The presence of polarity may also be advantageous.

Liquid crystalline systems can be either lyotropic or
thermotropic. In the case of lyotropic systems, liquid crystalline
characteristics are found in solution above a critical
concentration, and are concentration as well as temperature
dependent. On the other hand, thermotropic systems show liquid
crystalline behavior in the melt and are temperature dependent with
respect to mesophase behavior.

197

Jackson and Kuhfuss[2] reported a thermotropic liquid
crystalline polymer which was a copolyester of poly(ethylene)
terephthalate (PET) and parahydroxybenzoic acid (PHB). In this
copolyester, PET has a flexible chain conformation while PHB has a
stiff, rod-like conformation and the polymer is indicated to be the
liquid crystal forming component. Schematically, this copolyester
can be represented as $\sim\!\!\sim\!\!\sim$——$\sim\!\!\sim\!\!\sim$———$\sim\!\!\sim\!\!\sim$, where $\sim\!\!\sim\!\!\sim$ represents
the flexible PET units or sequences and — represents the stiff
PHB units. A series of copolymers containing different levels of
PHB were synthesized by these researchers and some unique properties
have been reported[2]. In general, the 60 mole % PHB-40 mole % PET
material had more desirable properties in terms of processing and
mechanical properties. While further rheological, morphological and
thermal studies on these PET/PHB copolymers have been performed by
us[3-6] and others[7-10], no work has been reported to our knowledge
on the blends of these liquid crystal systems with other polymers.

In general, when a rigid rod polymer and a flexible polymer are
mixed together, one would expect them to be incompatible. However,
since the liquid crystalline PET/PHB polymer referred to in this
paper is viewed as semi-flexible and has PET as one of its
components, it was felt that if this material was blended with pure
PET, some degree of compatibly might result. We also suspected that
this might be possible since we believe the PET/PHB chains are not
purely a random copolymer, i.e., we[4-6] and others have suggested
that these systems appear to display non-random character based on
various thermal, mechanical and microscopy studies[7,8]. The
purpose of this paper is therefore to report the results of initial
thermal, morphological and mechanical studies conducted on blends of
one of these liquid crystalline polymers (60 mole % PHB-40 mole %
PET) with PET. The 60 mole % PHB - 40 mole% PET copolyester will be
referred to as the liquid crystal polymer (LC) while the second
component will be denoted as PET homopolymer. These two resins were
blended using a solution casting technique, mechanically mixing in
the melt and, by injection molding. The resulting solid materials
were then analyzed with respect to their thermal and structural
characteristics.

EXPERIMENTAL

Materials

The two resins, LC polymer and PET homopolymer, were obtained
in pellet form from Tennessee Eastman Company. The initial resins
and the blends were identified as PET/LC-0, PET/LC-30, PET/LC-70 and
PET/LC-100 where PET and LC represent the polymer components
utilized to prepare the blend and the numbers indicate the amount of
LC used in weight percent. For example, PET/LC-0 represents PET
homopolymer while PET/LC-50 represents a blend containing 50 weight

percent LC polymer. In preparing the solution cast materials, trifluoroacetic acid (TFA) was used as the solvent and films were cast at room temperature in a Teflon mold. Blended materials were also obtained by mechanically mixing in the melt. The two resins were mixed in a Maxwell mixer (also known as a melt elastic extruder) at ∿ 290°C and then pressed into thin films at 280°C. These films were then quenched in an ice water bath. Injection molded plaques were obtained by mixing the two resins in pellet form at room temperature and then injection molding at 275°C using an Arburg injection molding machine. The plaque dimensions were 4"x4"x1/6".

Techniques

An ISI Super III-A scanning electron microscope was utilized to study the top surfaces of the final solution cast materials while light microscopy studies were carried out using a Zeiss polarizing optical microscope in conjunction with a Mettler FP2 hot stage. A Perkin-Elmer model DSC-4 with a SYSTEM-4 microprocessor was used for thermal studies. A heating rate of 20°/min. was utilized and the materials were scanned from 30°C to 300°C. Following the first scan, the samples were quickly quenched to room temperature and reheated using the same conditions. The crystallization behavior from the melt was monitored by heating the materials past the melting temperature in the DSC and then cooling down at 20°/min. Bending modulus measurements on the samples obtained from injection molded plaques were made using an Instron tensile tester equipped with two aluminum/stainless steel pieces. A schematic of the apparatus is given in Figure 1 and a detailed description is given elsewhere[11]

RESULTS AND DISCUSSION

Solution Cast Blends

Polarized optical micrographs of the solution cast films are shown in Figure 2 and they suggest the presence of a spherulitic morphology due to the appearance of a maltese cross structure. The optical micrograph of pure LC polymer, i.e. PET/LC-100 (not shown) surprisingly displays little depolarization and, not surprisingly, no sign of spherulitic texture. In order to confirm or disprove the occurence of this apparent spherulitic morphology, the techniques of small angle light scattering (SALS) and scanning electron microscopy (SEM) were utilized. The SALS patterns of the solution cast films are given in Figure 3 and it is evident that PET/LC-0, PET/LC-30 and PET/LC-70 show four-leaf clover scattering patterns characteristic of a spherulitic morphology. Also, the size of the scattering patterns decreases with LC content. This clearly

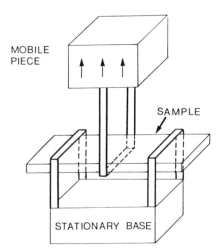

MOBILE
PIECE

SAMPLE

STATIONARY BASE

Fig. 1. Schematic diagram of the flexural test device which shows sample, rod positions and motion of the rods during loading (from ref. 11).

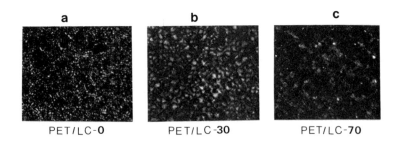

a   b   c

PET/LC-0   PET/LC-30   PET/LC-70

Fig. 2. Polarized optical micrographs of (a) PET/LC-0, (b) PET/LC-30 and (c) PET/LC-70 films cast from TFA. Magnification 125x.

indicates that the size of the spherulites increases with the addition of LC polymer since there is an inverse relationship between the scattering patterns and the size of spherulites. In the case of the PET/LC-100 material, no distinct scattering pattern is observed. Scanning electron micrographs of the solution cast free surfaces of PET/LC-0, PET/LC-30 and PET/LC-70 are shown in Figure 4(a)-(c). All three micrographs clearly display the presence of a spherulitic superstructure with the size increasing from approximately

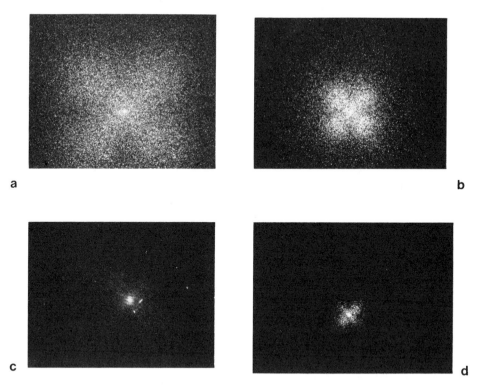

Fig. 3.   Small angle light scattering patterns obtained at the same
sample to film distance for films cast from TFA.   (a) PET/
LC-0, (b) PET/LC-30, (c) PET/LC-70 and (d) PET/LC-100.

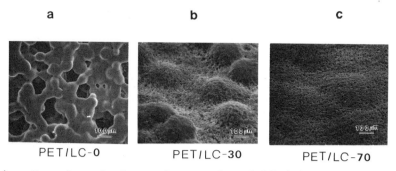

PET/LC-0   PET/LC-30   PET/LC-70

Fig. 4.   Scanning electron micrographs of PET/LC blends cast from
TFA (top surfaces).   (a) PET/LC-0, (b) PET/LC-30 and (c)
PET/LC-70.

201

10 microns to about 70 microns as the LC polymer content is
increased. This observation is in agreement with results obtained
by SALS measurements although it is clear that as LC content
increases, the texture of the spherulite superstructure is somewhat
atypical. Specifically, the SEM micrographs show that while the
surface texture of PET/LC-0 is smooth, a porous texture is present
in the PET/LC-30 and PET/LC-70 materials. It should be pointed out
that the mechanical properties of the solvent cast blends are
extremely poor which likely can be attributed to the porous nature
of the films. The porous nature also appears to be more uniform in
the higher LC polymer blend. In order to obtain further information
about this porous texture, the solution cast films were fractured in
liquid nitrogen and analyzed by SEM. The fracture surfaces are
shown in Figure 5. Clearly two distinct textures are observed; one

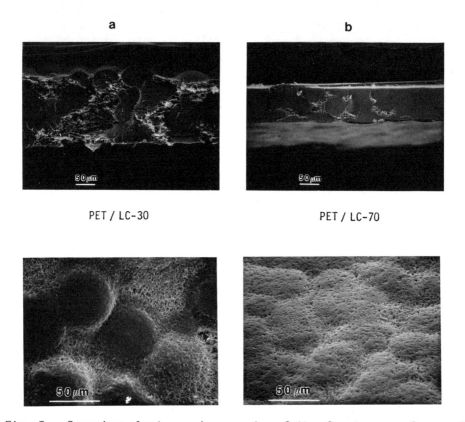

a   b

PET / LC-30   PET / LC-70

Fig. 5.   Scanning electron micrographs of the fracture surfaces of
          PET/LC blends cast from TFA along with corresponding top
          surfaces. (a) PET/LC-30, (b) PET/LC-70.

being solid like while the other is porous. As LC polymer content is increased the solid region increases, however, it is not possible to associate the solid "spherulitic" regions with either specific component of the blend. It is speculated that while the porous texture may be coupled to the solvent evaporation process, the solid regions consist of both PET homopolymer and LC polymer with one component dispersed in the other but not at the molecular level. That is, this dispersement is present at the macroscopic level but is not at the microscopic level for reasons to be discussed below.

In an attempt to verify the level of dispersion of one component in the other, hot stage microscopy studies were carried out. Polarized optical micrographs of PET homopolymer (i.e. PET/LC-0) obtained at 25°C and 250°C are given in Figure 6. From these it is clear that the depolarization at room temperature can be attributed to the presence of PET crystallinity, for at 250°C the depolarization is no longer present due to the melting of the PET crystals which is later confirmed by DSC analysis. Polarized optical micrographs of the PET/LC-30 blend obtained during a similar heating experiment are given in Figure 7. Again, the initial depolarization that is present can be attributed to PET crystallinity. However, the micrograph obtained at 275°C shows a decrease in depolarization but it also clearly shows that depolarization still remains in random regions. This lends some support to our earlier speculation that the LC polymer is dispersed throughout the PET matrix but not uniformly.

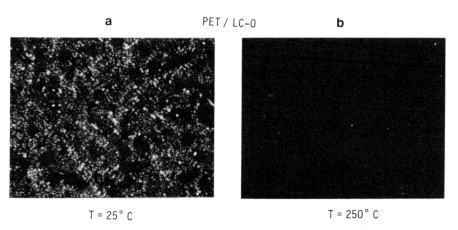

a   PET / LC-0   **b**

T = 25° C   T = 250° C

Fig. 6. Polarized optical micrographs of the solution cast PET/LC-0 film obtained at (a) 25°C and (b) 250°C.

PET / LC-30

T = 25° C   T = 120° C   T = 135° C

T = 180° C   T = 275° C

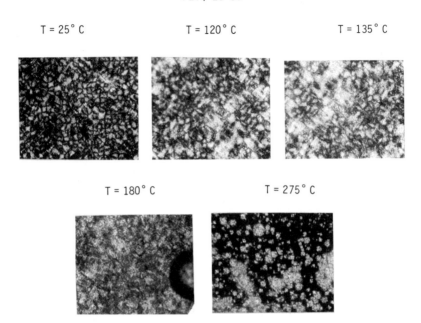

Fig. 7.   Polarized optical micrographs of the solution cast PET/LC-30
film obtained at different temperatures during a hot stage
experiment.

The thermal properties of the solution cast films were
investigated by differential scanning calorimetry and the DSC curves
are given in Figure 8.   An endothermic peak associated with the
melting of PET is observed in the PET/LC-0, PET/LC-30 and PET/LC-70
films.   However, a distinct exothermic (crystallization) peak that
is seen when amorphous PET is heated in the DSC is not seen in the
above mentioned scans.   This clearly indicates that PET
crystallinity is well developed in the solution cast films prior to
the DSC scan and that very little further crystallization occurs
during the scan.   As the amount of LC polymer is increased, the
melting temperature (T_m) remains nearly unchanged and the heat of
fusion (ΔH_f) per gram of sample decreases in line with the amount
of PET homopolymer present in the blend (see Table 1).   However,
when ΔH_f is normalized on the mass of PET homopolymer present in
the blend, the values of ΔH_f increase with LC polymer content
suggesting that a higher amount of PET crystallinity is present.
Specifically, crystallinity values increase from 45 percent to 54
percent as the LC polymer content increases from 0 wt.% to 70 wt.%.
These values which seem surprisingly high for the LC containing
systems are based on a heat of fusion value of 30 cal/gram for 100%
crystalline PET, and was obtained from the literature[12,13].   Further

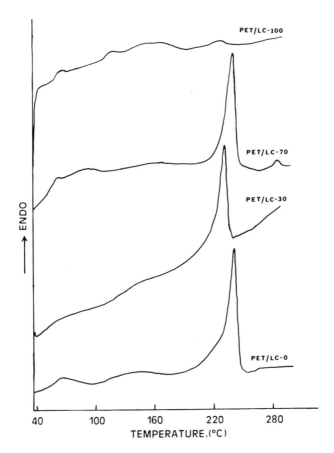

Fig. 8. Initial differential scanning calorimetry scans of the PET/LC blends cast from TFA.

Table 1. Differential Scanning Calorimetric Results of PET/LC Blends Cast from TFA (Initial heating experiment)

	T_g (°C)	T_c (°C)	T_m (°C)	ΔH_f (cal/gr of sample)
PET/LC-0	--	98	244	13.57
PET/LC-30	--	110	240	8.89
PET/LC-70	--	113	244	4.82
PET/LC-100	--	--	--	--

comment on this enhanced apparent crystallinity will be made later.

In order to erase all previous thermal history, the samples were heated up to 300°C at 20°/min in the DSC and quickly quenched. Then the same heating experiment was performed. The resulting DSC scans are given in Figure 9. The films that contain PET homopolymer now show a glass transition temperature (T_g), a distinct crystallization exotherm followed by a melting endotherm. Various parameters obtained from these scans are given in Table 2. As the amount of LC polymer is increased, the T_g remains unchanged while the crystallization temperature (T_c) and T_m are lowered. A T_g of 73°C is observed for the PET/LC-0, PET/LC-30 and PET/LC-70 samples and it is associated with the glass transition of the PET component.

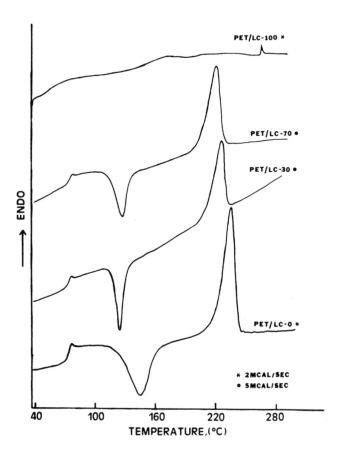

Fig. 9. Differential scanning calorimetry scans of the PET/LC blends from TFA that have been melted and then quenched in the DSC.

Table 2. Differential Scanning Calorimetric Results of
PET/LC Blends Cast From TFA (Second heating
experiment)

	T_g	T_c	T_m	ΔH_f	ΔH_c	$(T_m - T_g)$
PET/LC-0	73	145	236	8.88	6.21	163
PET/LC-30	73	128	234	6.88	5.00	161
PET/LC-70	74	128	223	4.17	2.12	150
PET/LC-100	--	--	--	--	--	--

The lowering of T_c in the blends may be explained by
considering the crystallization of the pure PET component. For a
given polymer, $(T_m - T_g)$ can be viewed as the "window for
crystallization". Pure PET has a $(T_m - T_g)$ value of approximately
163°C. As the amount of LC polymer is increased, the experimentally
observed T_m is distinctly decreased (which is in contrast to the
first thermal scan), while T_g appears unchanged thereby causing
$(T_m - T_g)$ to decrease. This observation in conjunction with the
lowering of T_c suggests that the LC component in the blend may
possibly be acting as a nucleating agent for the crystallization of
PET. Also supporting this concept are the values for $(\Delta T)_c$, which
represent the width of the crystallization temperature region.
Values for $(\Delta T)_c$ decrease from approximately 42°C to 24°C as the
liquid crystal polymer content increases from 0 wt. % to 70 wt. %
respectively. This indicates that as the level of the possible
nucleating LC species is increased, the crystallization rate is
apparently enhanced and hence a narrower $(\Delta T)_c$ is observed. This
speculation is somewhat similar to what Takayanagi et al. recently
proposed for the effect of Aramid surfaces on the crystallization
of Nylon 6[14]. In addition, the melting endotherm associated with
pure PET and, the associated observed depression of T_m (especially
in the PET/LC-70 material) appears to be due to the fact that the LC
polymer may be acting as a diluent. However, it is somewhat
surprising that this effect is present only after the first melting
of the solution cast films. No further changes in the transition
temperatures occur when cycled between 25°C and 295°C in the DSC.

As stated earlier the porous nature of the films gives rise to
poor mechanical properties. However, since the thermal properties
show interesting behavior (for example the possible nucleating
effect caused by the LC polymer) new blends were prepared by
mechanically mixing in the melt and these were then investigated in
terms of their morphology and the thermal behavior as discussed
within the next section.

Mechanically Mixed Blends

As mentioned earlier in the MATERIALS section, the two resins were mechanically mixed in a CSI-Max Mixing Extruder and then pressed into thin flims and quench cooled in ice water. The films were fractured in liquid N_2 and the SEM micrographs of these fracture surfaces are given in Figure 10. A non-homogeneous morphology is present in all three blends with the size of the spherical regions increasing from 0.5 microns to approximately 20 microns as LC polymer content increases form 30 wt % to 70 wt %. The spherical regions are associated with the LC Polymer and the micrographs may suggest that little compatibility exists between the two polymers. However, when the spherical regions are analyzed at higher magnification there appears to be signs of the disruption of adhesion between the LC polymer and PET homopolymer thereby suggesting that the two components may well indeed "wet" one another. Further investigations of this point is underway.

In order to verify if the thermal properties observed in the solution cast films were altered within the mechanically mixed blends, DSC measurements were performed with the results given in Figure 11. It is clear that a T_g, crystallization exotherm and, a melting endotherm due to PET are present in the initial scan. (Recall that for the solution blends, PET crystallinity was well developed within the cast films.) Some of the parameters obtained from the DSC scans have been tabulated and are given in Table 3. It is evident that the T_g of PET again remains unchanged with the addition of LC polymer content; however, there is a depression of T_m especially in the blend of highest LC content. Again, we believe the melting point depression of PET is due to a partial

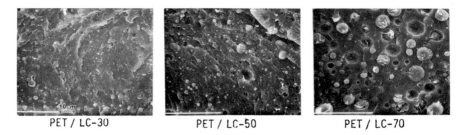

PET / LC-30   PET / LC-50   PET / LC-70

Fig. 10. Scanning electron micrographs of the fracture surfaces of the PET/LC blends that were mechanically mixed in the melt. (a) PET/LC-30, (b) PET/LC-50 and (c) PET/LC-70.

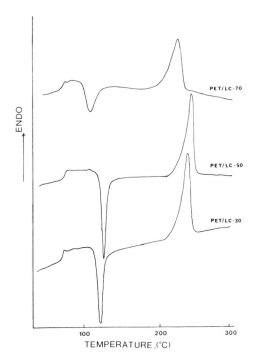

Fig. 11. Differential scanning calorimetry scans of the PET/LC blends mechanically mixed in the melt.

Table 3. Differential Scanning Calorimetric Results of Mechanically Mixed PET/LC Blends (Initial heating experiment)

	T_g (°C)	T_c (°C)	T_m (°C)	$(T_m - T_g)$ (°C)	ΔH_f (Cal/Gr. of PET Homopolymer)
PET/LC-0	78	148	250	172	8.69
PET/LC-30	74	122	242	168	12.37
PET/LC-50	74	125	243	169	19.48
PET/LC-70	74	106	225	151	20.63
PET/LC-100	76	96	188	112	0

diluent effect caused by the LC polymer although we do not believe this "diluent" is uniformily distributed. In terms of crystallization, the "window for crystallization" for a given blend, i.e.

(T_m-T_g), is again narrowed caused by the constant value of T_g and a lowering of T_m as LC content increases. This is particularly evident in the PET/LC-70 blend. One would expect that when (T_m-T_g) is narrowed, the rate of crystallization would decrease since it would be more difficult for the crystallization process to proceed. However, when one observes values for the breadth of the crystallization temperatures (ΔT_c) of the blends, it is apparent that upon the addition of LC polymer, there is a decrease in the ΔT_c values. Since ΔT_c can be viewed as a indicator of the rate of crystallization, the results suggest that the crystallization rate of PET homopolymer has somehow been enhanced upon the addition of LC polymer even though the apparent "window for crystallization" is narrowed. Again, one possible explanation for this behavior is that the LC polymer is acting as a nucleating agent for the crystallization of PET. Scanning electron micrographs of the fracture surfaces (Figure 10) show a random dispersion of LC polymer in PET and are in agreement with this speculation.

In an attempt to confirm this nucleating effect of the LC component, the blends were heated to 295°C in the DSC and then cooled back to room temperature at 20°/minute. The crystallization process was this time monitored during the cooling experiment and the corresponding DSC scans are given in Figure 12. Quantitative information on the crystallization behavior were obtained from these scans by two different parameters, namely, the area under the crystallization exotherm and, the width of the crystallization peak. The area under the crystallization exotherm is a measure of the amount of PET crystallinity, while the width of the crystallization peak can be used as an indicator for the time for crystallization to occur during the cooling cycle. These two parameters have been calculated and are shown in Figures 13 and 14. As LC polymer content increases, the heat of crystallization (which has been normalized per gram of PET homopolymer) increases linearly with a 200% increase seen in the PET/LC-50 blend! We believe these results occur due to an increase in the amount of nucleating species (LC polymer) which in turn enhances the crystallization of the PET homopolymer. At a LC polymer content of 50 weight percent, a critical level is reached where upon further addition of LC polymer, no significant increase in PET crystallinity is observed. The time required for crystallization shown in Figure 14 indicates that there is a linear decrease with the addition of LC polymer. In other words, as LC polymer content increases, the rate of crystallization of PET homopolymer increases by approximately a factor of 2.

The heat of crystallization values that are normalized per gram of PET homopolymer and obtained from DSC scans during the cooling cycle (Figure 13) show an increase with increasing LC polymer content. While this may be explained as due to the higher amount of

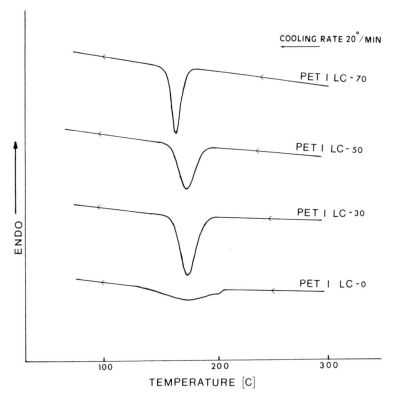

Fig. 12. Differential scanning calorimetry scans of the PET/LC
materials during the cooling experiment from the melt.
Cooling rate = 20°/min.

LC component present and hence a higher degree of nucleation, one
might expect that the amount of crystallinity based on PET
homopolymer would be approximately the same in all PET/LC blends if
annealed at a higher temperature. In other words, if all the PET/LC
blends were given appropriate conditions for crystallization, the
amount of PET crystallinity in all cases might be the same. In an
attempt to verify this, the PET/LC blend series were annealed at
150°C for 4 hours so that the crystallization level of PET
homopolymer might be enhanced. Differential scanning calorimetric
studies on these annealed materials show that as LC polymer content
increases, the heat of fusion (and hence the amount of
crystallinity) normalized per gram of PET homopolymer also

211

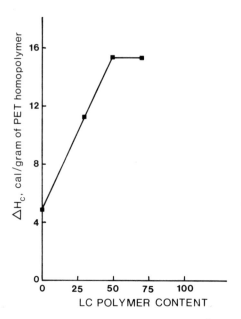

Fig. 13. Plot of enthalpy of crystallization (based on per gram of PET homopolymer) versus liquid crystal polymer content for the PET/LC materials mixed in the melt.

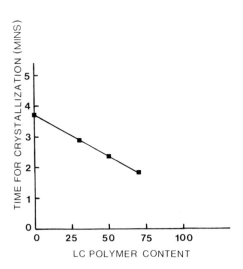

Fig. 14. Plot of "time for crystallization" versus liquid crystal polymer content for the PET/LC materials mixed in the melt.

212

increases. Specifically, the degree of crystallinity per gram of PET homopolymer increased from ∿ 36% for PET/LC-0 to ∿ 72% for PET/LC-70. This result is rather unusual in that as the LC content of the blend increased, the level of apparent PET crystallinity increased above the usual upper levels of pure PET under the same thermal conditions. Again, no good explanation is available for this behavior at this time. (Recall that similar results were denoted in the solution cast films discussed earlier although the final crystallinity levels were somewhat lower).

Injection Molded Plaques

It has been shown by Jackson and Kuhfuss that when the pure LC polymer is injection molded, materials with high flexural modulus can be obtained[2]. It was also reported that these injection molded plaques were highly anisotropic at small thicknesses (e.g. 1/16"), with the anisotropy decreasing as thickness is increased. An attempt was made to verify if similar high modulus materials can be obtained when a physical blend of PET homopolymer and LC polymer were injection molded into plaques of 1/16" thickness. The following materials were injection molded: PET/LC-30, PET/LC-50 and PET/LC-70.

Again, the injection molded plaques used in this study were 1/16" in thickness. Bending modulus measurements were made both along the flow direction and across the flow direction on the PET/LC materials and the results are given in Figure 15. As LC polymer content increases, the "along the flow" modulus of the PET/LC materials increases and at a LC polymer content of 50 weight percent this value is approximately three times the value of PET homopolymer[2]. However, this modulus value is ∿ 33% lower than the "along the flow" modulus value obtained for the pure LC polymer injection molded under similar conditions.

The morphology of these injection molded plaques were analyzed by fracturing specimens across the flow and along the flow, and then observed using SEM. The fracture surfaces of the PET/LC-50 material are shown in Figure 16 with the SEM micrographs representing the skin and core regions. The micrographs indicate the presence of a skin-core morphology that is clearly non-homogeneous in nature. In the skin region, the LC phases are elongated and appear as macroscopic "rods" oriented along the flow direction but are dispersed within the PET homopolymer matrix. This skin morphology is believed to give rise to the high bending modulus values when the material is fractured across the flow. The core regions on the other hand appear to have very little orientation but the dispersion of LC polymer in PET is evident. This lack of orientation of the LC phases in the core may be due to the relaxation of the flexible PET molecules caused by thermal effects,

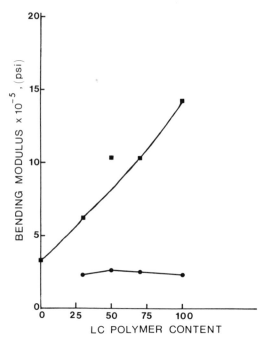

Fig. 15. Plot of bending modulus (along and across the flow
direction) versus liquid crystal content. Specimen
thickness = 1/16" and molding temperature = 275°C.

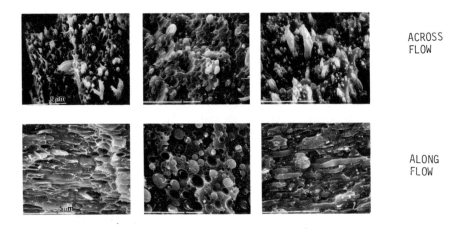

ACROSS
FLOW

ALONG
FLOW

Fig. 16. Scanning electron micrographs of the fracture surfaces of
a PET/LC-50 injection molded plaque. The sample was fractured along and across the flow direction and, the micrographs represent skin and core regions in the plaque.

i.e. slower cooling within the core. Further work is continuing to understand the effects of process variables on materials produced from these blends.

CONCLUSIONS

Morphological studies show that a non-homogeneous morphology is present in the mechanically mixed blends with the size of the spherical LC polymer phase increasing with LC polymer content. Differential scanning calorimetry studies indicate that the addition of LC polymer to PET enhances the rate of crystallization of PET homopolymer implying that it acts as a nucleating agent. This is observed in both the solution cast blends and the blends mixed mechanically in the melt. It is also very clear that the degree of crystallinity is increased significantly upon the addition of LC polymer although the reason for this behavior is not very well understood at present. Crystallinity values for PET ranging from 36% to 72% have been observed as LC content is increased from 0 wt. percent to 70 wt. percent. When PET and the LC polymer are injection molded at levels of 50 wt. percent LC polymer, the bending modulus value is approximately 3 times higher than that obtained for injection molded PET[2]. Morphological studies performed on the same injection molded plaques show the presence of non-homogeneous skin regions in which the LC polymer appears as oriented macroscopic rod-like regions dispersed in the PET matrix. Very little orientation of the LC phase is present in the non-homogeneous core.

ACKNOWLEDGEMENT

The authors would like to acknowledge the Army Research Office (Grant No. DAAG 29-80-K-0093) for their financial support for this project.

REFERENCES

1. L. A. Utracki, Polym. Eng. Sci., 20:17, 1166 (1982).
2. W. J. Jackson, Jr. and H. Kuhfuss, J. Polym. Sci., Polym. Chem. Ed., 14:2043 (1976).
3. R. E. Jerman and D. G. Baird, J. Rheol., 25:2, 275 (1981).
4. E. G. Joseph, G. L. Wilkes and D. G. Baird, Polym. Preprints, Am. Chem. Soc., Div. Polym. Chem., 22:2, 259 (1981).
5. E. G. Joseph, Ph.D. Dissertation, Virginia Polytechnic Institute and State University, Blacksburg, VA (1983).
6. E. G. Joseph, G. L. Wilkes and D. G. Baird, submitted to Polymer.
7. W. Meesiri, J. Menczel, U. Gaur and B. Wunderlich, J. Polym. Sci., Polym. Phys. Ed., 20:719 (1982).

8. A. E. Zachariades, J. Economy and J. A. Logan, J. Appl. Polym. Sci., 27:2009 (1982).

9. D. Acierno, F. P. LaMantia, G. Polizzotti, A. Cifferi and B. Valenti, Macromolecules, 15:6 1455 (1982).

10. C. Viney, A. M. Donald and A. H. Windle, J. Mat. Sci., 18:1136 (1983).

11. Y. Mohajer, E. Yorkgitis, G. L. Wilkes and J. McGrath, Proceedings of the Critical Review; Techniques for the Characterization of Composite Materials, Army Materials and Mechanics Research Center, May 1982.

12. F. Van Antwerpen, Ph.D. Dissertation, Technical University of Delft, Holland (1971).

13. D. J. Blundell, D. R. Beckett and P. H. Willcocks, Polymer 22:705 (1981).

14. M. Takayanagi, M. Kajiyama, F. Kumamaru and T. Ohno, Polym. Preprints, Am. Chem. Soc., Div. Polym. Chem., 24:1, 211 (1983).

STRUCTURE AND PROPERTIES OF RIGID AND SEMIRIGID LIQUID CRYSTALLINE

POLYESTERS

Gia Huynh-Ba and E. F. Cluff

Polymer Products Department
E. I. Du Pont De Nemours & Co., Inc.
Wilmington, Delaware 19898

SUMMARY

One class of liquid crystalline polymers consists of rigid, rod-like materials containing predominantly symmetrical aromatic components. Their typically high melting points are lowered to permit melt processability by any of several methods which disrupt their highly regular, crystalline structures. Several thermal transitions occur in these materials: glass transition, crystalline to liquid crystalline, and finally clearing point to isotropic melt.

Polymers with mesogenic side chains exhibit liquid crystalline behavior which is influenced by the nature of the polymeric backbone and its proximity to the mesogenic group in the side chain.

These materials exhibit mechanical properties (e.g., modulus, tensile) which are significantly greater than those exhibited by conventional thermoplastic resins. In fabricated articles these properties are highly anisotropic. This anisotropy is caused by the sensitivity of these stiff molecules to the imposed elongational flow fields which cause molecular orientation in the longitudinal direction.

Melt processable resins with outstanding mechanical properties have been synthesized from a variety of aromatic structures. Substituent structure appears to affect mechanical properties more than flow temperature. Molding temperature and molecular weight also significantly affect properties. Post-forming heat treatment causes marked improvement in rupture properties. Fibers tend to exhibit significantly higher tensile modulus and strength than do injection moldings due to the greater degree of molecular orientation achieved.

217

INTRODUCTION

For the last 15 years, there has been significant interest in a unique class of polymers which exhibit liquid crystalline behavior. Results of different types of structures have been published widely. We wish to review some of the structure-property relationships which have been reported for these materials.

Many polymers exhibiting liquid crystalline properties have been reported. From this work have come generalizations concerning the nature of the structural components or mesogenic groups necessary to impart liquid crystalline behavior to polymeric materials. These polymers are frequently rigid rods, and the groups which contribute

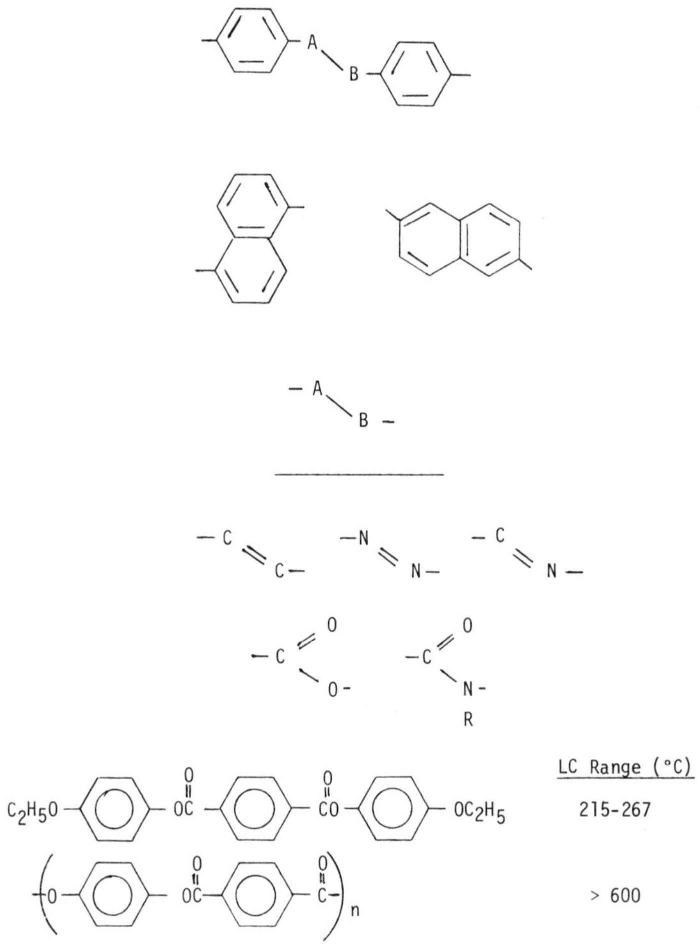

Fig. 1. Liquid Crystalline Groups.

to this rigidity are generally symmetrically substituted aryl, such as p-phenylene, 4,4'-biphenylene, 2,6 or 1,5-substituted naphthylene or 1,4-transcyclohexylidene. They can be connected directly by olefinic, azo, azomethine, ester or amide groups (Figure 1).[1,2,3]

A typical example of a small liquid crystalline molecule, derived from terephthalic acid and the ethyl ether of hydroquinone, has a liquid crystalline range between 215 and 267°C. However, the liquid crystalline polymer, hydroquinone terephthalate, is so high melting (above 600°C) that it decomposes before exhibiting a transition from the crystalline to the liquid crystalline state.[1]

METHODS FOR DEVELOPING MELT PROCESSIBILITY

In order to reduce the melting point of such high melting, highly symmetrical materials, several techniques to interrupt the crystalline order are available, as summarized in Figure 2.

1. Copolymerization of several mesogenic monomers such as p-hydroxy-benzoic acid (PHB) or 2-hydroxy-6-naphthoic acid produces random copolymeric structures with depressed melting points.[4]

2. Use of monomers with bulky side groups, such as phenylhydro-quinone, prevents close packing in the polymer crystals.[1,5]

3. Use of bent comonomers which contain the 1,3-disubstituted phenylene structure, such as isophthalic acid, (which are not inherently liquid crystalline precursors) interrupts crystalline order.[6]

4. Flexible spacers, such as alkylene groups, decrease polymer rigidity.[7,8] Polymers with the mesogenic group in the side chain are included in this class.[9,10]

Reduction of the melting point by copolymerization is illustrated by the two systems shown in Figure 3, in which symmetrical polyesters are modified by copolymerization with parahydroxybenzoic acid (PHB). In the case of poly(hydroquinone naphthalene-2,6-dicarboxylate), its melting point is reduced from 580°C to about 325°C by copolymerization with 70% PHB. More PHB increases the melting point until it reaches 600°C for pure poly-PHB.[1]

A similar effect is observed for copolymers of PHB and hydroxynaphthoic acid.[4]

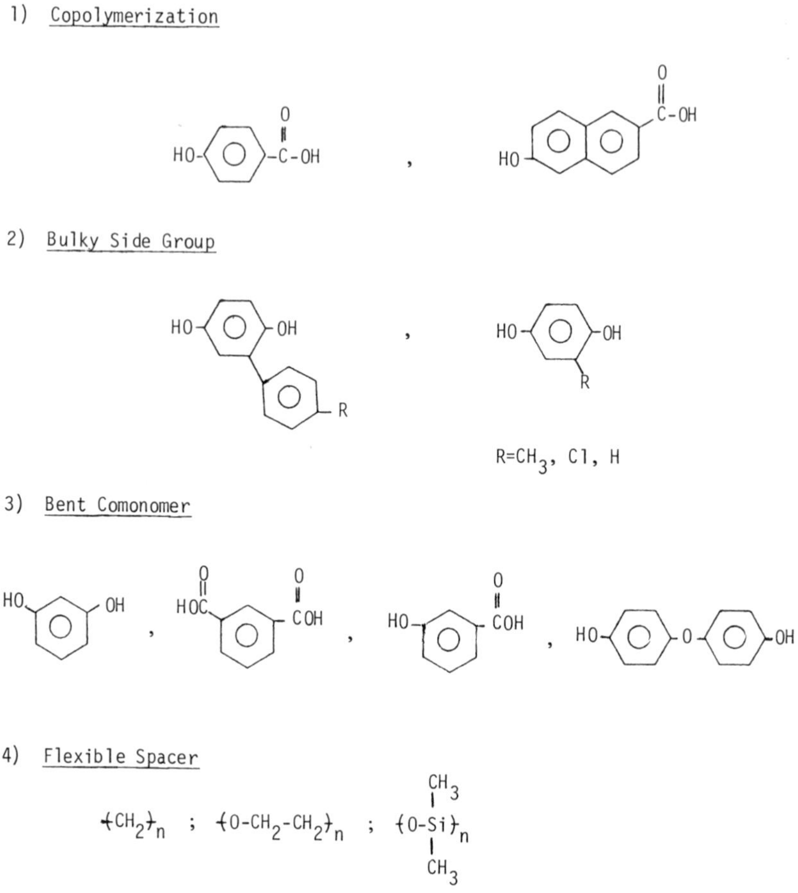

1) Copolymerization

2) Bulky Side Group

R=CH$_3$, Cl, H

3) Bent Comonomer

4) Flexible Spacer

$+CH_2+_n$; $+O-CH_2-CH_2+_n$; $+O-Si+_n$ with CH$_3$ groups

Fig. 2. Methods to Lower Melting Point.

The effect of bulky side groups is exemplified by the large melting point lowering observed when phenylhydroquinone is substituted for hydroquinone in terephthalic polyesters,[5] as shown in Figure 4.

Use of even 10 mole percent of a 1,3-disubstituted phenylene monomer (commonly called a bent comonomer), such as isophthalic acid, in typical polyesters also causes significantly lower melting points (Figure 4).[1,6]

By inserting a flexible group in the chain, the melting point of poly(methylhydroquinone terephthalate) was lowered from 400°C to about 210°C (Figure 4).[1]

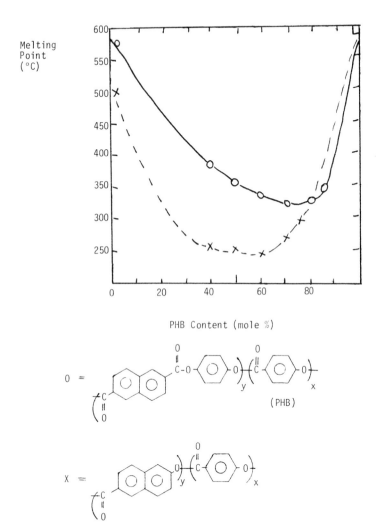

Fig. 3. p-Hydroxybenzoic Acid Copolymers.

Bulky Side Group   T (Melting) °C

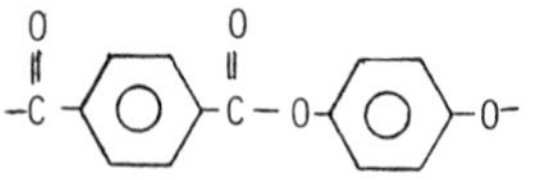

> 600

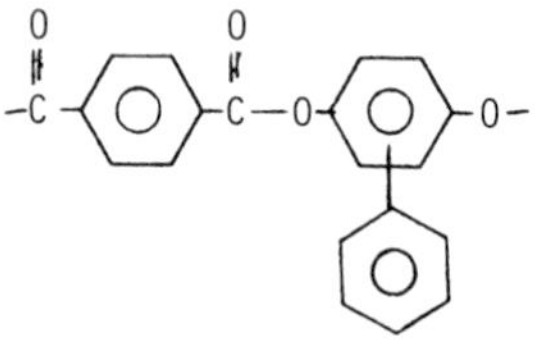

≈ 340

Bent Monomer   T (Melting) °C

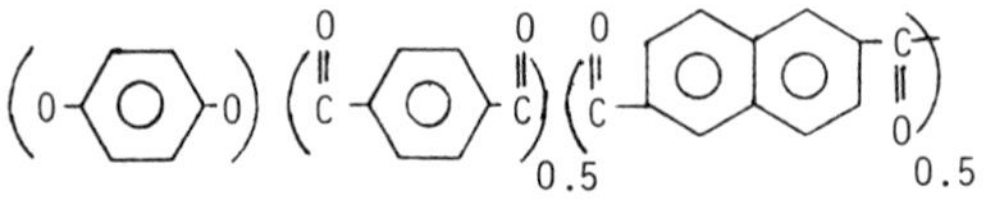

≈ 400

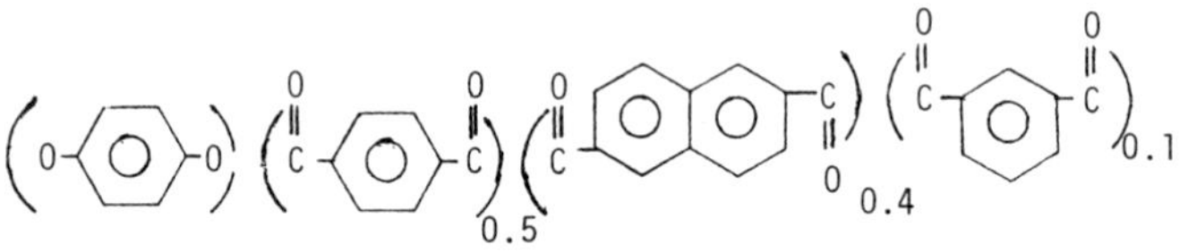

≈ 350

Flexible Group   T (Melting) °C

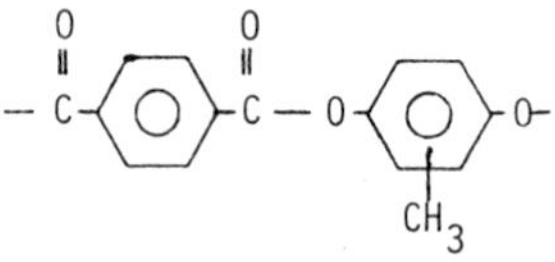

> 400

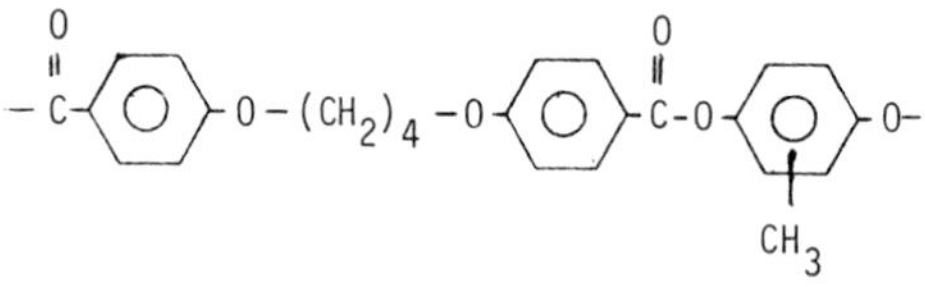

≈ 210

Fig. 4.   Melting Point vs Structure.

THERMAL CHARACTERIZATION

The unique morphologies of liquid crystalline polymers cause them to undergo numerous thermal transitions which can be observed in a Differential Scanning Calorimeter (DSC) scan. Figure 5 shows a typical scan with descriptions of the transitions and interpretations of the structural arrangements in the several phases.

Below the glass transition temperature (Tg), long range motions are frozen out. The materials behave like semicrystalline plastics. Above the Tg, the amorphous phase becomes liquid, but structure is imparted to the melt by whatever crystallinity exists.

Crystal lattice changes may occur, represented by the T_{K1-K2} transition.

At the melting point (Tm), three-dimensional order is lost, and a fluid melt with a loose two-dimensional structure is formed. This can be a smectic phase (LC_1). The polymer is now liquid crystalline and possesses all the unique features associated with this mesophase.

At still higher temperatures more disorder occurs, and transition from a smectic to a nematic (LC_2) arrangement is sometimes observed.

Finally, a transition to an isotropic phase occurs, and all liquid crystalline order is lost.

LCPs WITH MESOGENIC SIDE CHAINS

Ringsdorf and his group have reported the synthesis and thermal characterization of liquid crystalline polymers with methacrylate and acrylate backbones in which the mesogenic groups are in pendant side chains.[2,9,11] Some of their results are shown in Figure 6.

1. The Tgs of LCPs with methacrylate backbones are 30 to 60°C higher than those of polymers with acrylate backbones; however, clearing point temperatures are much less sensitive to changes in backbone structure.

2. As the length of the flexible group in the side chain increases, the Tg becomes lower. The clearing point shows a slight dependence on the length of the flexible group. An odd number of methylene groups (n) often prevents the formation of LCP, probably due to a disruption of order in chain packing.

3. Increasing the length of the "tail" (R_2) lowers the Tg.

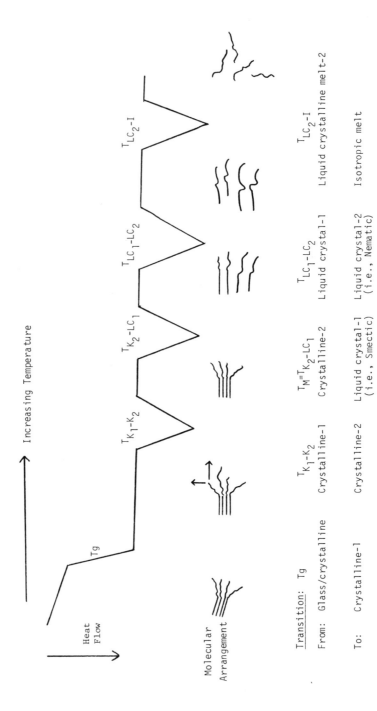

Fig. 5. Thermal Behavior.

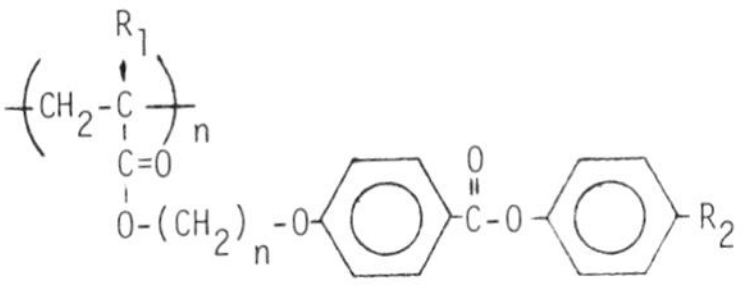

$$\left[CH_2-\underset{\underset{0-(CH_2)_n-O-\langle\phi\rangle-\overset{O}{\overset{\|}{C}}-O-\langle\phi\rangle-R_2}{\overset{|}{\underset{|}{C=O}}}}{\overset{R_1}{\underset{|}{C}}}\right]_n$$

n	R_2	$R_1=CH_3$; Methacrylate		$R_1=H$; Acrylate	
		Tg (°C)	T_{LC-I} (°C)	Tg (°C)	T_{LC-I} (°C)
2	OCH_3	101	121	62	116
6	OCH_3	95	111	35	123
6	OC_6H_{13}	60	115	28	130
6	 (phenyl)	132	184	--	--
2	OC_3H_7	120	129	--	--
3	OC_3H_7	No LCP		--	--
Ethyl ester		66°C	--	-24°C	--
Hexyl ester		-5°C	--	-57°C	--

Fig. 6. Mesogenic Group in Side Chain.

4. When the mesogenic group is lengthened by another phenyl group, both the Tg and the clearing point increase.

5. These materials are noncrystalline and, therefore, do not exhibit a crystalline-LC transition. Thus below the Tg the morphology can be described as a frozen liquid crystalline phase or a glass with local short range order.

The flexible spacer was shown to be necessary to at least partially decouple the motions of the mesogenic side chain from the motions of the flexible backbone, and thus to achieve liquid crystal-line behavior. Polymers in which there is a direct linkage between the mesogenic side group and the main chain usually do not reversibly maintain their liquid crystalline character through the clearing point (i.e., they are not enantiotropic). This effect is presumably due to the strong influence of the backbone conformation in ste-rically interfering with side chain ordering. Even with spacers present, some interaction between side chain and main chain exists, as evidenced by the large differences in Tg between the liquid crystalline polymers and their unsubstituted analogs such as the ethyl ester (Figure 6).

225

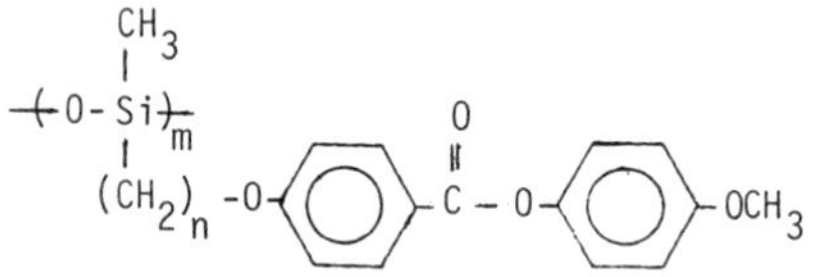

$$\left(O-\underset{\underset{(CH_2)_n}{\overset{CH_3}{|}}}{Si}\right)_m$$

the structure shows a polysiloxane with CH₃ and (CH₂)ₙ-O-⟨benzene⟩-C(=O)-O-⟨benzene⟩-OCH₃.

n	Tg (°C)	T_{LC-I} (°C)
3	15	61
6	5	112
11	*	133
PDMS	-130	None

* : $T_{K-LC} = 53\,°C$

Fig. 7. Mesogenic Group in Side Chain.

　　Further evidence of limited decoupling between flexible main chains and mesogenic side chains is seen in the case of substituted polydimethylsiloxanes, as illustrated in Figure 7.[10,12] The same trends with spacer length are observed in this system as with the corresponding acrylates. The Tgs are lower than the analogous acrylic-based polymers but not as low as might be expected considering the -130°C Tg of the unsubstituted siloxane. Again only limited decoupling occurs.

　　As described earlier, thermotropic, liquid crystalline polymers can be obtained by inserting flexible units in the main chain. As shown in Figure 8, these units can be hydrocarbon[8,13] (such as polymethylene), polyether or polysiloxane.[7] Liquid crystalline polymers with hydrocarbon spacers still exhibit relatively high transition temperatures. Significant, additional lowering is imparted by the very flexible, polydimethylsiloxane groups.

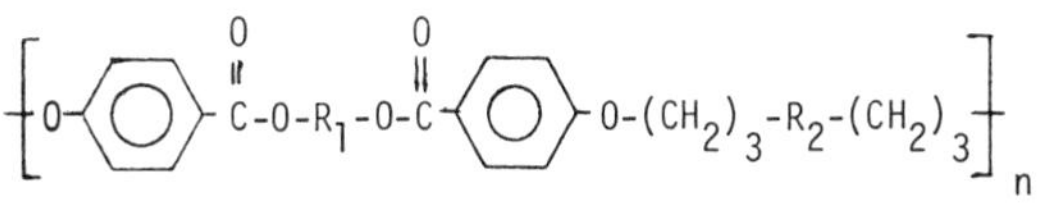

$$\left[O-\bigcirc-\overset{O}{\overset{\|}{C}}-O-R_1-O-\overset{O}{\overset{\|}{C}}-\bigcirc-O-(CH_2)_3-R_2-(CH_2)_3 \right]_n$$

R_1	R_2	T (Melting) °C	T_{LC-I} °C
 $\bigcirc$	$-(CH_2)_5-$	125	250
	$\{Si-O\}_2 Si-$ (with CH_3 groups)	35	130
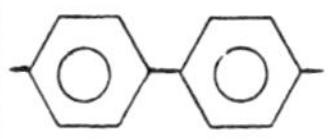 $\bigcirc\bigcirc$	$\{CH_2\}_5$	285	390
	$\{Si-O\}_2 Si-$ (with CH_3 groups)	95	274

Fig. 8. Flexible Spacer in Main Chain.

ORIENTATION PROCESSES IN RIGID MAIN CHAIN POLYMERS

Figure 9 shows tensile properties of common, commercial engineering plastics from nylon 6,6 to polyethylene. Tensile strengths range from 5 to 12 thousand psi, while tensile moduli range from 150 to 500 thousand psi.[14] Typical liquid crystalline polymers have significantly higher values - 30 and 3000 thousand psi, respectively; more than twice the strength and six times the modulus.

Why is this? Because of their uniquely rigid, rod-like structure, liquid crystalline polymers differ significantly from isotropic polymers in the response of their fluid melts to the forces impinging on them during processing. They undergo orientation to a much greater extent and relax more slowly. It is the high degree of retained orientation which imparts their exceptional mechanical properties. Figure 10 illustrates the differences between LCPs and isotropic polymers in this regard. An isotropic polymer, such as polyethylene terephthalate (PET), undergoes very little orientation and exhibits average strength and modulus, 10 and 500 Kpsi, respectively. On the other hand, LCPs with their

	Tensile Strength $(10^3$ psi)	Tensile Modulus $(10^3$ psi)
Nylon 6,6	12	500
Polyethylene Terephthalate	10	500
PET + Glass Fiber (30-40%)	23	1500
PEEK 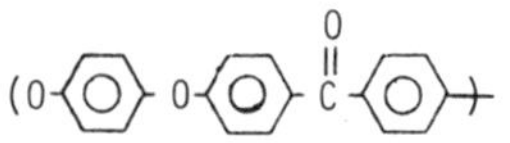 (0—⬡—0—⬡—C(=O)—⬡—)	10	350
Polyethylene	5	150
Polycarbonate	10	350
Polyarylate (Bisphenol A/T/I)	10	300
Typical LCP	30	3000

Fig. 9. Mechanical Properties of Engineering Plastics.

existing orientation in ordered domains need very little additional energy input to become very highly oriented. An elongational flow field is sufficient. Exceptional mechanical properties can be achieved. For instance, the liquid crystalline polymer shown in Figure 10 has typical tensile properties several times those of PET.

Articles fabricated from liquid crystalline polymers have highly anisotropic properties because of molecular orientation in a preferred direction. This orientation is caused by the imposed flow field which can be separated into its basic components - shear and elongation.[15,16] Ide and Ophir established the contributions of each flow field component to orientation, as shown in Figures 11a and b.

Figure 11a shows mechanical properties as a function of shear rate for rods extruded in a capillary rheometer. Mechanical properties, such as tensile modulus, are only 1.5 to 2X those of PET. They are essentially independent of shear rate.[15]

However, in other experiments, extruded filaments were drawn down and thereby subjected to elongational flow. Figure 11b shows that mechanical properties are very strong functions of the elongational strain-expressed as draw down ratio. Strength increases five fold and modulus ten fold as the melt is subjected to increasingly larger elongational strain.[15]

228

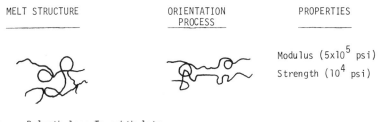

MELT STRUCTURE	ORIENTATION PROCESS	PROPERTIES

Modulus (5×10^5 psi)
Strength (10^4 psi)

e.g., Polyethylene Terephthalate

Modulus (25×10^5 psi)
Strength (3×10^4 psi)

e.g., P-Hydroxybenzoic Acid/
2,6 Dihydroxynaphthalene/
Terephthalic Acid

Fig. 10. Rheological and Orientation Processes–Melt Structure of LCP and Non-LCP.

The authors conclude that "in anisotropic, liquid crystalline melts where domains of local orientation are formed, shearing will not orient molecules if domains are stable. However, in elongational flow, each domain will be stretched out and all the molecules become aligned to the flow direction, yielding indistinct boundaries between domains."

Injection molding of LCPs produces molded pieces which exhibit a characteristic skin and core morphology, as shown in Figure 12. Jackson and Kuhfuss, for example, report data showing the significant effect of part thickness on the anisotropy of mechanical properties.[17] Ide and Ophir explain this skin and core morphology on the basis of LCPs' sensitivity to orientation under elongational flow. "The molten polymer injected into the cavity reaches the flow front and goes through 'fountain flow,' where it experiences a strong elongational flow; the oriented melt is deposited near the wall. The portions at the center are injected late and cannot

229

MECHANICAL PROPERTIES VS SHEAR RATE

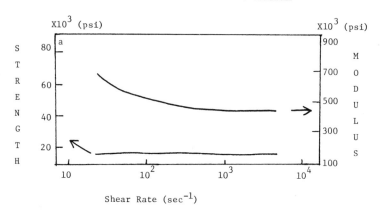

Shear Rate (sec^{-1})

MECHANICAL PROPERTIES VS DRAWN DOWN RATIO
(ELONGATIONAL FLOW)

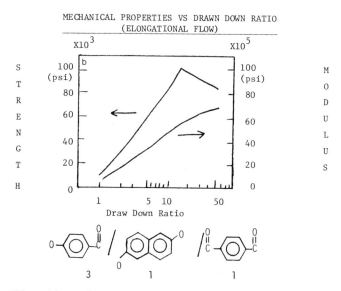

Fig. 11a & b.   Orientation vs Flow Field.

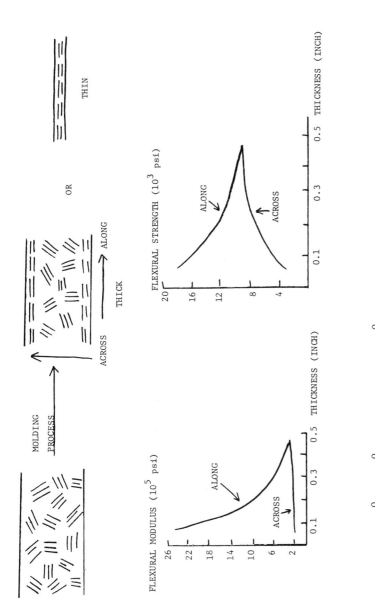

Fig. 12. Orientation by Elongational Flow.

reach the flow front. Thus, the portions near the wall are highly oriented (skin), but the middle portions are not oriented (core)."[15] In very thick moldings, the oriented skin is a small fraction of the total, and its anisotropic properties are almost totally diluted by the larger core.

MECHANICAL PROPERTIES OF MELT PROCESSIBLE LCPs

As discussed earlier, in order to reduce melting temperature to allow melt processibility, sometimes flexible groups are inserted in the main chain. However, their presence has an adverse effect on mechanical properties (Figure 13). As the length of the flexible unit increases, mechanical properties decrease significantly, as does heat resistance and melting point. The rigidity of the chain, which is responsible for these outstanding properties, is significantly impacted by a small change in flexible spacer length.[1]

Thermotropic liquid crystalline polyesters can be prepared by the use of meta-oriented modifiers to reduce the melting points of para-substituted aromatic polyesters to the processible range. Resorcinol, isophthalic acid and m-hydroxybenzoic acid are commonly used. Figure 14 shows the effect of copolymerized m-hydroxybenzoic acid on the physical properties of melt processable, modified polyesters based on terephthalic acid and methylhydroquinone. As the m-hydroxybenzoic acid (MHB) content increases

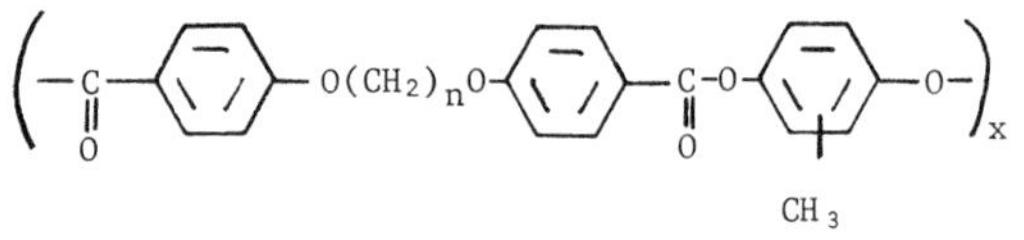

Value of n	2	3	4	5
TENSILE STRENGTH, PSI	37,600	26,700	10,800	5600
FLEXURAL STRENGTH, PSI	29,200	20,200	15,000	6700
FLEXURAL MODULUS, 10^5 PSI	16.6	9.3	12.0	3.9
ROCKWELL HARDNESS, L	102	81	88	34
HEAT-DEFLECTION TEMP. °C	177	-	148	74
POLYMER MELTING POINT, °C	316	-	212	-

Fig. 13. Properties of Injection Molded 4,4' (Alkylenedioxy) Dibenzoic Acid Polyesters.

232

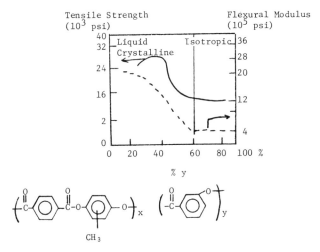

Fig. 14. Mechanical Properties Modified by Bent Comonomer.

above 60%, the polymers are no longer liquid crystalline but are isotropic, and the mechanical properties characteristic of LCPs deteriorate to significantly lower values. However, at low MHB concentration, exceptional physical properties can be obtained along with processibility.[1,18]

When bulky side groups are used to reduce the melting point, the packing ability (or arrangement) of the chain is an important factor in determining mechanical properties.[3,19] Figure 15 shows the effect of various substituted hydroquinones in mixed terephthalate/naphthalene dicarboxylate polyesters. While flow temperatures or melting points are lowered from those of the intractable, unsubstituted polyester, the flow temperatures are not greatly dependent on substituent structure. However, both tensile modulus and tensile strength are significantly affected by the nature of the substitutent. Chloro is better than methyl and ethyl. Symmetrical dimethyl shows a high modulus without a correspondingly high tensile strength.

In Figure 16, results of Jackson and Kuhfuss show the effect of molding temperature and polymer morphology weight on the mechanical properties for poly(ethylene terephthalate) modified with p-hydroxybenzoic acid. "The temperature at which the PHB copolyesters are injection molded affects the orientation of the 'liquid-crystal' polymer chains and, therefore, affects the mechanical properties."[17]

233

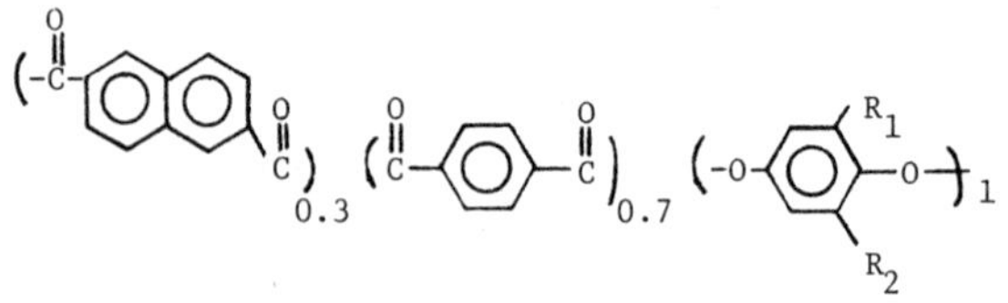

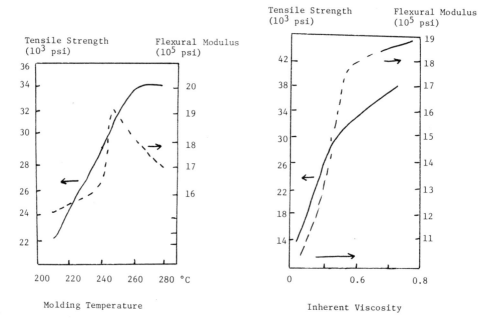

R_1	R_2	Flow Temperature °C	Fiber or Yarn (Heat Treated)	
			Tensile Modulus 10^3 psi	Tensile Strength 10^3 psi
H	H	>450	Melt >450°C	Melt >450°C
H	Cl	302	9500	540
H	CH_3	301	6600	360
H	CH_2-CH_3	292	6300	210
CH_3	CH_3	306	8600	280

Fig. 15. Effect of Bulky Side Groups.

Fig. 16. Control of Physical Properties.

234

"For PET modified with 60% PHB, as the temperature was increased, the melt viscosity decreased, the speed of the polymer melt injected into the mold increased, and the orientation of the polymer chains therefore increased."

"The inherent viscosity affects the mechanical properties of the molded copolyesters. The tensile strength, stiffness and notched Izod impact increased as the inherent viscosity increased up to about 0.6-0.7 (number average MW about 16,000-20,000)."

A significant increase in the level of mechanical properties is obtained when shaped articles such as fibers, films or molded parts are subjected to a high temperature, heat treatment process.[5,20] As shown in Figure 17, there is in general an improvement in rupture properties, such as breaking strength and impact resistance. Modulus changes are variable, sometimes increasing and sometimes remaining unchanged. Probably additional polymerization is taking place under these severe conditions to produce higher molecular weight which is responsible for the correspondingly better mechanical properties, although defect healing could be important.

Figure 18 summarizes the physical properties of several different chemical structures including both a lyotropic LCP[21] as well as several thermotropic LCPs.[1,15,22] Those materials fabricated as fibers exhibit tensile moduli and strengths much higher than materials fabricated by injection molding. The reason for this difference is probably the larger contribution of elongational flow in fiber spinning and draw down compared to injection molding where the

	Modulus $(10^3$ psi)	Break Strength $(10^3$ psi)	Max Strain To Break (%)	Energy To Break (psi)	Impact Strength (ft-lb/in)
As-Molded	980	13.2	1.9	125	1.7
Heat Treated	1500	24.7	2.4	296	3.1
% Change	+53	+87	+26	+136	+83

Heat Treatment: 2 hrs/260°C + 3 hrs/275°C + 16 hrs/300°C

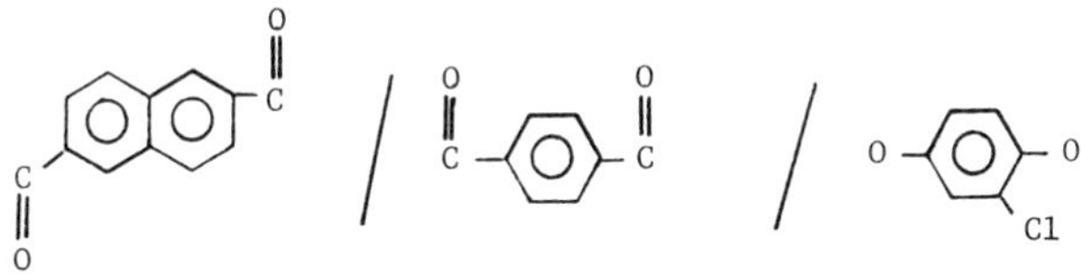

(0.3) (0.7)

Fig. 17. Effect of Heat Treatment.

235

MONOMER STRUCTURE

A–A	B–B	A–B	Type	T_m °C	Modulus (10^3 psi)	Strength (10^3 psi)
 –NH–◯–NH–	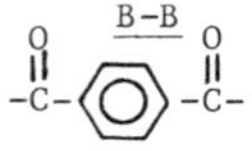 ⟨benzene diacid chloride⟩	—	Fiber	>700	9000	300
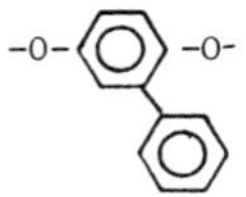 –O–◯–◯–O– (biphenyl)	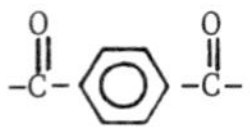 ⟨benzene diacid chloride⟩	—	Fiber	340	15000	480
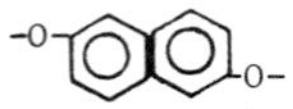 –O–◯◯–O– (naphthalene)	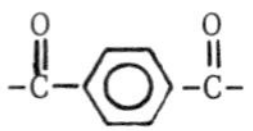 ⟨benzene diacid chloride⟩	3 –O–◯–C=O 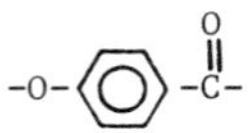	Fiber	340	6000	100
Ibid	Ibid (naphthalene diacid chloride)	Ibid	Molded	340	2900 (tensile) 2200 (flex)	25
 –O–◯–O–	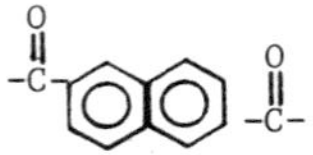 Ibid	3 –O–◯–C=O 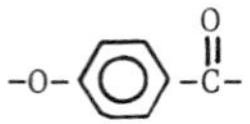	Molded	330	2200 (flex)	30

Fig. 18. Physical Properties of Fibers and Molded Parts.

skin/core effect in a much thicker sample dilutes the average degree of orientation.

Interestingly, the properties of the two resins which were injection molded are very similar even though their compositions are significantly different.

Probably all rigid, rod-like molecules have a significant degree of structural similarity, and it should not be surprising that they are remarkably similar in physical properties. It is the strong sensitivity of LCPs of whatever structure to elongational flow that provides the greatest change in mechanical behavior.

CONCLUSION

We have attempted in this brief review to summarize some basic principles relating the structure and properties of rigid and semirigid liquid crystalline polyesters. The references used were chosen to be illustrative of the principles involved, and no attempt was made to be comprehensive. Many workers in both academic and industrial laboratories have contributed a great deal of fine work which we were not able to include in this limited review.

REFERENCES

1. W. J. Jackson, Brit. Pol. J., 12, 154 (1980).
2. H. Ringsdorf, M. Happ, and H. Finkelmann, German Patent 2,722,589.
3. J. Preston, Synthesis and Properties of Rod-Like Condensation Polymers, in: "Liquid Crystalline Order in Polymers," A. Blumstein, ed., Academic Press, 1978.
4. W. G. Calundann, US Patent 4,161,470 (1979).
5. R. C. Payet, US Patent 4,159,365 (1979).
6. W. J. Jackson and J. C. Morris, US Patent 4,181,792 (1980).
7. H. Ringsdorf, C. Aguilera, J. Bartulin, and B. Hisgen, Makromol. Chem., 184, 253 (1983).
8. L. Strzelecki and D. V. Luyen, Eur. Poly. J., 16, 299 (1980).
9. H. Ringsdorf, M. Portugall, and R. Zentel, Makromol. Chem., 183, 2311 (1982).
10. H. Ringsdorf and A. Schneller, Brit. Pol. J., 13, 43 (1981).
11. H. Ringsdorf, H. Finkelmann, W. Siol, and J. H. Wendorff, ACS Symp. Series, 74, 22 (1978).
12. H. Finkelmann and G. Rehage, Makromol. Chem. Rap. Comm., 1, 733 (1980).
13. L. Strzelecki and D. V. Luyen, Ibid, 16, 303 (1980).
14. Modern Plastics Encyclopedia, 58 (10A), 1981-1982.
15. Y. Ide and Z. Ophir, Pol. Eng. Sci., 23 (5), 261 (1983).
16. Y. Ide, US Patent 4,332,759 (1982).

17. W. J. Jackson and H. F. Kuhfuss, <u>J. Poly. Sci.</u>, <u>14</u>, 2043 (1976).
18. W. J. Jackson and E. J. Morris, US Patent 4,146,702 (1979).
19. J. J. Kleinschuster, T. C. Pletcher, J. R. Schaefgen, and R. R. Luise, Belgium Patents 828,935 and 828,936 (1975).
20. R. R. Luise, US Patent 4,247,514 (1981).
21. J. Preston, Ang. Makromol. Chem., 109, 1 (1982).
22. Z. Ophir and Y. Ide, Antec (1982).

STRUCTURE-PROPERTY RELATIONS IN FLEXIBLE THERMOTROPIC
MESOPHASE POLYMERS: I. PHASE TRANSITIONS IN POLYESTERS
BASED ON THE AZOXYBENZENE MESOGEN

R.B. Blumstein, O. Thomas, M.M. Gauthier,
J. Asrar and A. Blumstein

Polymer Science Program
Department of Chemistry
University of Lowell
Lowell, MA   01854

ABSTRACT

Phase transitions in thermotropic nematic and cholesteric main chain polyesters were investigated by polarizing microscopy and differential scanning calorimetry.
The polymers were formed by condensation of 4,4'-dihydroxy-
benzene and 2,2'-dimethyl-4,4'dihydroxyazoxybenzene with
various diacid chlorides acting as flexible spacer groups.
Polydisperse homopolymers and copolymers, sharp fractions
of homopolymers and mixtures of polydisperse polymers with
a low mass mesogen were investigated. Supercooling at the
mesophase-isotropic and solid-mesophase transitions, sharpness of the nematic-isotropic transition (range of N+I biphase), polymer crystallization from the mesophase melt,
and enhancement of crystallinity upon addition of a low
mass nematic, were studied.

INTRODUCTION

Thermotropic Polymeric Liquid Crystals (PLCs) which
are formed by regularly alternating mesogenic elements
and flexible spacer groups in the main chain are currently
the focus of intensive investigation[1]. The standard mesogens, such as biphenyl, stilbene, azo or azoxybenzene derivatives, which form the core of low molecular mass liquid crystals (LMLCs), are used in synthesis of PLCs as
well.

239

Herein, we present a discussion of phase transitions
in nematic and cholesteric main chain polyesters formed
by condensation of 4,4'-dihydroxyazoxybenzene (mesogen 8)
and/or 2,2'-dimethyl-4,4'-dihydroxyazoxybenzene (mesogen
9) with various diacid chlorides which are used as flexible spacer moieties. Polarizing microscopy and differential scanning calorimetry (DSC) results are presented for
polydisperse homopolymers and copolymers, well fractionated samples, and mixtures of polydisperse polymers with
a LMLC. Supercooling at the mesophase/isotropic phase an
solid/mesophase transitions, extent of the nematic-isotropic (N+I) biphase, polymer crystallization from the mesophase melt and enhancement of crystallinity upon addition
of a LMLC were investigated.

The qualitative description of the nature of phase
transitions in main chain mesophase polymers which emerge
from these data will serve as a basis for more quantitative treatments in subsequent publications.

EXPERIMENTAL SECTION

Sample synthesis was carried out as previously described[2], using either solution or interfacial polycondensation. Characterization was as in reference 3; DSC
and polarizing microscopy were carried out using control]
sample thermal history as described in references 4,5.

Two polymers based on mesogen 9 were fractionated.
They are designated as polymer AZA-9 (n=7) and polymer
DDA-9 (n=10):

$$\left[O-\!\!\left\langle\bigcirc\right\rangle\!\!-\!\!\underset{\underset{CH_3}{}}{\overset{\overset{O}{\uparrow}}{N}}\!\!=\!\!N\!\!-\!\!\left\langle\bigcirc\right\rangle\!\!-\!\!\underset{CH_3}{}\!\!O-\!\!\underset{\underset{O}{\|}}{C}-(CH_2)_n-\underset{\underset{O}{\|}}{C}\right]$$

Polymer AZA-9 was fractionated using the solvent-
nonsolvent fractional precipitation method (DMF-ethanol)
to yield 10 separate fractions. Polymer DDA-9 was fractionated by means of the temperature gradient method from
a solution in DMF to yield 11 separate fractions. The
molecular weights were measured using a combination of
vapor pressure osmometry, dilute solution viscometry, DSC
and NMR end-group analysis[5]. Molecular weight distributions were established by gel permeation chromatography
(GPC) in chloroform spiked with 10% absolute ethanol
(vol/vol). Sample concentration was 0.25% by weight.
The instrument was calibrated with model compounds, oli-

240

gomers and fractions of polymers DDA-9 and AZA-9.

RESULTS

Figure 1 shows some of the symbols used to characterize the phase transitions. The DSC peak maximum of a typical nematic-isotropic (N/I) transition is represented by T_{NI} (or T_{IN}). The DSC peak width is measured by the interval $|T_{IN}"-T_{IN}'|$, on heating or cooling. Supercooling of peak maximum and of peak end is represented by $\Delta(T_{IN})$ and $\Delta(T_{IN}")$, respectively. Similarly, supercooling at the N/K or Ch/K transitions is symbolized by $\Delta(T_{NK})$ or $\Delta(T_{ChK})$.

Symbols T_2 and T_c on Figure 1 refer to nomenclature used for phase transitions observed by microscopy. A homogeneous nematic phase is observed between temperatures designated at T_N and T_2. The first isotropic droplet appears on heating at T_2 and the last trace of birefringence disappears at T_c. A N+I biphase is displayed between T_2 and T_c, and the same sequence of symbols is used in reverse to characterize the N+I biphase on cooling. It should be noted that T_c is a purely empirical clearing temperature based on microscopy observations under specified scanning rates.

On Figure 1 are also plotted the values of the nema-

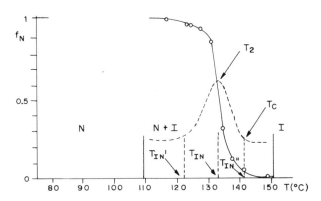

Fig. 1. The Nematic-Isotropic Biphase. Nematic fraction f_N measured by NMR on slow cooling from the isotropic phase (sample thermal history as in ref. 5). Comparison with $|T_{IN}"-T_{IN}'|$; the DSC peak width and T_c-T_2 or biphase observed by microscopy (scanning rate 10 °C/min.; thermal history as in ref. 5). Sample: DDA-9; $\overline{M}_n=4,000$.

tic fraction f_N, measured throughout the N+I biphase on cooling from the isotropic phase. The nematic fraction was deduced from proton NMR spectra or from the FID signals, taking as reference the intensity in the isotropic phase, under slow cooling as described in ref. 5,6. Consequently, the range of the N+I biphase on Fig. 1 is delineated: i) by the NMR experiment (interval between the two solid vertical lines); ii) by polarizing microscopy (interval $|T_2-Tc|$); and iii) by the width of the DSC peak (interval $|T_{IN}''-T_{IN}'|$). Note that the biphase as observed by microscopy is well approximated by the interval $|T_{IN}-T_{IN}''|$ comprised between the DSC peak maximum and peak end. Although Fig. 1 illustrates the case of a specific sample (DDA-9, $\overline{M}_n=4,000$), it is representative of the behavior of the other samples as well.

Phase transitions of homopolymers $\left[R-\underset{O}{C}-(CH_2)_n-\underset{O}{C} \right]$,

where R is mesogen 9 ($\left[O-\bigcirc-N\overset{\uparrow}{=}N-\bigcirc-O \right]$),
$\qquad\qquad\qquad\qquad CH_3 \quad\; CH_3$

are summarized in Table I. Synthesis of this homologous series was first reported in ref. 7. Supercooling at I/N is reported in Column 4 and supercooling at T" in Column The glass transition temperature Tg is reported in the last column.

The polymers listed in Table I were prepared by interfacial polycondensation and are polydisperse, with a polydispersity index of approximately 1.5.

Influence of molecular weight and molecular weight distribution on the sharpness and reversibility of phase transitions was investigated in the case of polymers DDA- and AZA-9 by comparing the behavior of unfractionated samples and of fractions ranging in molecular weight from oligomers to about 20,000. The fractions studied were fairly sharp, with a polydispersity index $\overline{M}_w/M_n \leqslant 1.1$. The range of molecular weights was as follows:

AZA-9: 15,000; 9,700; 7,300; 5,500; 5,200; 4,200; 3,700; 3,000; 2,700; 2,500

DDA-9: 19,000; 17,200; 13,000; 9,500; 7,000; 5,600; 4,800; 2,500; 2,300; 1,900.

Phase transition temperatures of fractions display

TABLE I. PHASE TRANSITIONS OF HOMOPOLYMERS

$$\left[R-\overset{\underset{\|}{O}}{C}-(CH_2)_n-\overset{\underset{\|}{O}}{C} \right]$$

n	Transitions[a] (on heating)	$\overline{M}_n$[b]	$(\Delta T)_{IN}$[a]	$(\Delta T)_{IN''}$[a]	T_g[c]
5	K135.8N173.8I	13,300	5	6.8	35
7	K98.3N164.6I	36,000	9.8	11.6	26
9	K121N140I	17,800	3	10.8	25
11	K100.8N132.8I	34,000	5	7.3	27
6	K160.3N230.3I	27,000	24.5	25.8	24
8	K136.8N188.3I	36,400	15.8	17.6	24
10	K118.2N163.5I	18,000	3.8	6.6	14
12	K110.8N143.8I	17,500	8	9.8	17
14	K121N134.5I	25,000	12.5	14.4	18

a   in °C; scanning rates 20 °C/min.

b   from ref. 7

c   in °C; measured on heating at 40 °C/min.

qualitatively the same molecular weight dependence as pr
viously investigated polydisperse samples[3,5] but, at a
given value of the molecular mass, the value of T_{NI} is
approximately 8-10° C. higher.

In Table II are listed some characteristic features
of supercooling at the I/N transition for various polydi
perse samples of DDA-9 and AZA-9. The corresponding val
obtained for p-azoxyanisole (PAA), a low molecular mass
nematic, are listed for comparison. Table II shows that
the phase transitions are broad in polydisperse samples
of relatively low molecular mass. The DSC peaks become
sharper and the biphasic range $|T_c - T_2|$ narrower with in-
creasing molecular weight as reported previously[5]. For
values of $\overline{M}_n > 12,000-15,000$, the transitions are sharp.
The transitions are sharper on cooling than on heating,
mostly due to supercooling at T_{IN}" or T_c. This is furth
illustrated on Fig. 2a and b, which shows the polarizing
microscopy phase diagram for mixtures of PAA with DDA-9

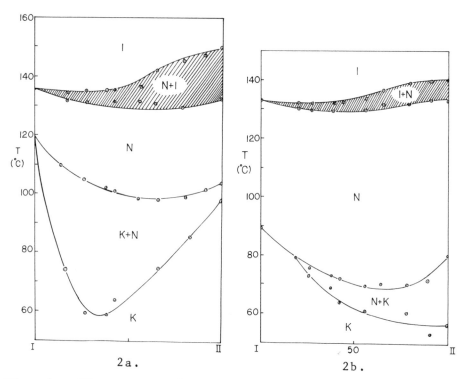

2a.   2b.

Fig. 2. Microscopy Phase Transition in Mixtures of PAA
         and DDA-9 ($\overline{M}_n$=4,000), on a) heating; b) cooling.
         Mixtures prepared by casting from chloroform.
         Thermal history as in 5. Scanning rates 10°C/mi

TABLE II. SUPERCOOLING AT THE I/N TRANSITION

SAMPLE	$\overline{DP}$	$T_{IN}-T_{IN'}$ Heat	Cool	$\Delta(T_{IN})$	$\Delta(T_{IN''})$	$Tc-T2$ Heat	Cool	$\Delta(Tc)$
PAA		2.3	2.2	2.9	3.7	0.4	0.1	1.7
DDA-9	2	*	58	*	*	*	32	*
	6	29	18	6	10	19	10	5
	9	23	17	6	10	15	8	6
	15	25	18	8	11	20	8	12
	26	17	13	7	10	6	3	6
	41	11	9	4	6	6	4	4
AZA-9	32	14.2	7.1	2.0	6.4	5.3	2.3	5.5
	88	10.6	6.9	9.6	11.4	4.8	2.1	7.8

$\overline{DP}$ Average number of repeating units per chain.

Scanning rates 10°C/min. All values reported in C.

*   Monotropic nematic (no mesophase on heating).

245

($\overline{M}_n$=4,000), on heating and cooling. The interval $|T_C-T_2|$ (I+N biphase, as observed by microscopy) is considerably narrower on cooling than on heating.

The transitions are sharp for all fractions, regardless of molecular weight; for AZA-9 fractions the values of $|T_{IN}''-T_{IN}'|$ on cooling range from 6 to 11, and for DDA-fractions from 9 to 13 (scanning rate: 10°C/min).

Supercooling at the mesophase/K transition is dependent on structure and molecular mass. The polymers listed in Table I all have a value of $\overline{M}_n$ sufficiently high to eliminate fluctuations due to the influence of molecular mass. Upon cooling at scanning rates reported in Table I polymers with spacer length n=5,6,7 and 11 fail to crystallize, while the remaining samples crystallize with values of $\Delta(T_{NK})$ of 30-70°C. All samples eventually develop some crystallinity upon annealing above T_g: a quenched mesophase cannot be permanently locked-in above room temperature in these homopolymers.

Table III, on the other hand, shows some copolymers in which the mesophase can be locked-in for use till temperatures of about 200°C. These are novel cholesteric compounds resulting from the copolymerization of mesogen 9 with the mixed spacer MAA*/TAA, where MAA* is the optically active 3-methyladipic acid and TTA is trans-traumatic acid, $HOCO-CH=CH-(CH_2)_8-COOH$.

We have previously reported synthesis of cholesteric PLCs based on mesogen 8 or 9 and MAA*-DDA[2,3]. Samples shown in Table III were prepared in a similar fashion. The mesophase is locked-in indefinitely by simple cooling from the isotropic phase. Supercooling at T_{IN} is from ∿10 to 20°C.

Table IV shows phase transitions for some nematic copolymers based on spacer DDA and the mixed mesogen 8-9. The polymers were prepared by polymerization in solution and have a molecular mass $\overline{M}_n$∿3,000-4,000. They show some of the features characteristic of polydisperse samples of homopolymer DDA-9 of relatively low molecular mass[5], such as broad transitions or supercooling of mesophase to room temperature and below, occasionally followed by cold crystallization (N/K) upon reheating. Supercooling at T_{IN} is 2-12°C.

DISCUSSION

The isotropic/mesophase, transition in PLCs discusse

TABLE III DSC PHASE TRANSITIONS OF SOME COPOLYMERS MAA*/TTA-9

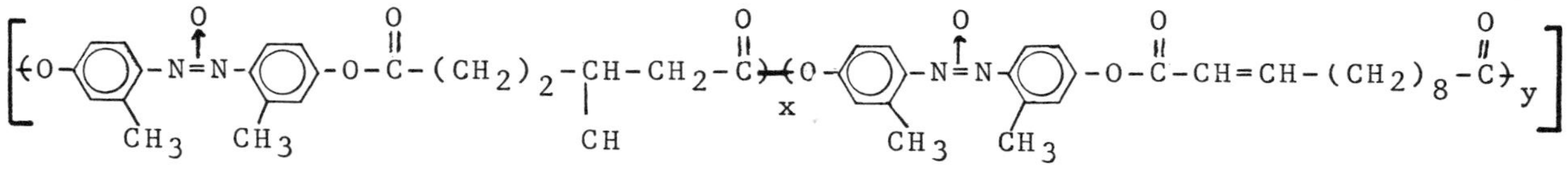

MAA*/TTA-9 (x/y)	$\overline{M}_n$	Transition Temp. (°C) Heating	Cooling	$\Delta(T_{IN})$ (°C)	$\Delta(T_{ChK})$ (°C)
66/44	11,000	Ch211I	I198Ch	13	a
49/51	15,500	Ch211I	I198Ch	13	a
36/64	17,500	Ch216I	I197Ch	19	a
23/77	14,000	Ch212I	I202Ch	10	a
0/100	9,000	K129N209I	I197N∼70K	12	59

Scanning rates 20°C/min

a = Locked-in mesophase

TABLE IV. DSC PHASE TRANSITIONS OF COPOLYMERS DDA-8/9

$$\left[O-\underset{\overset{|}{N=N}}{\bigcirc} -O-\underset{\overset{\|}{O}}{C}-(CH_2)_{10}-\underset{\overset{\|}{O}}{C}-O-\underset{CH_3}{\bigcirc}-N=N-O-\underset{CH_3}{\bigcirc}-O-\underset{\overset{\|}{O}}{C}-(CH_2)_{10}-\underset{\overset{\|}{O}}{C}\rightarrow \right]_y$$

x/y	Transition (°C)		Δ(T_IN) (°C)	Δ(T_KN) (°C)
	Heating	Cooling		
0/100	N57K96N139I	I128N	11	a
11/89	N140I	I128N	12	b
25/75	N163I	I156N57K	7	c
35/65	N169I	I166N69K	3	c
43/57	N173I	I167N73K	6	c
50/50	K105N169I	I167N75K	2	

Scanning rate 20°C/min.

a: Supercooling of mesophase, followed by cold crystallization characteristic of DDA-9 of molecular mass <∿5,000.5

b: Locked-in mesophase.

c: K/N transition very broad; difficult to pinpoint.

here differs from that of LMLCs in several respects:
i) drastic conformational changes can occur at the I/N
transition; ii) a nematic-isotropic biphase (I+N) is present; iii) supercooling of the N/I transition may be significant; iv) extent of supercooling at the mesophase/
K transition depends on structure and molecular mass;
the mesophase can be locked-in in some instances.

Summary of previous results

We know from investigation of the isotropic melt[8]
of several polymers that the chains are randomly coiled,
with a persistence length roughly that of the repeating
unit. On the other hand, at the I/N transition of DDA-9,
measurements of magnetic birefringence show drastic intramolecular conformational changes from randomly coiled
chains to chains with a high degree of alignment. High
degree of chain alignment and extension of spacer in the
nematic phase have been confirmed by other methods as
well[9], including proton[10] and deuterium NMR spectroscopy.

The extent of correlation between repeating units is
dependent on molecular mass: entropy of isotropization[4]
and nematic order parameter at the I/N transition[5] both
increase rapidly with molecular mass, before leveling off.
The N/I transition temperature T_{NI} also follows a similar
trend, leveling off at $\overline{M}_n \gtrsim \sim 10,000$. As a result of this
molecular mass dependence of T_{NI}, a N+I biphase is observed in polydisperse samples.

In addition, extent of correlation between repeating
units is drastically influenced by structural factors,
such as spacer length. In the homologous series in
Table I, polymers in which spacer length n is an even number display a micellar cybotactic nematic organization,
with high values of isotropization entropy and nematic
order parameter[7,12]. This odd-even effect in supramolecular organization is distinctly different from that found
in LMLCs. In the homologous series of n-alkoxysubstituted
azoxybenzenes, for example, all members except n=1 and 2
are cybotactic nematics[13].

In what follows, we discuss some additional features
characteristic of phase transitions in this series of polymers.

The Nematic-Isotropic Transition

Supercooling at the I/N transition. Supercooling

of the isotropic melt with respect to the nematic phase
of a typical LMLC is illustrated in Table II for the case
of PAA. The values of $\Delta(T_{IN})$, $\Delta(T_{IN}")$ and $\Delta(T_c)$ are
small, as expected. Indeed, the only known example of
significant supercooling of the I/N transition in LMLC
was reported for unsymmetrical systems of the type[14].

RO-◯-CH=N-◯◯-N=N-◯

In the case of PLCs, it is also generally assumed
that the I/N transition is one of near equilibrium and
should not be significantly supercooled[15]. Tables I
through IV, however, show that $\Delta(T_{IN})$ (supercooling of
DSC peak maximum) can be significant. Although molecular
weight and molecular weight distribution undoubtedly play
a part in supercooling (compare AZA-9 DP=32 and 88 in
Table II), structure appears to be the primary determin-
ant. This is illustrated in Table I, where phase transitions are listed for the homologous series based on
mesogen 9 in the case of molecular mass sufficiently high
to minimize fluctuations which might be due to mass[4].
Supercooling of T_{IN} varies from 3 to 25°C and supercooling
at T" varies from 7 to 26°C, with an even-odd oscillation
(except for n=10).

N+I biphase. The nematic-isotropic intervals as observed by microscopy provide a very inadequate representative of the extent of the N+I biphase. This is clearly
illustrated on Fig. 1, where the $|T_2-T_c|$ interval measured
by microscopy is considerably smaller than the biphase
observed by NMR. Microscopic observation under slow rates
of scanning with measurement of the transmitted light intensity by means of a photomultiplier, to eliminate obser-
ver subjectivity, gives substantially the same results.
Fig. 1 shows that the N+I biphase extends well into what
is seemingly a "homogeneous" nematic phase. In polymer
DDA-9, for example, the shortest chains become isotropic
approximately 90° before chains whose mass is $\overline{M}_n \geqslant \sim 10,000$
and act as isotropic "contaminants" of the polydisperse
mesophase.

Preliminary NMR investigation of polydisperse DDA-9
samples does indeed suggest that selective partitioning
according to chain length takes place at the I/N transition[5]. More detailed analysis of undulations observed
in the variation of nematic order parameter with temperature further supports the conclusion that the longest

250

molecules are the first ones to be transferred into the anisotropic phase at the I/N transition[6].

Presence of small amounts of isotropic material within the mesophase can have important practical implications in processing of polydisperse samples. We have observed, for example, that upon fiber extrusion from the nematic melt, optimum mechanical strength is not developed until extrusion temperatures are below the I/N transition temperature of the shortest components[16]. Similarly, measurement of order parameter shows that presence of a minor isotropic component deters optimum overall alignment until a temperature some 25-30° C below the transition to a "homogeneous" nematic phase, as observed by microscopy[6].

Fig. 1 and Table II show that $|T_{IN''}-T_{IN'}|$, the width of the DSC peak at the N/I transition, is a much better gauge of the biphasic interval than the microscopic determination of $|T_c-T_2|$. Furthermore, comparison of GPC and DSC data shows a reasonably good correlation between width of DSC peak and polydispersity index $\overline{M}w/\overline{M}n$.

Consequently, we use the width of the N/I DSC transition peak as an approximation of the width of the biphase and the polydispersity of the sample (occasional shoulders in the DSC N/I peak can be correlated with shoulders in the GPC chromatograms).

Quenching of the nematic phase. Because of the sequential incorporation into the anisotropic phase of decreasing chain lengths, quenching of a biphasic sample in a magnetic field should lead to (semi)macroscopic domains. Formerly anisotropic domains would be rich in long chains and homogeneously aligned, while formerly isotropic domains would be rich in short chains and unaligned.

This raises the question of the homogeneity of the nematic phase; namely, an apparently homogeneous phase might actually be formed by a distribution of domains of (sub)macroscopic size. If this distribution is dependent on sample thermal history, quenching of the nematic phase or of the biphase might lead to mesophase glasses of varying morphologies.

We are currently investigating the homogeneity of the nematic phases of DDA-9 polymer. Preliminary NMR data[6] suggest the intriguing possibility that the "homogeneous" nematic phase might also be composed of domains segregated by chain size.

Mesophase -- Solid Transition

Crystallization of an oriented mesophase. The crystallization process of oriented nematic melts can lead in certain cases to highly oriented crystalline fibers while, in other cases, the crystallite orientation is totally or partly disrupted by the process of crystallization. Thus, DDA-8, MAA-8 and its copolymers MAA-DDA-8 lead to strongly oriented crystalline fibers on crystallization from nematic melts oriented in a magnetic field of 12 Tesla[17]. On the other hand, DDA-9 and its copolymers are much less oriented under the same circumstances. From preliminary observation of magnetically induced birefringence, it appears that the poor orientation of crystallites is related to the width of the mesophase and or kinetic factors governing the cooling experiment.

Fibers obtained by extrusion from the nematic melt form either oriented crystals (n=8,9,10,12 and 14) or supercooled oriented nematic glass fibers with a high orientation function, $f_c \sim 0.9$ (n=3,5,7,9,11 and copolymers)[12].

Crystallization from the unaligned melt. Samples of relatively high molecular weight ($\overline{M}_n > \sim 5,000$) tend to crystallize on cooling from the melt more readily than samples of low molecular weight[5]. For high molecular weight DDA-9 for example, the maximum heat of crystallization developed under scanning rates of 10° C/min. was 19.5 J/g. Under the same conditions, a sample with $\overline{M}_n = 4,000$ crystallized very poorly ($\Delta H_{NK} = 3.2$ J/g). However, addition of 10% PAA (w/w) enhanced the crystallization of the low mass polymer to the maximum of 19.5 J/g of polymer.

Enhancement of crystallinity of the polymer grown on cooling from the mixed nematic melt was even more pronounced for a polymer such as AZA-9, which by itself either does not crystallize at all, or crystallizes very poorly, regardless of molecular mass. Thus, in AZA-9 the heat of crystallization increased from nil to 22.6 J/g of polymer ($\overline{M}_n = 4,000$) upon addition of 10% PAA (no annealing; scanning rates 10°/min).

Crystals obtained from main chain nematic polymers appear to be very poorly ordered. This has already been pointed out previously by us and other authors[18,19], and it is logical to expect that if we are to realize an ordered melt (low ΔS_M) at reasonably low transition temperatures, as is the case here, it must be at the expense of the enthalpy of fusion; that is, at the expense of the degree of order in the crystal.

In the homologous series based on mesogen 9, the total entropy of fusion ($\Delta S_{KN} + \Delta S_{NI}$, after extrapolation to 100% crystallinity) was found to range from 3.3 to 4.3 J/°K mole of rigid backbone group. These are indeed values reflecting a poor level of order in the completely crystalline material and can be compared with values of 10-12 J/°K mole of rigid backbone group found in well-ordered crystals, such as PAA or 4,4'-acetoxy-2,2'-dimethylazoxybenzene, a model compound of mesogen 9.

In LMLC the ratio $\Delta S_{NI}/\Delta S_{total}$ is traditionally used to visualize the partitioning of order between the mesophase and the crystalline phase. For LMM nematogens this ratio is typically ∼0.03-0.05. In the case of nematic main chain polymers, ratios of ∼0.07 to 0.17 have been reported[20,21], and these relatively high values have been interpreted to indicate a relatively high level of order in the polymeric nematic phase.

In the case of the homologous series based on mesogen 9, the values of $\Delta S_{IN}/\Delta S_{total}$ range from ∼0.13-0.16 for spacers containing an odd number of methylene units and from ∼0.33-0.43 for even membered spacers (a discussion of odd-even effects will be presented in the next paper of this series). For PAA this same ratio is 0.03, although the degree of order as measured by the nematic order parameter is about the same in the odd-membered spacer polymers as in PAA[22]. While the even-membered spacer lengths do indeed lead to nematic phase with a high degree of order[12,22], the relatively high value of $\Delta S_{IN}/\Delta S_{total}$ observed for the odd-membered series probably just reflects the poor quality of crystals of mesophasic polymers.

CONCLUSION

In conclusion, we see that polyesters based on flexible spacers and substituted azoxybenzene moieties form a versatile family of stable mesophases with moderate to low phase transition temperatures. Nematic or cholesteric glasses can easily be formed from supercooled, nematic or cholesteric mesophases of polymers with proper choice of spacer. Homogeneous alignment of the nematic phase can be achieved by shear or by application of electric or magnetic fields. Because of sequential incorporation of chain lengths into the anisotropic phase, optimum sample alignment in a seemingly homogeneous mesophase or mesophase glass may actually be jeopardized by the presence of small amounts of short chains with low isotropization temperatures. Consequently, in polydisperse samples, the extent of the N+I biphase should be carefully determined. The width of the DSC peak at the N/I transition appears to be a much better measure of the biphasic interval than polarizing microscopy observations.

Although there exist two distinct levels of molecular order in the nematic phase, crystals grown from the mesophase all appear to be poorly ordered. Crystallization can be nucleated by addition of a LMM mesogen.

ACKNOWLEDGEMENT

The authors acknowledge the support of the National Science Foundation's Polymer Program under Grant No. DMR-8303989.

REFERENCES

1. For a review article. see A. Blumstein, J. Asrar, and R.B. Blumstein in Liquid Crystals and Ordered Fluids, Vol. IV, J.F. Johnson and A.C. Griffin, Eds., (Plenum Press, N.Y., 1984), pp. 311-346.

2a. S. Vilasagar and A. Blumstein, Mol. Cryst. Liq. Crys (Letters), 56, 263 (1980).

 b. A. Blumstein and S. Vilasagar, Mol. Cryst. Liq. Crys (Letters), 72, 1 (1981).

3. A. Blumstein, S. Vilasagar, S. Ponrathnam, S.B. Clou, and R.B. Blumstein, J. Polymer Sci. (Polymer Physics), 20, 877 (1982).

4. R.B. Blumstein, E.M. Stickles and A. Blumstein, Mol. Cryst. Liq. Cryst. (Letters), 82, 205 (1982).

5. R.B. Blumstein, E.M. Stickles, M.M. Gauthier, A. Blu stein and F. Volino, Macromolecules, 17, 2, 177 (1984).

6. F. Volino, J.M. Allonneau, A.M. Giroud, R.B. Blumstein, E.M. Stickles and A. Blumstein, Mol. Cryst. Liq. Cryst. (Letters), in print.

7. A. Blumstein and O. Thomas, Macromolecules, 15, 1264 (1982).

8. A. Blumstein, G. Maret and S. Vilasagar, Macromolecules, 14, 1543 (1981).

9. A. Blumstein, S. Vilasagar, S. Ponrathnam, S.B. Clou, and R.B. Blumstein, J. Polymer Sci. (Polymer Physics, 20, 877 (1982).

10. A.F. Martins, L.B. Ferreira, F. Volino, A. Blumstein and R.B. Blumstein, Macromolecules, 16, 279 (1983)

254

11. E.T. Samulski, M.M. Gauthier, R.B. Blumstein and
    A. Blumstein, _Macromolecules_, 17, 479 (1984).
    R.B. Blumstein, A. Blumstein, E.M. Stickles,
    M.D. Poliks, A.M. Giroud and F. Volino, _Polymer
    Preprints_, 24, 2, 258 (1983).
12. A. Blumstein, O. Thomas, J. Asrar, P. Makris, S.B.
    Clough and R.B. Blumstein, _J. Polymer Sci. (Letters)_
    22, 13 (1984).
13. I.G. Chistyakov and Q.M. Chaikovsky, _Mol. Cryst.
    Liq. Cryst._, 7, 269 (1969).
14. H. Kelker and R. Hatz, _Handbook of Liquid Crystals_,
    Vg. Chemie, 1980, p. 364.
15. J. Grebowicz and B. Wunderlich, _J. Polymer Sci.
    Polymer Physics)_, 21, 141 (1983).
16. A. Blumstein et al., work in preparation.
17. G. Maret and A. Blumstein, _Mol. Cryst. Liq. Cryst._,
    88, 295 (1982).
18. D.J. Blundell, _Polymer_, 23, 359 (1982).
19. B. Wunderlich, _Macromolecular Physics_, Vol. 3, "Crystal Melting", (Academic Press, N.Y., 1980).
20. P. Iannelli, A. Roviello and A. Sirigu, _Polymer J._,
    18, 759 (1982).
21. A.C. Griffin and S.J. Havens, _J. Polymer Sci. (Poly-Physics)_, 19, 951 (1981).
22. A. Blumstein, R.B. Blumstein, M.M. Gauthier, O. Thomas
    and J. Asrar, _Mol. Cryst. Liq. Cryst. (Letters)_,
    92, 87 (1983).

LIQUID CRYSTAL POLYMERS: 16 POLAR SUBSTITUENT EFFECTS

ON THERMOTROPIC PROPERTIES OF AROMATIC POLYESTERS

Qi-Feng Zhou* and   Robert W. Lenz
Chemical Engineering Department
University of Massachusetts
Amherst, MA   01003, U.S.A.

Jung-Il Jin
Chemistry Department
Korea University
1 Anam Dong, Seoul 132, Korea

SUMMARY

A series of polyesters containing monosubstituted p-phenylene
bis-terephthalate mesogenic groups and decamethylene flexible
spacers was prepared and characterized for the effect of substituent polarity and size on the transition temperatures and on the
thermal stability and type of liquid crystalline phase formed. No
simple correlation between the polar effects of the lateral substituents and the thermal stabilities of the polymer mesophases was
observed. The polar effects of the substituents appeared to be less
important than their steric effects in controlling the thermotropic
transitions of these polymers and of essentially no importance in
controlling their glass temperatures and melting points. Optical
textures of the polymer melts did not provide clear conclusions as
to the nature of the liquid crystalline phases formed by these
polymers, but their textures resembled those of low molecular weight
nematics.

* Present address: Chemistry Department, Beijing University,
  Beijing, China

INTRODUCTION

There are only a few reports on the effects of lateral substituents attached to the mesogenic units on the liquid crystal
behavior of semiflexible, main chain polymers having an alternating
sequence of rigid mesogenic elements and spacers.[1-4] In contrast,
many studies have been reported on the effects of introducing lateral
substituents into low molecular weight liquid crystalline compounds.[5]
In one of our earlier studies we showed that the clearing temperatures of a series of polymers having the structure I shown below
decreased continuously with increasing size of the substituent, X,
when the substituent was either H, CH_3, Cl or Br. This result was
taken as an indication that the predominant steric effect of the
substituents was to increase the separation of the mesogenic units
and, thereby, that of the long axes of polymer chains in the same
manner as has been reported for low molecular weight mesogens. All
of the polymers in this series were nematic.

$$\{O-\langle O \rangle-\overset{\overset{O}{\|}}{C}-O-\langle \overset{O}{\underset{X}{}} \rangle-O-\overset{\overset{O}{\|}}{C}-\langle O \rangle-O\{CH_2\}_{10}\}$$

I

We also described, in a previous report,[4] the preparation and
properties of a series of main chain, liquid crystal polyesters
containing n-alkyl substituents on the central aromatic ring of the
mesogenic group which varied in length from the methyl to the decyl
group. The polymers in this series, with the following structure,
II, were liquid crystalline for the methyl to hexyl substituents,
n = 1-6, but not for the octyl and decyl substituents, n = 8 or 10:

$$\{O-\overset{\overset{O}{\|}}{C}-\langle O \rangle-\overset{\overset{O}{\|}}{C}-O-\langle \underset{C_nH_{2n+1}}{O} \rangle-O-\overset{\overset{O}{\|}}{C}-\langle O \rangle-\overset{\overset{O}{\|}}{C}-O\{CH_2\}_{10}\}$$

II

For the members of the series which were thermotropic, the
glass transition temperatures, T_g, melting points, T_m, and clearing
temperatures, T_c, decrease in a uniform manner, for the most part,
with increasing alkyl group size. These polymers were all nematic,
but the temperature range over which they were liquid crystalline,
ΔT, did not follow a simple relationship with substituent size. The

temperature range ΔT is usually taken as an indication of the thermal stability, in a physical sense, of the mesophase.

All of the studies described above involved only either alkyl or halogen substituents. It was of interest, therefore, to examine whether the effects of polar substituents in general were similar for polymers and for low molecular weight compounds showing liquid crystalline behavior, so a closely related series of substituted polymers, of the same basic structure as that of II, was prepared. That is, the same mesogen and flexible spacers were used, but the substituent was either the Br, NO_2, CN or OCH_3 group. It is known from the properties of related model compounds that changes caused by a substituent in the molecular polarizibility and polarity of the mesogen, as well as in steric factors, strongly influence their thermotropic behavior.

RESULTS AND DISCUSSION

The polymers of this study were prepared by the reaction of 1,10-bis(p-chloroformylbenzoyloxy)decane, III, with the appropriate 2-substituted hydroquinone, IV, as follows:

$$Cl-C(=O)-\langle O \rangle-C(=O)-O-(CH_2)_{10}-O-C(=O)-\langle O \rangle-C(=O)-Cl \; + \; HO-\langle O \rangle(X)-OH \longrightarrow II$$

III   IV

In the latter and in the resulting polymer, X was Br, NO_2, CN or OCH_3. The polymers in which X was H, CH_3 or C_2H_5 were prepared in our previous study. The polycondensation reaction was carried out at room temperature in 1,1,2,2-tetrachloroethane (TCE) with pyridine as the catalyst. The polymers were all soluble in TCE, so that polymerizations proceeded homogeneously. The solution viscosities of the polymers obtained after precipitation in acetone are listed in Table 1. The structures of the polymers were confirmed by their IR and ^{13}C-NMR spectra. The thermal properties given in Table 1 and the types of thermotropic behavior were determined by differential scanning calorimetry (DSC) and by visual observation of a sample placed on a hot stage on a polarizing microscope.

Four of the monosubstitued hydroquinone units in the mesogenic groups in this series of polymers are not much different in size, those in which X is CH_3, Br, NO_2 or CN, as indicated by their molecular diameters listed in Table 1,[6,7] while the diameters of the mesogenic units with X as C_2H_5 and OCH_3 are slightly larger than others. Hence, it is expected that any differential effects of these two sets of substituents on the thermotropic properties of the polymers of the present series should be caused by electronic effects rather than steric effects, assuming that molecular weight differences of the polymers were not important. That is, while the inherent viscosities of the polymers prepared suggest that their molecular weights are relatively low, it is believed that the transition temperatures observed were approximately those of their high molecular weight polymers because gel permeation chromatography studies on related polymers indicate that the inherent viscosity values observed represent molecular weights in the range of approximately 10,000 to 20,000.[4]

All of the polymers exhibited melting endotherms in their DSC thermograms, and the melting points, T_m, of all of the substituted polymers were much lower than those of the parent polymer. For all substituted polymers, T_m values were close to $150^\circ C$ regardless of the structure of the substituents. This result indicates that the depression of the melting point of the polymers arises mainly from the steric effects by the substituents which reduce the molecular packing efficiency, with very little contribution by polar effects. Indeed, a comparison of melting points and glass temperatures of the polymers with CH_3, Br, CN, and NO_2 substituents clearly suggests that the polarity of the substituent has almost no effect on these properties, and the controlling importance of the steric effect is also apparent from the very low value of T_m for the polymer with the C_2H_5 substituent.[4]

The effect of substituent size and polarity on the clearing temperatures, T_c, and the thermal stabilities of the mesophases, ΔT, of the polymers is less clear. The polymer containing the CN group, which has the lowest value of η_{inh} in Table 1, has the highest value of both T_c and ΔT, while the nitro polymer has the lowest ΔT. These two substituents should have the strongest polar effects of the six groups of substituents shown in Table 1. So it is apparent that there is no simple correlation between polar effects and thermotropic properties in the present series. Nevertheless, the clearing temperatures of the three liquid crystalline polymers substituted with polar groups were higher than that of the polymers with the methyl group, so in that sense, the polarity of the substituent appears to contribute to the stability of the mesophase. The results reported by Dubois and Beguin[8] on the influence of lateral substitution on mesomorphic properties of phenyl 4-benzoyloxybenzoate derivatives also show a measurable degree of polar effects induced by strongly polar substituents such as nitrile group.

In general a lateral polar substituent can influence the thermotropic behavior of a polymer by two oppositely operating effects: steric separation of neighboring molecules or polymer chains and increased dipolar attraction between them. Increased polarizability, which tends to stabilize the mesophase, can, of course, result from the presence of the polar substituents, but no quantitative measures have been developed to relate the degrees of contribution of each of these factors, steric and polar, to the T_m and T_c transition temperatures. Even for well-defined low molecular weight compounds a quantitative evaluation of those factors has not been attempted, although it is well known that steric factors play the predominant role in controlling the thermal stability of mesophases of such compounds.[5] Dewar and Riddle[9] arrived at that conclusion from their study on the thermotropic properties of a series of hydroquinone di-p-methoxybenzoates with different substituents on the central phenyl ring.

One surprising observation in the present study was that the methoxy-substituted polymer did not show a clearing transition in the heating cycle of the DSC thermogram, although a small shoulder was observed on the crystallization exotherm in the cooling cycle. The melt phase of this polymer showed strong stir-opalescence and birefringence on cooling to a temperature just above the crystallization point, which suggests that this polymer may be monotropic, but the evidence for monotropic behaviour is not unequivocal. Except for the methoxy polymer, all of the others showed birefringence upon melting and only became isotropic liquids at a substantially higher temperature. Thread-like textures were observed for these polymers on a hot-stage of a polarizing microscope, and it is believed from these qualitative observations that all of these polymers formed nematic mesophases. In the previous report of this investigation,[4] it was also concluded that the methyl and ethyl polymers formed nematic phases.

EXPERIMENTAL

Preparation of Monomers

Cyanohydroquinone, nitrohydroquinone and methoxyhydroquinone were prepared by following the literature procedures and were purified either by distillation under reduced pressure or by recrystallization.[10-13] Bromohydroquinone was purchased from Aldrich Chemical and purified by vacuum sublimation. 1,10-Bis(p-chloroformyl-benzoyloxy)decane, III, was synthesized by a multi-step sequence of reactions following the literature methods[14] and purified by recrystallization from n-heptane. All of the compounds were identified by melting point, elemental analysis, infrared spectra with a Perkin-Elmer IR-283 spectrometer, ^{13}C-NMR with a Varian FT-20 spectrometer, and ^{1}H on a Varian S-60T spectrometer.

Table 1. Inherent viscosities and thermal properties of unsubstituted and substituted polymers, I.

Hydroquinone Residue X	Size, Å [a]	η_{inh} [b] dl/g	Transition Temperatures, °C [c]			
			T_g	T_m	T_c	ΔT
H	6.7	0.68	67	231	267	36
CH_3	7.8	0.41	44	154	190	36
Br	8.1	0.33	44	140	196	56
NO_2	8.3	0.56	42	161	194	33
CN	8.3	0.29	47	157	219	62
OCH_3	8.9	0.63	50	158	--d	--d
C_2H_5	9.0	0.55	35	71	127	56

[a] Defined as the diameter of the narrowest cylinder through which the monosubstituted hydroquinone ring would pass.
[b] Inherent viscosity measured at a concentration of 0.5 g/dl in tetrachloroethane at 40°C.
[c] Transition temperatures determined by DSC: T_g--glass temperature observed in the first heating cycle; T_m--melting temperature taken at peak of endotherm; T_c--clearing temperature taken at peak of endotherm; ΔT--temperature range of nematic phase.
[d] No clearing endotherm present in DSC thermogram.

Preparation of Polymers

A representative procedure used for the preparation of polymers is as follows: a solution of monomer III (3.40 g; 6.7 mmole) dissolved in 20 ml dry TCE was added dropwise to a solution of the substituted hydroquinone monomer IV (6.7 mmole) dissolved in a solvent mixture of 20 ml TCE and 6 ml of dry pyridine under a nitrogen atmosphere with vigorous stirring. The mixture was stirred for 20-24 hours at room temperature, after which the polymer was precipitated into 400 ml of acetone with stirring. The product was filtered and thoroughly washed with acetone, then with water and acetone, and dried at room temperature under vacuum. Polymer yields ranged from 89 to 97% by weight. The structures of the polymers were confirmed for their IR and ^{13}C-NMR spectra.

Characterization of the Polymers

The inherent viscosity measurements of the polymers were made at 40°C using a Ubbelohde viscometer. Solutions of 0.5 g/dl in TCE were used for the measurements. The thermal transitions of the polymers were measured using a Perkin-Elmer DSC-II at the heating and cooling rates of 10°C/min under a nitrogen atmosphere. The peak maximum positions were taken as those of the transition points. Optical textures of the polymer melts were observed on the hot-stage (Mettler FP-2) of a polarizing microscope (Leitz, Ortholux).

ACKNOWLEDGEMENT

The support of this research by the NSF-sponsored Materials Research Laboratory of the University of Massachusetts, by the Air Force Office of Scientific Research and by the Korea Science and Engineering Foundation is gratefully acknowledged. Q.-F. Zhou wishes to express his appreciation to the government of the People's Republic of China for the support of his study in the U.S.A.

REFERENCES

1. J.-I. Jin, S. Antoun, C. Ober and R. W. Lenz, Brit. Polymer J., 12, 132 (1980).
2. S. Antoun, J.-I. Jin and R. W. Lenz, J. Polymer Sci., Polymer Chem. Ed., 19, 1901 (1981).
3. A. Blumstein, S. Vilasaga, S. Ponrathnam and S. B. Clough, J. Polymer Sci., Polymer Chem. Ed., 20, 877 (1982).
4. Q.-F. Zhou and R. W. Lenz, J. Polymer Sci., Polymer Chem. Ed., in press.
5. G. W. Gray and P. A. Winsor (Ed.), "Liquid Crystals and Plastic Crystals, Vol. 1," Halsted Press, New York, 1974, pp. 126-136.

6. G. W. Gray and B. M. Worrall, J. Chem. Soc., 1545 (1959).

7. G. W. Gray, B. Johns and F. Marson, J. Chem. Soc., 393 (1957).

8. J. C. Dubois and A. Beguin, Mol. Cryst. Liq. Crystl., $\underline{47}$, 193 (1978).

9. M. J. S. Dewar and R. M. Riddle, J. Amer. Chem. Soc., $\underline{97}$, 6658 (1975).

10. K. Wallenfels, D. Hofmann and R. Kern, Tetrahedron, $\underline{21}$, 2231 (1965).

11. W. Baket and N. C. Brown, J. Chem. Soc., 2303 (1948).

12. J. Fouest and V. Petrow, J. Chem. Soc., 2441 (1950).

13. A. Chatterjee, D. Gangnly and R. Sen, Tetrahedron, $\underline{32}$, 2407 (1976).

14. C. Ober, J.-I. Jin and R. W. Lenz, Polymer J., $\underline{14}$, 9 (1982).

PREPARATION AND PROPERTIES

OF THERMOTROPIC POLYAMIDOESTERS

Leszek Makaruk, Hanna Polanska
and Barbara Wazynska*

Department of Chemistry, Technical University
Noakowskiego 3, 00-662 Warsaw, Poland
*Institute of Materials Science & Engineering
Technical University, Narbutta 85
02-524 Warsaw, Poland

INTRODUCTION

Up to date, quite a great number of low-molecular weight segments have been incorporated either in the main or in the side-chains of the macromolecules allowing the polymer to form a mesophase upon heating. There are, however, such polymers, for example aromatic polyamides, that do not turn liquid crystalline and start decomposing without melting. Usually mesophases in such systems can be observed only after dissolution of the polymer. As an example, melt spinning of high-modulus Kevlar fibres from the solution in concentrated sulphuric acid can be given. To eliminate this commonly used, though very aggressive solvent, the related but thermotropic polyamides are widely studied.

Aharoni (1,2) has attempted the synthesis of random and alternating rigid-flexible polyamides and copolyamides with the mesogenic elements in the main chain. The resulting polymers gave, however, lyotropic liquid crystals. Mc Intyre et al. (3) have recognized that thermotropic polyesteramides could be obtained from p-aminophenol with several monoalkoxyterephthalic acids or p-carboxyphenoxyacetic acid. According to them this is due to the fact that each of these compounds can be arranged in the macromolecule successively either head-

to-head or head-to-tail. Therefore, the polymers are
expected to exhibit lower crystallinity and melting
points compared to these with alternation of amide and
ester groups. Adduci et al.(4) prepared the series of
polymers in which the basic structure of Kevlar was
maintained but the p-phenylenediamine units were replaced with various semi-flexible diamines capable of exhibiting limited internal rotational mobility. Those
diamines contained either an ether linkage or a combination of ether and sulphone linkages. As a result the
thermotropic liquid crystalline polymers with the melting points lower than the decomposition temperatures
and substantially decreased as compared to Kevlar, were
obtained.

In our recent paper (5) we have reported the preparation and properties of thermotropic liquid crystalline
polyesters and copolyesters derived from dicarboxylic
acid dichlorides and bisphenols containing ketone groups
between the aromatic rings. Partial replacement of ketone
groups by alternating them with amide groups seemed to us
a way to produce the polyamidoesters with the increased
transition temperatures and broadened mesophasic range.
In the present paper it is shown that polymers under the
study behave indeed as it was expected, as well as the
upper limit of stability of the anisotropic phase in
such copolyamidoesters is lower than their decomposition
temperatures.

EXPERIMENTAL

Materials

4,4'-Dihydroxybenzophenone (DHBP) a comonomer in
polycondensation reaction and a substrate for the preparation of its amide, was obtained from Merck-Schuchardt
and was not further purified. It had the melting point of
214 °C.
Suberyl and sebacyl dichlorides were prepared and
purified according to the procedures previously described (5).
All solvents were purified and stored according
to the details given in (6).

Synthesis of Monomers

Amides are usually obtained from their oximes by the
Beckmann rearrangement. We found it very difficult to
prepare the DHBP oxime with the reasonable yield, accord-

ing to the methods given by Vogel (7), Spiegler (8), Zigeuner and Ziegler (9). We have succeeded in 87% of the theoretical yield following the method of Bender et al. (10), but using $NH_2OH \cdot HCl$ instead of $NH_2OCH_3 \cdot HCl$. DHBP-oxime of the following chemical formula:

$$HO-\bigbit\!\!-\!C\!-\!\bigbit\!\!-OH$$
$$\underset{NOH}{\overset{\|}{}}$$

was crystalline, beige precipitate with the melting point of 270 °C. Treatment of this oxime with tionyl chloride in ether solution (7) resulted in crystalline DHBP-amide of light-beige color and a melting point of 273 °C. Both DHBP-oxime and DHBP-amide (DHBP-A) structures were confirmed by elemental analyses and IR spectra.

Polymer Preparation

Two series of polyamidoesters, having the same molar ratios of DHBP and DHBP-A in the rigid part of the macromolecule (i.e. 100/0, 75/25, 50/50, 25/75 and 0/100) and different number of methylene groups (the residues of either sebacyl or suberyl acids) in the flexible part, were synthesized following the general procedure for the interfacial polycondensation (5).

Polymer Characterization

Elemental analyses on new compounds were performed on Perkin-Elmer Model 240 microanalyzer.

Melting points for all crystalline monomers were determined using a Boetius hot-stage microscope (PHMK VEB Analytik Dresden).

Inherent viscosities of polymers were measured with Ubbelohde type viscometer for 0.5 g/dl solutions in o-chlorophenol as a solvent at 25 °C.

DSC measurements were carried out on Perkin-Elmer DSC-2 apparatus under an argon atmosphere with a heating or cooling rate of 10 °/min. Peak areas were determined by planimetry. The results of DSC measurements were connected with the visual observations of the texture of the samples by means of hot-stage microscope equipped with a pair of crossed polarizers.

Small-angle light scattering (H_v) measurements were carried out at room temperature with a He-Ne (λ=6328 AU) laser.

For the detailed description of samples preparation and the procedures refer to our earlier paper (5).

RESULTS AND DISCUSSION

Polymers obtained were white-brownish powders, insoluble in virtually all common solvents at room temperature, although their molecular weights were rather low as witnessed by their inherent viscosities ($\eta_{inh} \simeq$ 0.1 dl/g).

Examination on a hot-stage polarized microscope, as well as SALS and DSC data let us to the conclusion, that all polymers derived from the mixture of DHBP-A and DHBP with suberyl dichloride (Series I) or sebacyl dichloride (Series II) behave as thermotropic liquid crystals. For most of them DSC scans revealed two endotherms, which are believed to correspond to solid-to-liquid crystalline (T_m) and mesophasic-to-isotropic (T_i) transitions. In few cases, i.e. for polyamidoester derived from DHBP-A with sebacyl dichloride and the copolyamidoesters of Series II in which DHBP-A content exceeded 40 mol%, the multiple melting endotherms were observed. Though the similar phenomena were demonstrated by several authors (11,12) their origin is still not clear.

It was also found for the copolyesters in which the amide group (from DHBP-A) content exceeded 40 mol%, that the endotherms encountered during the first DSC scan differ from those obtained on subsequent heating run. Therefore it can be concluded that during the heating of the polymer in the DSC to approximately 30-40 °C above T_i (what is the usual procedure) the partial decomposition occurs. However, when the first heating was terminated just after T_i was reached, the reproducible transition temperatures and enthalpies were obtained.

The curves presented in Figs 1 and 2 compare the relationships of T_m and T_i against molar concentration of amide containing bisphenol in the rigid part of the polymer for two series of copolyamidoesters differing in the length of the flexible segment. As expected, the copolyamidoesters in which the ketone groups were partially or fully alternated with amide ones (capable of forming the strong hydrogen bonds) had higher both T_m and T_i temperatures. The reverse behavior was observed for the samples in which the shorter aliphatic chains were replaced by longer ones (compare data for polymers based on DHBP or DHBP-A with suberyl and sebacyl dichlorides).

It is also worth noting that two members of the Series I with 50 and 75 mol% of DHBP-A content in the

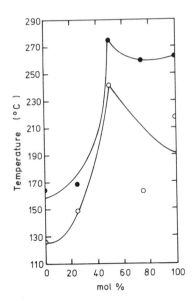

Fig. 1. T_m (●) and T_i (○) dependence on molar concentration of DHBP-A in the mixture of bisphenols
used for the synthesis of Series I copolyamidoesters.

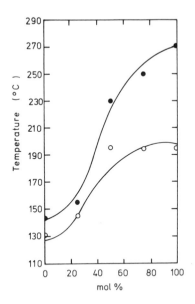

Fig. 2.   Relationship between T_m (●), T_i (○) and molar
concentration of DHBP-A in the mixture of bisphenols used for the synthesis of Series II copolyamidoesters.

rigid part of the macromolecule, exhibited very unique properties. The former had the highest T_i (271 ⁰C) and the latter - the broadest temperature range of liquid crystallinity (97⁰C) out of all polymers of this work (Fig.1). We must stress as well, that the copolyamido-ester obtained from suberyl dichloride with DHBP-A/DHBP mixture of 75/25 did not show any traces of decomposition after heating much above T_i.

One of the most important features characterizing the thermotropic liquid crystalline polymers is the temperature range of their mesophases. The results of this work show, that the incorporation into the rigid part of the macromolecule of groups capable of packing in an ordered arrangement due to the strong intermolecula interactions (in this case hydrogen bonds), broaden the temperature range of liquid crystallinity ($\Delta T = T_i - T_m$). This relationship is particularly pronounced for Series II (Fig.2). For these polymers we observed the upward trend in ΔT with the increase of the concentration of the amide containing bisphenol in the rigid part of the macromolecular chains.

Another observation we have made was a rather low enthalpy change involved in the solid-to-mesophase tran-sition (ΔH_m) as compared to the isotropisation enthalpy (ΔH_i), especially for the members of Series II (Fig.3) The dependence of ΔH_m and ΔH_i on the composition of polymer (Fig.3 and 4) shows a regular behaviour within the Series revealing a minima for approximately equimolar content of amide and ketone groups incorporated into the rigid part of the macromolecules (with the exception for ΔH_m of Series I, in which minimum reaches 75 mol% of DHBP-A in the macromolecule).

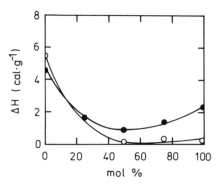

Fig. 3. ΔH_m (0) and ΔH_i (o) changes with the compo sition of copolyamidoesters of Series II.

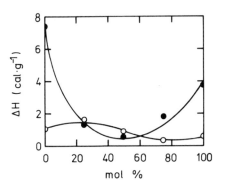

Fig. 4. Plot of ΔH_m (●) and ΔH_i (o) versus composition of copolyamidoesters of Series I.

Such behaviour of ΔH_m may be the evidence of lower crystallinity or that the observed transition are transitions from the crystalline state to highly ordered mesomorphic state. Both facts could explain the decrease in ΔH_m as well as its relatively low values. It is known (13, 14) that for polyesters the clearing temperature (T_i), the clearing enthalpy (ΔH_i) and consequently the clearing entropy are little affected by the thermal history of the sample, while the melting parameters are highly dependent on the thermal treatment of the sample. Assuming that crystallinity of the samples does not change drastically with the composition, one may expect the observed effect on ΔH_m being mainly connected with the ordering of the mesophase.

Since ΔH_i is more directly related to the degree of order of the mesophase, its changes (at least qualitatively) can be attributed to the description of the state of mesophase.

It is seen from the curves that within one series, with the increasing of DHBP-A content in macromolecules, the ΔH_i increases, while ΔH_m changes very slightly. It is very likely then, that it corresponds to higher ordering of polymer macromolecules in mesophase due to the stronger intermolecular interactions (namely hydrogen bonds) in the system.

We have also attempted to determine the superstructure of the scattering elements in the melt. SALTS investigations revealed four-clover type scattering patterns of different sizes and intensities for all members of both series, however any regular behaviour could not be detected. These results can suggest then, that the scattering elements of the mesomorphic melts were generally spherulitic in shape.

We are now carrying out X-ray studies, which we hope will give us additional information concerning the degree of crystallinity of the prepared polymer samples and the type of the mesophases they form.

In view of the results of this work it can be concluded, that partial replacement of ketone groups by alternaing them with amide groups in the rigid part of the macromolecules, produces thermotropic liquid crystalline copolamidoesters of increased transition temperatures and broadened mesophasic ranges.

REFERENCES

1. S. M. Aharoni, Rigid Backbone Polymers.XVIII. Strictly Alternating Rigid-Flexible Polyamides, J.Polym. Sci.Polym. Phys.Ed., 19:281 (1981).
2. S. M. Aharoni, Rigid Backbone Polymers.XVI. Random Copolyamides, J.Appl.Polym.Sci., 25:2891 (1980)
3. J. E. Mc Intyre and A. H. Milburn, Thermotropic Liqui Crystalline Polyesters and Polyesteramides, Br. Polym.J., 13:5 (1981).
4. J. Adduci, L. L.Chapoy, G. Jonsson, J. Kops and B. M. Shinde, Semi-Stiff Chain Aromatic Polyamides: New Candidates for Thermotropic Liquid Crystalline Polymers, Polym.Eng.Sci., 21:712 (1981).
5. L. Makaruk and H. Polanska, Effect of Molecular Structure of Rigid-Flexible Polyesters on Their Thermotropic Liquid Crystalline Properties, Liq.Cryst. Ordered Fluids, in press.
6. J. A. Riddick and E. F. Toops, Organic Solvents - Physical Properties and Methods of Purification, in: "Techniques of Organic Chemistry", vol.VII, A. Weissberger, ed., Interscience Publishers, Inc. New York (1955).
7. A. I. Vogel, "A Textbook of Practical Organic Chemistry", Longmans (1959), pp. 345, 721 and 741.

8. E. Spiegler, Zur Kenntniss der Euxanthongruppe, <u>Monatsh.</u>, V: 195 (1884).
9. G. Zigeuner and E. Ziegler, Spaltung mittels Diazonium-verbindungen. IV. Mitteilung: Über Oxybenzophenone, <u>Monatsh.</u>, 80:359 (1949).
10. M. Bender, Z. Buczkowski and J. Plenkiewicz, NMR Study of Certain Aliphatic Azomethines. II. Ketoxime O-methyl Ethers and Their Complexes with Trimethyl-aluminium, <u>Bull. Acad. Pol. Sci. Ser. Sci. Chim.</u>, 17:643 (1969).
11. A. G. Griffin and S. J. Havens, Mesogenic Polymers. 3. Thermal Properties and Synthesis of Three Homologous Series of Thermotropic Liquid Crystalline "Backbone" Polyesters, <u>J. Polym. Sci. Polym. Phys. Ed.</u>, 19:951 (1981).
12. L. Strzelecki and L. Liebert, Influence de la Structure sur le Proprietes Mesomorphes de Polyesters- III, <u>Europ. Polym. J.</u>, 17:]271 (1981).
13. A. Roviello and A. Sirigu, poly[oxytetradecanedioyloxy-],4-phenylene-(2-methylvinylene)-1,4-phenylene],M Makromol. Chem., 180:2543 (1979).
14. A. Roviello and A. Sirigu, A Mesophasic Polymer: Poly[-oxydodecanodioyloxy-1,4-phenylene-(2-methylvinylene)-1,4-phenylene],<u>Makromol. Chem.</u>, 181:1799 (1980).

PHOTOELASTIC BEHAVIOR OF LIQUID CRYSTALLINE POLYMER NETWORKS

H.J. Kock, H. Finkelmann, W. Gleim, and G. Rehage

Institute of Physical Chemistry
Technical University Clausthal
Clausthal-Zellerfeld, FRG

1. INTRODUCTION

The linkage of low molar mass liquid crystals as side groups
to a polymer main chain leads to a new class of substances which
are important with respect to theoretical aspects as well as
future technological application. They combine the liquid crystalline (LC) behavior with specific polymer properties [1,2]. To get
an almost complete decoupling of the motions of the polymer backbone and the mesogenic moieties they are connected by covalent
bonds to the backbone by a flexible spacer. Investigations on the
state of order of LC-polymers compared with the state of order of
low molar mass LCs indicate that the linkage to the polymer chain
by sufficiently long flexible spacers does not essentially influence the state of order [3]. Under these conditions the mesogenic
side chains can spontaneously form an anisotropic orientation and
thus build up a LC phase [4]. There are some indications that the
conformation of the polymer chain is changed by the formation of the
anisotropic LC phase. Instead of a statistical coil conformation
it rather has to be assumed that the coil is more or less aniso-
tropically deformed.

In the last years polymeric side chain LCs with nematic,
smectic or cholesteric LC phases have been synthesized by varying
the chemical constitution of the mesogenic side chains [2,5]. By
lowering the temperature the macroscopic texture as well as the
order parameter freeze in and one gets an anisotropic glassy state.

If the polymer chains are crosslinked to form a network, the
micro-Brownian motion of the chain segments is not essentially
influenced [6]. The macro-Brownian motion, that are the motions

of entire molecules, is prevented by crosslinking. The LC side
groups possess nearly the same freedom as in linear polymers. Therefore the LC properties of the crosslinked system should be very
similar to those of the linear LC polymers. This is proved by DSC
measurements. There is only a shift of the glass transition temperature and transformation temperature (liquid crystalline - isotropic) to lower temperatures caused by the plasticizing effect of
the high flexible non-mesogenic crosslink agent. It has to be
pointed out that higher amounts of crosslink agent will completely
disturb the formation of the LC phase and only a common optical
isotropic polymer network is maintained. Therefore the content of
crosslinking agent is reduced to the minimum value necessary for
building up a stable three-dimensional network. By the crosslinking
process the viscous melt of the linear LC polymer is changed now
into an elastomer and the LC properties are combined with rubber
elasticity. Especially the stability of shape and the reversible
elastic deformation by mechanical forces as well as the glass transition are of great importance for future application of LC-elastomers.

Conventional low molar mass LCs as well as linear LC-polymers
can be macroscopically ordered by external electric or magnetic
fields, which is widely applied in optoelectronics in the case of
low molar mass LCs. For LC-elastomers it is very important to
know whether a macroscopic mechanical deformation of the polymer
network influences the liquid crystalline side groups and whether
a mechanical stress or strain produces similar effects as observed
for conventional LCs by external fields.

From the statistical theory of rubber elasticity follows for
the case of uniaxial deformation of a network of independent
Gaussian chains the relation [7]:

$$\sigma = \nu k T \frac{\langle r^2 \rangle}{\langle r^2 \rangle_0} (\lambda^2 - \lambda^{-1}) \qquad (1)$$

σ is the retractive force per area of the deformed sample, λ the
relative deformation, ν the number of elastically active chains
per volume unit, $\langle r^2 \rangle$ the mean square end-to-end distance of the
network chains in the undeformed sample and $\langle r^2 \rangle_0$ the same quantity if the chains are free and otherwise under the same conditions
kT has the usual meaning. This relation follows from the model
case of a phantom network where the chains can freely intersect.
The crosslinking points are assumed not to fluctuate and subjected
to an affine deformation.

The macroscopic deformation of the networks leads to an orientational birefringence which has been correlated by Kuhn und Grün [8]
to the relative deformation:

$$\Delta n = \frac{2\pi}{45} \; \nu \; \frac{\langle r^2 \rangle}{\langle r^2 \rangle_0} \; \frac{(\bar{n}^2 + 2)^2}{\bar{n}} \; \Delta\alpha \; (\lambda^2 - \lambda^{-1}) \qquad (2)$$

Δn is the difference in the refractive indices n_1 in the plane of the drawing direction and n_2 perpendicular to it. $\bar{n} = (n_1 + 2n_2)/3$ is the mean refractive index of the sample and $\Delta\alpha$ the difference of the polarizabilities of the statistical segment ($\Delta\alpha$ = optical anisotropy) parallel and perpendicular to the axis of the segment. The quotient of orientational birefringence and stress is defined as the stress optical coefficient C. From the simple model of the network chains the product CT (equation 3) should be independent of T if a slight temperature dependence of $\bar{n}$ is neglected. CT is proportional to the optical anisotropy of the statistical segment.

$$C \, T = \frac{2\pi}{45 \; k} \; \frac{(\bar{n}^2 + 2)^2}{\bar{n}} \; \Delta\alpha \qquad (3)$$

The optical anisotropy calculated from this equation enables the discussion of the orientation of the optical anisotropic meso-genic side groups with respect to the mechanically oriented polymer backbone. Additionally the question of the decoupling of the motions of the polymer chains and the mesogenic side groups by a flexible spacer can be treated qualitatively. This is possible by comparing the stress optical coefficients of LC-networks and analogous net-works without mesogenic side groups. In the case of LC-elastomers with a polysiloxane backbone the results have to be compared with usual PDMS networks.

For the discussion of the results it has to be emphasized that the Kuhn-Grün relation is based on the application of the Lorenz-Lorentz-equation. It has already been discussed earlier that the assumption of a spherical internal field could not be valid for extended non-spherical statistical segments of polymer chains [9]. The strong influence of different internal fields on the sign of the stress optical coefficient has been described by Pietralla [10]. The same problem arises for liquid crystalline phases and has been discussed in detail for low molar mass LCs [11] whereby the validity of the Lorenz -Lorentz equation in the isotropic state still has been assumed [12].

2. SYNTHESIS AND PHASE BEHAVIOR

Starting with linear polysiloxanes the networks were prepared by addition reaction of the mesogenic molecules and crosslinking agent via the following scheme (Fig. 1.) Because of the similar reactivity of the crosslinking agent and the mesogenic molecules, a statistical addition to the backbone has to be expected. The LC-elastomers investigated in this paper and their phase behavior are listed in Table 1.

Table 1: Phase behavior of liquid crystalline elastomers
(m, x, y, z: see Fig. 1)

Spacer Length m	Composition x	y	z	Phase Behavior (DSC)	
3	0	108	12	g 285 n 332 i	
3	0	120	0	g 288 n 334 i	uncrosslinked
3	60	48	12	g 263 n 283 i	
4	0	108	12	g 275 n 343 i	
4	0	120	0	g 288 n 368 i	uncrosslinked
4	60	48	12	g 261 n 301 i	

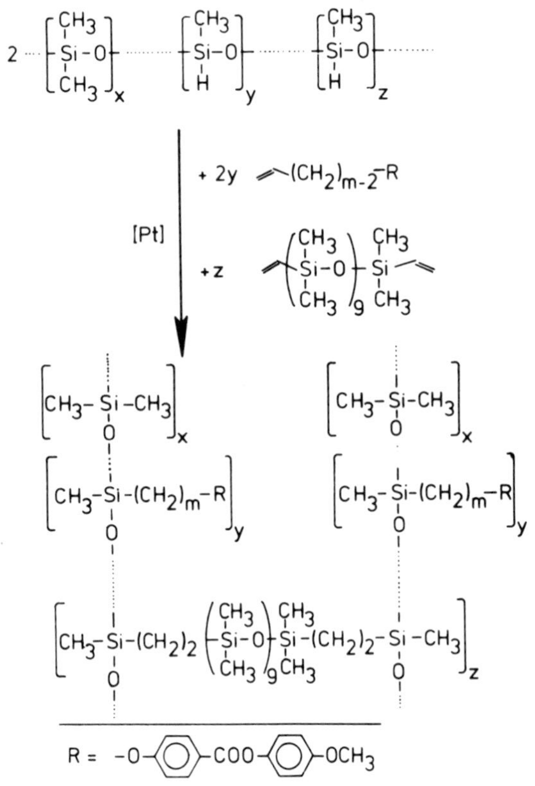

Fig. 1: Synthesis of liquid crystalline polymer networks (x,y,z
are the numbers of monomer units of one polymer chain;
x+y+z = 120; m is the number of methylene units of the
flexible spacer)

It has been found by DSC measurements that the phase behavior of the elastomers is analogous to the phase behavior of the linear polymers[6]. Starting at high temperatures in the isotropic rubbery state a transformation to the liquid crystalline rubbery state by decreasing temperature is observed. By further cooling down the elastomers become glassy at T_g.

3. EXPERIMENTAL

The stress-birefringence measurements were performed with an experimental arrangement as described by Treloar[7] (Fig. 2). Different measuring cells for uniaxial compression of cylindrical samples and uniaxial elongation of long strip form samples, respectively, have been used. The retractive force of a deformed sample has been measured as a function of the relative deformation λ($\lambda = l/l_o$ with l and l_o the lengths of the deformed and the undeformed sample) by means of a load cell. λ was obtained by using a cathetometer. The birefringence of the deformed network was determined with a Babinet compensator by measuring the phase shift of polarized laser light parallel and perpendicular to the axis of deformation.

The temperature dependence of birefringence and stress were measured under the conditions of constant force or constant deformation of the sample. Both experimental techniques lead to the same results of the quotient $\Delta n/\lambda^2-\lambda^{-1}$. The stress - strain and birefringence - strain measurements do not differ from these generally received for common rubbers, where σ and Δn linearly depend on the deformation factor $(\lambda^2-\lambda^{-1})$. With respect of the formation of the anisotropic phase at low temperatures especially the temperature dependence of Δn and σ are of interest. Therefore in order to obtain results which are comparable with results of amorphous rubbers having no mesogenic side chains we plotted the product CT versus temperature.

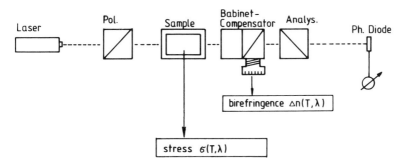

Fig. 2: Experimental arrangement for stress-birefringence measurements

Unfortunately because of strong light scattering in the LC state of the elastomer, stress optical measurements are restricted to the isotropic state. Therefore conclusions taken from the results above the clearing temperature T_c to orientational effects in the anisotropic liquid crystalline state have to be treated guardedly. To get further and better information about the orientation of the mesogenic side chains below T_x X-ray experiments on elongated liquid crystalline elastomers and polarizing microscopic measurements on thin films have been performed.

4. RESULTS

4.1. LC-Elastomers with the Spacerlength m = 3

4.1.1. Mechanical behavior of uniaxial elongated and compressed LC-elastomers

In Fig. 3 we will demonstrate the thermoelastic behavior of the LC-elastomer with m = 3 (x=0) at constant load (mean relative deformation $\bar{\lambda}=0,78$). Above T_c in the isotropic state σ and the modulus $\sigma/\lambda^2-\lambda^{-1}$ respectively increase linearly with increasing temperature corresponding to the behavior of common elastomers.

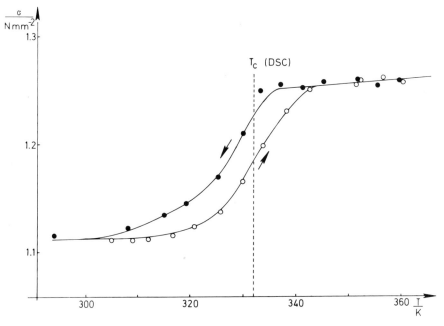

Fig. 3: Temperature dependence of true stress $\sigma(T)$ at constant load ($\bar{\lambda}=0,78$); experiments performed at constant length show the same relation ($T>T_c$: isotropic rubber; $T<T_c$: anisotropic (liquid crystalline) rubber)

Experiments performed at lower elongations (or compressions) below the so-called thermoelastic inversion point lead to decreasing o-values with increasing temperature because the thermal expansion of the samples predominates the effect of the retractive force.

In the vicinity of T_c a drastic change in the mechanical behavior is observed. In the case of constant load (Fig. 3) the true stress rapidly decreases. This is caused by a spontaneous change of the geometrical dimensions and can be only explained by an anisotropic ordering of the mesogenic side chains below T_c. Besides in the vicinity of T_c there are strong hysteresis effects for cooling and heating experiments, a phenomenon observed in a large variety of systems (ferromagnetism, ferroelectricity a.s.o.). An explanation of this on a molecular basis is not possible at present.

The temperature dependence of the sample length of an elongated strip at constant load is shown in Fig. 4. It is remarkable that the length of the sample increases by decreasing the temperature below T_c. By further cooling in the liquid crystalline state the length increases continuously until reaching the glass transition temperature T_g. This unusual temperature dependence is caused by rising orientation of the mesogenic groups in the nematic phase at lower temperatures and consequently by a higher anisotropic deformation of the polymer chains. This means that the thermal linear expansion coefficients $\beta_\parallel$ parallel and $\beta_\perp$ perpendicular to the axis of deformation are different. This effect predominates the normal decrease of the length by cooling. Below T_g the length decreases as usual because the degree of orientation does not change by further cooling.
In compression experiments the length of the cylindrical sample decreases by cooling below T_c while the diameter increases.

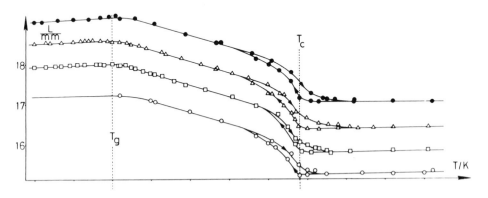

Fig. 4: Temperature dependence of sample length at constant load F
(O : F=0,   □ : F=0.058 N, △ : F=0.121 N, ● : F=0.183 N;
LC-elastomer with m=3 (x=0); uniaxial elongation)

In both cases, elongation and compression, the lengthening happens
in the direction of the elongated polymer backbone. It is noteworthy that these effects still exist in the undeformed state of
the networks with always the same tendencies of lengthening and
shortening. This could be explained on a molecular basis by our
first X-ray measurements on crosslinked and uncrosslinked samples
which indicate that the possible direction of orientation in the
LC phase is fixed by crosslinking of the polymer chains.

4.1.2. <u>Stress optical measurements in the isotropic state</u> $(T > T_c)$

As mentioned before, the stress optical measurements could
be performed only above T_c. But the results especially near T_c
still allow to discuss the orientation of the LC side chains with
respect to the elongated main chain. It is of interest whether
the orientation of the mesogenic moieties in the pretransformational region enables us to predict their orientation in the LC
state. Fig. 5 and 6 show the CT(T)-curves of the LC-elastomers
with m=3 (x=0) and m=3 (x=60) (see Table 1) measured in uniaxial
compression and elongation. Both types of deformation lead to
the same results within the experimental error. The deviations at
higher values of CT arise from the difficulty of preparing totally
identical samples with the same degree of order and from different
limits of inaccuracy in both experiments.

The CT(T)-curves of all samples can be divided into two
regions, one is high above T_c, showing constant values of birefringence, and the other near to T_c with significant temperature
dependence.

At high temperatures the product CT is independent of the
temperature as is predicted by the theory and the network behaves
similar to conventional elastomers. For comparison the CT-values
of natural rubber (NR) are plotted in both figures. The negative
values of CT (Fig. 5) indicate that the mesogenic side chains
orient more or less perpendicular to the deformed network chain.
Only in this case the polarizability perpendicular $(\alpha_\perp)$ to the
axis of the polymer segment is greater than parallel $(\alpha_\parallel)$ to this
axis and therefore $\Delta\alpha(\sim CT) = \alpha_\parallel - \alpha_\perp < 0$.

This does not mean that all mesogenic side groups of a
statistical segment, which contains a certain number of monomeric
units, are oriented perpendicular to the segment axis. There may
be always some mesogenic side groups which are oriented more or
less parallel to the segment axis with a positive contribution
to the total optical anisotropy of the segment. Therefore the
results only show that the perpendicular position of the side
groups is preferred and as a consequence the mean optical anisotropy of the statistical segment is negative. The sign of $\Delta\alpha$
does not change if the plane of the mesogenic groups adopts

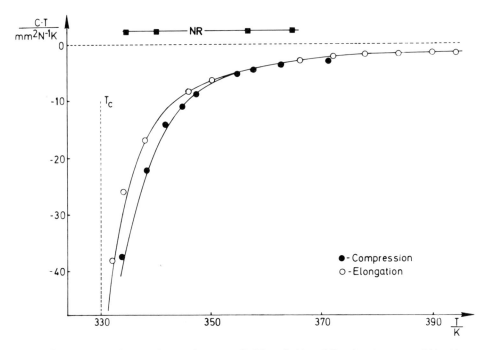

Fig. 5: Temperature dependence of CT of the LC-elastomer with the spacerlength m=3 (x=0) (see Table 1) in the isotropic state $(T>T_c)$; NR=natural rubber; optical negative, uniaxial

different orientations. With respect to the polymer main chain, only the value changes.

The anisotropic deformed LC-elastomer above the clearing temperature is uniaxial (optical axis=axis of deformation) and optically negative. This result does not differ from other conventional elastomers with optically anisotropic side groups, e.g. polystyrene. Besides the high negative values indicate even in the isotropic state $(T>T_c)$ the mesogenic side chains are influenced by the anisotropic matrix of the network. This can be explained by the assumption that the elongation of the network chain also reduces the number of conformations of the side groups resulting in a preferred mean position. If complete decoupling of the mesogenic side groups from the motion of the network chains is assumed only the oriented network chains should contribute to the birefringence. The distribution of the mesogenic side groups should remain statistically unordered. Contrary to this statement the sign of the stress optical coefficient of polydimethylsiloxane (PDMS) networks is positive and the value is substantially smaller than for LC-elastomers in the optical isotropic state For example the product CT of a PDMS-network at 360 K is $+ 1,8 \cdot 10^{-2}$ mm^{2}N^{-1}K

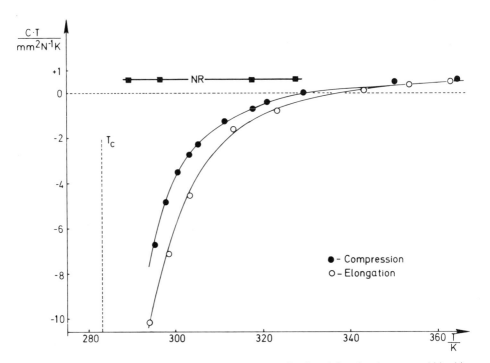

Fig. 6: Temperature dependence of CT of the LC-elastomer with the
spacer length m=3 (x=60) (see Table 1) in the isotropic
state (T>T$_c$); NR =natural rubber; <u>optical negative, uniaxial</u>

while the LC-elastomer with m=3 (x=0) shows a value of -5 mm^{2}N^{-1}K
at the same temperature. This comparison proves that there is a
correlation between the deformation of the polymer network and the
orientation of the side chains. The orientation of the mesogenic
side groups is more pronounced if the LC-elastomers are deformed
in the isotropic state and then cooled below T$_c$ by fixing the
length of the sample. The deformation below T$_c$ leads not only to
smaller degrees of orientation but also to longer periods for
reaching the final value of the degree of orientation.

The shape of the CT(T)-curve of the LC-elastomer with m=3
(x=60) does not differ markedly from the LC-elastomer with m=3
(x=0), only the values are smaller and the sign of CT changes
to positive values at higher temperatures. A comparison of the
CT-values is only efficient if they are determined at the same
reduced temperature T$^+$ (T$^+$ = T/T$_c$). If this is performed in
the pretransformational region near T$_c$ it is remarkable that
the values only differ slightly. At a reduced temperature T$^+$ = 1.04
the CT-values are -9 mm^{2}N^{-1}K for the LC-elastomer with m=3 (x=0)
and -9.2 mm2N-1K for the LC-elastomer with m=3 (x=60).

Lowering the temperature the CT values increase rapidly to

higher negative values, indicating strong deviations from the
Kuhn-Grün-theory. This effect is first observed about 30 K above
T_c. The strong deviations from Eq. 3 can be easily understood by
pretransitional effects which are well known for low molar
mass LCs [13-16]. The pretransition associated with short range
order effects above the transition temperature T_c is successfully
interpreted by the so-called Landau-de Gennes-theory [17].

From these measurements the following aspects can be concluded: In the same way as the deformation of the network influences the orientation of the mesogenic side chains, the transition to the LC-state by lowering the temperature influences
the situation of the polymer network. In the first case the deformation leads to a partially orientation of the mesogenic side
chains. In the second case the orientation of the mesogenic side
chains forming the LC-state is associated with a deformation of
the polymer matrix as is indicated by a change of the geometrical
dimensions of the sample (see Fig. 4).

4.1.3. Conoscopic polarizing microscopic measurements in the LC-state ($T < T_c$)

Drawing conclusions from the stress optical results with
respect to the orientation of the mesogenic groups in the isotropic state to their orientation in the LC-state is not possible
because of the first order transformation, which causes a jump
of the CT-values at the transformation temperature T_c. To get
an information of their orientation in the LC-state additional
microscopic and X-ray experiments have been performed.

A thin film of the LC-elastomer with m=3 (x=0) has been
stretched approximately 20 % at room temperature ($T < T_c$) in the
LC-state and then observed in the polarizing microscope using the
conoscopic method.

The axis of deformation is always perpendicular to the
direction of observation. Two different positions of the axis of
deformation with respect to the polarizer are shown in Fig. 7.

In the first case ($\vartheta = \pi/2$) polarizer and axis of deformation
are in a perpendicular position. The conoscopic picture is schematically shown in Fig. 7 ($\vartheta = \pi/2$). The dark cross is typical for
monaxial crystals between crossed polarizers if the optical axis
is parallel to the incident light. This also holds for biaxial
crystals, if the plane of the polarized light (plane of oscillating
electric field vector) is perpendicular to the plane formed by
the two optical axis of a biaxial crystal.

This means that the mesogenic side groups are oriented perpendicular to the second optical axis, that is the axis of defor-

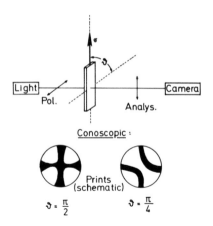

Fig. 7: Polarizing microscopic measurements on stretched LC-
elastomers with the spacerlength m=3 (x=0) (see Table 1)
in the LC-state $(T<T_c)$; optical positive, biaxial

mation. Only in this case the optical axis of the LC-phase is
parallel to the incident light. The orientation can be designated
as a homeotropic orientation between the surfaces of the film.
The use of a quarter wave plate shows that the LC-state corresponds
to an optically positive crystal.

This result is not contrary to the conclusions following from
the stress optical measurements at temperatures above T_c, where
the LC-elastomer is optically negative. In this case the optical
axis of the network corresponds to the axis of deformation. In
the LC-state the characterization "optically positive" is refered
to the optical axis of the LC-phase, which must not necessarily
coincide with the axis of deformation.

For monaxial crystals the picture of the dark cross has to be
invariant of changing ϑ when the microscopic stage is turned. This
is not the case for the LC-elastomer as to be seen in the second
position $(\vartheta =\pi/4)$. One finds a picture which is comparable with
those received for biaxial crystals in the diagonal position.
Further turning to $\vartheta=0$ leads to the same picture (dark cross)
as shown for $\vartheta= \pi /2$. The second optical axis of the stretched
LC-elastomer is parallel to the axis of deformation. The use of
a quarter wave plate finally indicates that the deformed LC-elastomer corresponds to an optically biaxial positive crystal.

Regarding the LC-elastomer (below T_c) as a "two phase-system"
one can assign one optical axis to the anisotropic LC-state of the
mesogenic side chains and one to the anisotropic orientation of
the network chains. The second one corresponds to the optical axis
as usually determined by stress optical measurements of conventional

rubbers. The angle between the optical axes is $\pi/2$.

It has to be pointed out that there is a fundamental difference between both optical axes. The optical axis of a LC-phase without crosslinking is a consequence of a specific orientation of the mesogenic molecules in the LC-state and is not caused by external effects. In opposite to this the networks only possess an optical axis if they are deformed by external forces which induce an ani-sotropy into the network. In this case the optical axis of the network and the axis of deformation are identical. Further infor-mation about the orientation of the mesogenic side groups are available by measurements of the linear dichroism which are per-formed at present.

The results of the microscopic measurements are in agreement with the conclusions taken from the stress optical measurements in the isotropic state which indicate the mesogenic groups being perpendicular to the axis of deformation. But there is one impor-tant question unsolved. Until now it cannot be decided whether the mesogenic groups are oriented homeotropically (perpendicular to the surface of the film), as found by conoscopic measurements, or whether thay are radially distributed with respect to the axis of the polymer chain, as indicated by the stress optical measure-ments on cylindrical samples. In the second case the liquid crystalline phase cannot be designated as a normal nematic phase, because it would correspond to an optically negative crystal with the long axis of the mesogenic groups perpendicular to the optical axis. As mentioned before it cannot be excluded that these different results are caused by different conditions of preparation of the networks. Further investigations of this problem are necessary.

4.1.4. X-ray measurements in the LC-state

The X-ray diffraction measurements have been performed with samples oriented above T_c and then cooled below T_c by fixing the length of the sample. The incident X-ray beam is always perpendi-cular to the axis of deformation as shown in Fig. 8, where σ in-dicates the axis of deformation. This axis corresponds to the meridian on the resulting X-ray photographs.
Two different positions (I,II; see Fig. 8) by turning the sample about the axis of deformation have been investigated. The resulting X-ray photographs are shown schematically in Fig. 8.

In position I the X-ray photograph is the same as found for nematic liquid crystals if the optical axis of the LC-phase is parallel to the incident X-ray beam (Fig. 9a). In this case no reflections can be observed and the diffraction diagram is very similar to that of the isotropic fluid state. The halo in the wide angle region is attributed to the short range positional order [18].

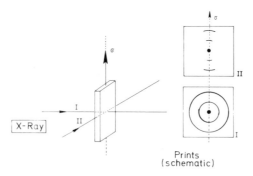

Prints
(schematic)

Fig. 8: X-ray diffraction measurements on stretched LC-elastomers
with the spacerlength m=3 ((x=0) (see Table 1) in the
LC-state $(T<T_c)$

The X-ray photograph of position II (turning of the sample
by $\pi/2$) corresponds to a nematic liquid crystal if the optical
axis is perpendicular to the indicent X-ray beam. The reflections
on the meridian indicate that the mesogenic side groups are
oriented perpendicular to the axis of deformation (Fig. 9b).

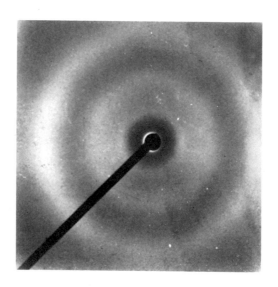

Fig. 9a: X-ray diagram of the LC-elastomer with m=3 (x=0) (see
Table 1); (position I; see Fig. 8)

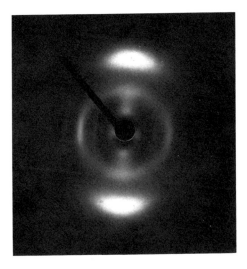

Fig. 9b: X-ray diagram of the LC-elastomer with m=3 (x=0) (see
         Table 1) (position II; see Fig. 8)

This verifies the results obtained by the polarizing microscopic measurements. In both cases it was found that the mesogenic
side groups are oriented homeotropically with respect to two
opposite surfaces (Position I) while they are oriented homogenously
with respect to other surfaces (Position II). Even the undeformed
sample heated above T_c and then cooled down shows the same results
but less marked. This leads to the assumption that the later
orientation of the mesogenic groups may be fixed at the time of
formation of the network.

4.2. LC-Elastomers with the Spacerlength m=4

The results of the LC-elastomers with the spacer length m=4
are in contrast to the results described above. This is shown by
stress optical- and X-ray diffraction measurements.

The CT-curves above T_c show the same shape but the values are
positive in the pretransformational region while they were negative
for the LC-elastomers with m=3 (see Fig. 5).

This means that the mesogenic side chains are oriented parallel
to the axis of deformation. The LC-elastomer is optically monaxial
because both optical axes, of the LC-phase and the anisotropic

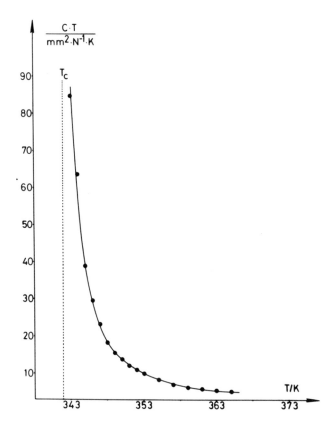

Fig. 10: Temperature dependence of CT of the LC-elastomer with
the spacerlength m=4 (x=0) (see Table 1) in the isotropic
state (T>T$_c$); <u>optical positive, uniaxial</u>

polymer network, possess the same direction. This orientation is
preserved by cooling below T$_c$ as indicated by X-ray measurements.

The X-ray measurements of the LC-elastomer with m=4 (x=0)
have been performed in the same way as described for the sample
with m=3 (x=0) (Fig. 8). In both cases (Positions I,II; analogous
to Fig. 8) the X-ray photographs show identical diffraction diagrams (Fig. 11). In opposite to the X-ray diagram of the LC-elastomer with m=3 (x=0; Position II, Fig. 8) the reflections are on the
equator. This means that the mesogenic side groups are oriented
parallel to the axis of deformation in the LC state as well as in
the isotropic state. An explanation why the orientation of the
mesogenic side chains changes by lengthening of the flexible
spacer by one methylene unit is not possible at present.

290

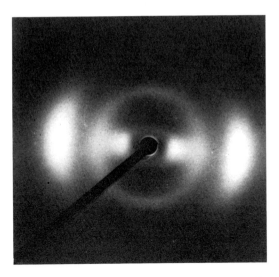

Fig. 11: X-ray diagram of LC-elastomer with m=4 (x=0) (Position I and II, see Fig. 8)

5. CONCLUSIONS

The investigations of LC-elastomers by stress optical-, X-ray diffraction- and polarizing microscopic measurements lead to the same results with respect to the orientation of the mesogenic side groups in the deformed network. The LC-elastomers with the spacer lengths m=3 and m=4 show different states of orientation. In the first case (m=3) the mesogenic side groups are oriented perpendicular to the axis of deformation and in the second case (m=4) parallel to this axis (Fig. 12).

This means that the LC-elastomer LC-1 (m=3) is optically biaxial (negative) while the LC-elastomer LC-3 (m=4) is optically uniaxial (positive).

In both cases the deformation of the networks and the orientation of the mesogenic groups are correlated:
- The deformation of the network causes a partial orientation of the mesogenic side groups. The degree of orientation depends on the degree of deformation.
- The formation of the LC-phase causes an anisotropic deformation of the network even without action of mechanical forces. The degree of deformation depends on the degree of order of the LC-phase.

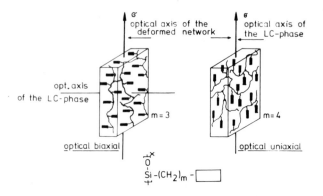

Fig. 12: Model of LC-elastomers in the LC-phase

Further investigations, especially the evaluation of the degree of order, are necessary for characterizing the LC-elastomers. The explanation of the orientation of the mesogenic side groups with respect to the direction of the deformed network will be the main subject of further investigations and theoretical considerations.

REFERENCES

1. V. P. Shibaev, A. N. Platé, Polym. Sci. USSR 19, 1065 (1978)
2. H. Finkelmann, H. Ringsdorf, J. H. Wendorff, Makromol. Chem. 179, 273 (1978)
3. H. Finkelmann, H. Benthack, G. Rehage, J. de Chim.Phys. 80, 163 (1983)
4. H. Finkelmann, G. Rehage, Makromol. Chem. Rapid Commun. 1, 31 (1980)
5. H. Finkelmann, G. Rehage, Makromol. Chem. Rapid Commun. 3, 859 (1982)
6. H. Finkelmann, H.J. Kock, G. Rehage, Makromol. Chem. Rapid Commun. 2, 317 (1981)
7. L. R. G. Treloar, "The Physics of Rubber Elasticity" Oxford (1975)
8. W. Kuhn, F. Grün, Kolloid-Z. 101, 248 (1942)
9. M. Fukuda, G.L. Wilkes, R.S. Stein, J. Polym. Sci. A-2 9, 1417 (1971)
10. M. Pietralla, J. Polym. Sci. Phys. Ed. 18, 1717 (1980)
11. H. E. J. Neugebauer, Can. J. Phys. 18, 292 (1950)
12. H. S. Subramhanyam, D. Krishnamurti, Mol. Cryst. Liq. Cryst. 22, 239 (1973)
13. T. W. Stinson, J. D. Litster, Phys. Rev. Letters, 25, 503 (1982)

14. R. S. Porter, J. F. Johnson, J. Appl. Phys. $\underline{34}$, 51 (1963)
15. A. Torgalkar, R. S. Porter, E. M. Barrall, J. F. Johnson, J. Chem. Phys. $\underline{48}$, 3897 (1968)
16. G. S. Attard, P. A. Beckmann, J. W. Emsley, G. P. Luckhurst, D. L. Turner, Mol. Phys. $\underline{45}$, 1125 (1982)
17. P. G. de Gennes, "The Physics of Liquid Crystals" London (1974)
18. J. H. Wendorff, H. Finkelmann, H. Ringsdorf, J. Polym. Sci. Polym. Symposium $\underline{63}$, 245 (1978)

STRUCTURE OF NEMATIC SIDE CHAIN POLYMERS

H. Finkelmann, and H.J. Wendorff*

Institut für Physikalische Chemie, TU Clausthal, FRG
*Deutsches Kunststoff Institut, Darmstadt, FRG

INTRODUCTION

Systematic investigations of the past years have proved that the structures of the liquid crystalline phases of liquid crystalline (l.c.) side chain polymers are similar to the structures of conventional low molar mass liquid crystals (l.lc). Nematic and cholesteric phases as well as S_A and S_C phases have been clearly identified [1],[2]. The systematic realization of l.c. side chain polymers, however, requires a flexible linkage of the rigid mesogenic moieties to the polymer backbone via a flexible spacer. The flexible spacer more or less decouples the motions of the rod-like side chains and the backbone. Two extremes are conceivable. With a long flexible spacer the main chain does not influence the anisotropic packing of the side chains. In case of no flexible spacer, motions of the mesogenic side groups are directly correlated with the motions of the chain segments, nor-mally preventing the l.c. state.

Following these considerations and considering polymers having linked mesogenic groups as side chains via flexible spacers of different length, a distinct length of the spacer is required for which the polymers will start to become liquid crystalline. For these polymers having a "minimum length" of the flexible spacer, strong interactions of side and main chain are expected. Actually some of these polymers exhibit some anomalous optical properties [3],[4],[5]. It could not definitely be established, whether these polymers can be appointed to nematic phases or whether the influence of the backbone induces another structure. Some new experimental results concerning the l.c. structure of these polymers will be presented in this paper.

RESULTS

1. Material

A typical series of polymers displaying the properties discussed above are the poly(methacrylates) :

$$CH_3 - \overset{\overset{\displaystyle CH_2}{|}}{\underset{|}{C}} - COO - (CH_2)_m - O - \bigcirc - COO - \bigcirc - OCH_3 \quad (1)$$

(1)	m	Phase transitions in K
a	2	g 374 n 394 i
b	6	g 368 n 378 i

For (1b), having the long flexible spacer of six methylene units
a normal nematic phase is observed displaying the typical well
known nematic textures under the polarizing microscope. Marbled
textures as well as Schlieren textures can be seen, and they
exhibit the same types of defects as ordinary l.-lc's. Conoscopic
observations clearly prove the optically uniaxial positive
character. The absolute value of the birefringence Δn as well as
the temperature dependence of Δn correspond to chemically similar
l.-lc's [4].
Polymer (1a) in principle has the same chemical constitution as
polymer (1b), however, the rigid mesogenic benzoic acid phenylester group is linked to the backbone by the short flexible
spacer of only two methylene units. Owing to their chemical
constitution strong interactions of the main chain and the
mesogenic side groups are to be expected. Actually this polymer
exhibits unusual optical properties [3,4]. No typical nematic
textures can be observed. Instead this polymer displays a more
or less unique birefringence in the polarizing microscope without the typical defects of nematic textures. Furthermore, birefringence measurements indicate an optical uniaxial negative
character ($\Delta n = n_{e,n} - n_{o,n} < 0$; $n_{e,n}$, $n_{o,n}$ are the extraordinary
and ordinary refractive indices of the nematic phase). On the
other hand, thermodynamic investigations prove the l.c. state of
this polymer. In order to clarify the structure of this polymer
detailed X-ray measurements and miscibility experiments with
l.-lc's will be described. The synthesis of polymers (1 a,b) is
described elsewhere [6].

2. X-ray Investigations

The small angle X-ray curve as obtained by means of the Kratky
camera is displayed in Fig. 1. It is obvious that the scattering
is independent of the absolute value s of the scattering vector s
($s = |s| = 4\pi/\lambda \, (\sin \theta)$, θ scattering angle, λ wavelength) up
to values of s of about 0,2 $\overset{\circ}{A}^{-1}$ [7]. A slight increase of the

296

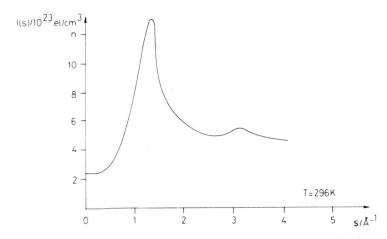

Fig. 1: Small angle X-ray curve of polymer (1a)

intensity with increasing s above this range can be attributed
to the fact that the amorphous halo in the wide angle regime
is approached as s increases. This can be seen from Fig. 2,
which shows the total scattering curve in the small angle as
well as in the wide angle region.

Two amorphous halos are observed in the wide angle region
which are mainly determined by the intermolecular packing,
characteristic of the fluid or glassy state. A detailed analysis
of the wide angle scattering requires the calculation of the pair
correlation function from the scattering curve by means of Fourier
transform. Due to the complex chemical structure of the polymer,
the pair correlation function cannot be interpreted in a straight-
forward way in terms of the intermolecular arrangement of the
molecular groups. Thus we will focus on the information, which
can be derived from the small angle scattering curve.

The small angle scattering curve is very similar to that
which is displayed by isotropic fluids or glasses. In the presence
of orientational order, as in the case of a nematic phase, the
scattering is in a first approximation determined by thermal density fluctuations [7] related to the spherical symmetric part of
the general pair correlation function. Recently we were able to
show [7,8] that an additional scattering component may arise from
orientation fluctuations in the case of nematic phases, which
leads to a total scattering curve that is independent of the
scattering vector in the small angle region. The scattering
curves obtained for low molar mass nematic liquid crystals (shape
and absolute values), which have been analyzed in detail[7,8],
agree with the scattering curve of the polymer (1a) considered
here. We thus conclude that the structure of (1a) is either

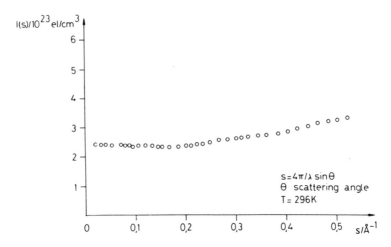

Fig. 2: Wide angle X-ray curve of polymer (1a)

that of an isotropic liquid or glass, which is ruled out by optical and thermodynamical studies, or that of a nematic phase. A smectic phase would exhibit additional scattering.

3. Miscibility Experiments

Following Arnold and Sackmann [9] only liquid crystals having the same phase structure are miscible over the whole concentration range. This rule can be also applied for mixtures of l.-lc's and l.c. polymers. It has been proved that normally nematic side chain polymers are miscible over the whole concentration range with low molar mass nematics [10]. For these mixtures, however, care has to be taken with respect to the incompatibility of some polymer/ monomer mixtures, which will be indicated by a more or less pro- nounced miscibility gap in defined concentration ranges [11]. The unmiscibility of two components, however, does not imply that the two components have to differ in their phase structure. The appearance of miscibility gaps, which are well known for isotropic polymer mixtures, have not been observed for l.-lc's so far. Nevertheless the rule of Arnold and Sackmann is applicable for l.c. polymer systems, presumig a miscibility is indicated over the whole concentration range. With this, miscibility experiments are a valuable method to clarify the l.c. strucutre of unknown l.c. polymers.

To ascertain whether the l.c. phase of polymer (1a) can be appointed to a nematic phase, two binary phase diagrams have been investigated of the polymer with the two selected, low molar mass nematic l.c. (2) and (3):

$$CH_3 - CH(CH_3) - (CH_2)_2 - O - \bigcirc - COO - \bigcirc - \bigcirc - OCH_3 \quad (2)$$
$$CH_2 = CH - CH_2 - O - \bigcirc - COO - \bigcirc - OC_6H_{13}$$

The l.-lc's deviate in their mesogenic groups (benzoic acid bi-
phenylester and benzoic acid phenylester).

The phase diagram of mixtures of polymer (la) and the l.-lc's
(2) is shown in Fig. 3. Actually we find a total miscibility over
the whole concentration range in the nematic state, which indicates
that polymer (la) also exhibits a nematic phase. By adding polymer
to the l.-lc the phase transformation line crystalline to isotropic
is poorly affected owing to the high molar mass of the polymer
($\bar{M} \approx 34000$). Above 55 % of polymer, increasing polymer concentration
suppresses the crystallization of the mixture under the experimental conditions and the temperature interval of the nematic phase
becomes markedly enlarged. Even the annealing of the samples for
some days gives no crystallization. Instead of low temperatures,
the mixtures freeze at the glass transition temperature. One important observation has to be mentioned with respect to the nematic texture of the mixtures. By adding small amounts of the l.-lc
to polymer (la), a strong decrease in the glass transition is
observed, because of the plasticizing effect of the low molar
mass component. For mixtures containing more than about 5 % of the
l.lc normal nematic textures are observed and the above described
anomalous optical behaviour of the pure polymer vanishes. This indicates that with increasing mobility of the system, which is
demonstrated by the the lower glass transition, the polymer is
able to form normal nematic textures. For the pure polymer the
formation of the nematic texture of the mesogenic side groups
is suppressed by the polymer backbone.

The same behaviour with respect to the formation of the nematic texture has been found for the mixtures of polymer (la) with
the low molar mass derivative (3), whose phase diagram is shown in
Fig. 4. Adding small amounts of (3) to (la) suppresses the glass
transition and normal nematic textures appear in the polarizing
microscope. For this system, however, the phase diagram strongly
differs from the previous diagram. Although we also find miscibility
of polymer and monomer over the whole concentration range, at low
temperatures a broad miscibility gap is observed in the nematic
state. In this regime two nematic phases coexist which differ in
their composition. Recently the appearance of such a miscibility
gap has been predicted theoretically by Brochard et.al.[12]. For
this phase diagram it has to be noted that the isotropic-nematic
biphasic regime is drastically enlarged in the concentration range
of the miscibility gap.

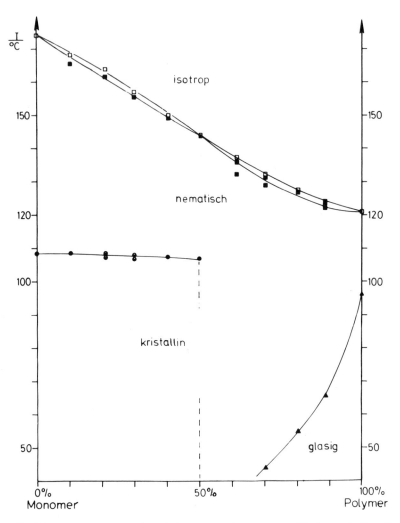

Fig. 3: Phase diagram of mixtures of polymer (1a) and the low
molar mass nematic l.c. (2); I=isotropic, N=nematic,
K=crystalline, G=glassy

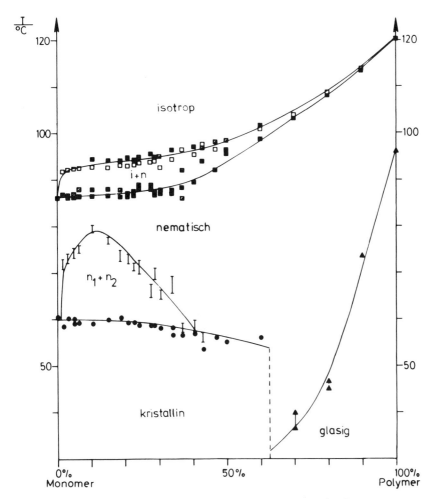

Fig. 4: Phase diagram of mixtures of polymer (1a) and the low
molar mass nematic l.c (3); I=isotropic, N=nematic,
K=crystalline, G=glassy

CONCLUSIONS

If nematogenic groups are linked via short flexible spacers to the polymer backbone, the macroscopic properties of nematic polymers may deviate from conventional nematic systems, owing to a disturbing effect of the polymer main chain. This is indicated e.g. by the absence of typical nematic textures and some anomalous optical properties [3,4]. The symmetry of the phase structure, however, corresponds to that of conventional nematic liquid crystals. The model proposed by Cser [5] (aperiodical helical packing) for these polymers is not applicable.

REFERENCES

1. H. Finkelmann, "Polymer Liquid Crystals", Eds. A. Ciferri, W.R. Krigbaum, R.B. Meyer, Academic Press Inc., pp 35 (1982)
2. S.G. Kostromin, V.V. Sinitzyn, R.V. Talroze, V.P. Shibaev, N.A. Platé, Makromol. Chem. Rapid Commun. 3, 809 (1982)
3. H. Kelker, U.G. Wirzing, Mol. Cryst. Liq. Cryst. Let 49, 175 (1979)
4. H. Finkelmann, D. Day, Makromol. Chem. 180, 2269 (1979)
5. F. Cser, K. Nyitrai, G. Hardy, Advances in Liquid Crystal Research and Applications, Ed. L. Bata, Pergamon Press, Oxford, pp 845 (1980)
6. H. Finkelmann, H. Ringsdorf, J.H. Wendorff, Makromol. Chem. 179, 273 (1978)
7. W. Kopp, J.H. Wendorff, Mol. Cryst. Liq. Cryst. 84, 63 (1982)
8. W. Kopp, J.H. Wendorff, Colloid Polym. Sci. 260, 1071 (1982)
9. H. Arnold, H. Sackmann, Z. Phys. Chem. 213, 137 (1960); 213, 145 (1960)
10. H. Finkelmann, H.J. Kock, G. Rehage, Mol. Cryst. Liq. Cryst. 89, 23 (1982)
11. C. Casagrande, M. Veyssie, H. Finkelmann, J. Physique, Letters 43, L671 (1982)
12. F. Brochard, J. Jouffroy, P. Levinson, to be published

CHOLESTERIC POLYMERS WITH MESOGENIC SIDE GROUPS

Ya. S. Freidzon, N.I. Boiko, V.P. Shibaev,
and N.A. Platé

Department of Polymer Chemistry
Moscow State University
Moscow, USSR

INTRODUCTION

Today it is a well known fact that polymers containing mesogenic side groups attached to the backbone via a flexible aliphatic spacer are capable of forming mesophases of smectic, nematic, and cholesteric types[1,2]. For a cholesteric structure of liquid crystals to be formed it is necessary that the macromolecules contain chiral groups. If a chiral group is mesogenic, homopolymers are liable to exhibit a cholesteric type of mesophase[3]. In the case of a non-mesogenic chiral group, a cholesteric polymer may be obtained by copolymerization of a chiral monomer with a nematic one[4]. A variety of cholesteric copolymers are produced by copolymerizing nematic and chiral monomers[5-7].

The work presented aims at the study of structure, optical properties and the formation of cholesteric mesophases in homopolymers as in copolymers, the monomers being of the following structural formulae:

$$CH_2=C-R \qquad CH_3 \qquad CH-(CH_2)_3-CH-CH_3 \qquad (I)$$
$$\underset{O}{\overset{C=O}{|}}$$

$CH_2=C-R$
|
$C=O$
|
$O-(CH_2)_n-\overset{O}{\overset{\|}{C}}-O-$ [cholesteryl steroid with CH_3, CH_3, and $CH-(CH_2)_3-CH-CH_3$ side chain] (II)

$CH_2=C-R$
|
$C=O$
|
$O-(CH_2)_n-\overset{O}{\overset{\|}{C}}-O-\bigcirc-\overset{O}{\overset{\|}{C}}-O-\bigcirc-O-CH_3$ (III)

$CH_2=C-R$
|
$C=O$
|
$O-(CH_2)_n-O-\bigcirc-\bigcirc-C\equiv N$ (IV)

The following notation for monomers will be in use in the subsequent discussion.

Table 1.

Monomer	R	n	Notation of Monomer
I a	H	–	ChA
I b	CH_3	–	ChMA
II a	H	5	ChA-5
II b	H	10	ChA-10
II c	CH_3	5	ChM-5
II d	CH_3	10	ChM-10
III	H	5	AM-5
IV	H	5	AC-5

The types of monomers chosen permitted us to elucidate the effect of the nature of the main chain (acrylic
or methacrylic), of the length of aliphatic spacer within
cholesterol-containing units, as well as of the nature of
nematogenic group, on the properties of cholesteric mesophase.

304

EXPERIMENTAL

The monomers were synthesized according to the procedure described earlier in[7-9]. All polymers were obtained by radical polymerization in solution (initiator - azobis-isobutyronitrile, solvent - benzene), and subsequently purified by triple reprecipitation from benzene solutions with hot methanol. The copolymer composition was determined by UV-spectroscopy of their dichloroethane solutions.

A polarizing microscopy MIN-8, equipped with a heating table and a camera, was used to investigate the textures and temperatures of phase transitions. X-ray diagrams were obtained with URS-55 apparatus, the sample-film distance being 60.6 mm. The wave-length of selective light reflection from polymer samples was determined by measuring the transmittance at various temperatures (thermostating accuracy was $\pm 0.1^\circ$ C) with a "Specord UV-VIS" spectrophotometer (Karl Zeiss, Yena). Circular dichroism spectra were registered with a "Dichrograph II" (Roussel Jouan, Paris). Selective light reflection and circular dichroism were examined for the planar texture of polymer samples placed between two quartz plates. The planar texture was produced by a slight shift of one plate relative to the other.

RESULTS AND DISCUSSION

Let us first examine the properties of cholesterol-containing homopolymers. As was already shown[8-10], polymers with cholesterol groups attached directly to the backbone (PChA and PChMA), are amorphous, glassy compounds displaying high glass transition temperatures. This is because the main chain retards the packing of large cholesterol groups, thus preventing the formation of a mesophase.

All homopolymers with cholesterol attached to the backbone via a spacer (PChA-n and PChM-n) exhibit liquid crystalline structure[11]. To detect the type of liquid crystalline structure of these polymers, let us examine the results of X-ray and microscopic studies. X-ray patterns of all PChA-n and PChM-n samples display three reflections in a small-angle region of scattering, their position being dependent on the length of the mesogenic side group[11]. Such a characteristic in the X-ray patterns indicates the presence of a layer ordering in the packing of mesogenic side groups and may thus produce a smectic type of mesophase. Microscopic studies revealed that at room temperature all homopolymers examined form a focal-conic texture only (Fig. 1a); the latter is characteristic of smectic, and also cholesteric liquid crystals. A prolonged annealing at a temperature above glass transition leads to the formation of the fan-like texture Fig. 1b) which is more common in cholesteric mesophases. Shifting the covering glass plate results in a texture, analogous to the planar one typical of cholesteric liquid crystals (Fig. 1c). Hence, a cholesteric mesophase may be attributed to PChA-n and PChM-n homopolymers.

A specific feature of cholesteric liquid crystals is selective light reflection and circular dichroism. Spectrophotometric study of the films of PChA-n and PChM-n homopolymers in the 20-150°C temperature range (the clearing temperatures of homopolymers are 180-220°C) has shown the existence of a broad reflection maximum in the 270-300 nm region. Circular dichroism is observed in the same wavelength region (Fig. 2). Thus PChA-n and PChM-n homopolymers produce a cholesteric mesophase with intrinsic selective UV-light reflection. In contrast to low molecular cholesterics, cholesteric polymers seem to display a tendency to layer packing which is promoted by the structuring effect of the main chain.

For cholesteric liquid crystals, the wavelength of selective light reflection λ_R is related to the pitch of the cholesteric helix P in the following manner: $\lambda_R = nP$, where n is the refractive index. It is evident that for a cholesteric mesophase with selective reflection of the

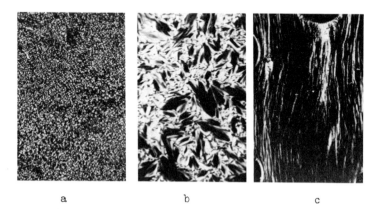

a   b   c

Fig. 1. Optical microphotographs of focal-conic (a), fan like (b) and planar (c) textures of polymers PChA-10(a), PCHA-5 (b) and PChM-10 (c).

Table 2. Properties of cholesterol-containing copolymers.

Number	Copolymer Mole % of cholesterol-containing units in copolymer	Glass transition temperature T_g, °C	Clearing temperature T_{cl}, °C	Wavelength of selective reflection λ_R, nm
	ChA/AM-5			
1	15	30	120	620[a]
2	21	40	117	530[a]
3	33	65	95	450[a]
4	40	80	90	–
5	52	120	–	–
	ChA-5/AM-5			
6	18	25	115	555[b]
7	35	35	103	495[b]
8	45	40	110	400[b]
9	65	45	140	UV-range
	ChA-10/AM-5			
10	17	20	120	740[a]
11	25	20	110	630[a]
12	32	20	115	540[a]
13	43	20	112	420[a]
	ChM-10/AM-5			
14	19	20	116	690[a]
15	22	20	118	590[a]
16	33	25	120	480[a]
	ChA-5/AC-5			
17	19	50	98	850[b]
18	28	50	102	660[b]
19	36	55	105	555[b]
20	52	55	150	500[b]

a - λ_R was measured at room temperature
b - λ_R was measured at $T=0.99T_{cl}$

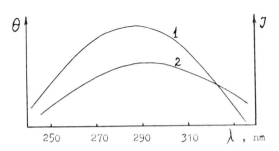

Fig. 2. CD-spectrum (1) and the curve of selective light reflection (2) for PChA-5.

visible light to be produced, it is necessary that the
pitch of the cholesteric helix be large. One of the ways
to regulate the pitch of the helix in polymers may be by
copolymerization of chiral and nematic monomers. This is
why we have synthesized and investigated copolymers of
various cholesterol-containing monomers with two nematic
monomers - AM-5 and AC-5, differing in the chemical natur
of mesogenic groups. The composition and some characteristics of copolymers synthesized are given in Table 2.

As is seen from the data presented in Table 2, copol-
mers of the nematic monomer AM-5 with cholesteryl acrylat
(ChA) with up to 33 mole % ChA content exhibit a cholesteric mesophase displaying selective reflection of the
visible light. Subsequent increase in ChA content leads
to gradual destruction of the mesophase. For instance,
copolymer 4, containing 40 mole % ChA, is only slightly
birefringent, while at 50 mole % content of ChA, the copolymer becomes totally amorphous. These data reveal
that if a cholesteric monomer gives an amorphous homopolymer, then a cholesteric copolymer may be obtained by copolymerization of the cholesteric monomer with a nematogenic one only at a relatively low content of the chiral
monomer. The cholesteric texture obtained is very imperfect, which leads to significant light scattering by films
of these copolymers and results in a broad peak of selective light reflection (Fig. 3, curve 1).

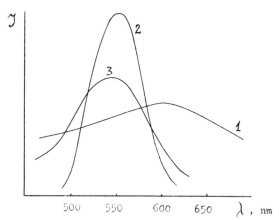

Fig. 3. Selective light reflection curves for copolymer 1
         at room temperature (1), for copolymer 6 at
         T=102°C (2) and for copolymer 6 at room tempera-
         ture (3).

308

Transparent films with perfect texture and a narrow
peak of selective light reflection are obtained on copolymerization of a nematogenic monomer with a cholesterol-
containing monomer that forms a mesomorphic homopolymer
(Fig. 3, curves 2 and 3). Such copolymers display liquid
crystalline properties for any content of their components.
The range of compositions for which the liquid crystalline
phase is cholesteric depends on the chemical nature of
the monomers. Let us compare the properties of copolymers
of one and the same chiral monomer and various nematogenic
monomers (copolymers series 6-9 and 17-20). On X-ray diagrams of all copolymers examined an amorphous halo corresponding to the distance between mesogenic side groups is
observed. The distance increases from 0.46 nm to 0.53 nm
when the fraction of cholesterol-containing units is increased. Small angle X-ray reflections appear for copolymers at 35 mole % content of cholesterol-containing units.
These reflections coincide with that of homopolymer PChA-5
and indicate a layer ordering of mesogenic side groups.
At the same time the films are characterized by selective
reflection of light. It is seen from Fig. 4 and Table 2
that for all copolymers the wavelength of selective light
reflection λ_R decreases when the fraction of chiral units
is increased; but for copolymers ChA-5/AM-5, the selective
light reflection is observed at lower values of λ than for
copolymers ChA-5/AC-5 of analogous compositions. Apparently, the helix is more difficult to form from the
nematic structure formed by more polar molecules (AC-5),
than from less polar molecules (AM-5).

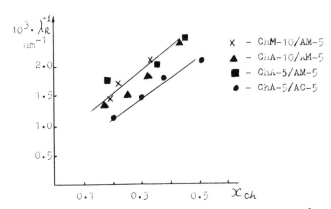

Fig. 4. Inverse wavelength of reflection λ^{-1}_R vs. mole
fraction X_{ch} of chiral units for a series of
copolymers 6-20.

If the same nematic monomer but various mesogenic cholesterol-containing monomers are used for copolymerization (copolymer series 6-9, 10-13, and 14-16), the optical properties of copolymers are only slightly affected The dependence of $\lambda^{-1}{}_R$ on the content of cholesterol-containing units is described by a single straight line for all copolymers of these series (Fig. 4).

At the same time, glass transition temperatures of these copolymers vary significantly. For instance, for copolymers with ChA-5, the T_g values are above the room temperature and they increase with the increase in ChA-5 content, while the films of ChA-10 and ChM-10 copolymers are highly elastic at room temperature. The selective reflection is retained for all cholesteric copolymers for a wide temperatire range. However, when the temperature is decreased, a certain widening of the peak is observed (Fig. 3, curve 3). Such a phenomenon is not common for low molecular cholesterics, but is characteristic of cholesteric polymers. The reason for this must be the presence of a polymeric chain with mesogenic groups attached via spacers. At high temperatures when the segmental mobility of the backbone is high enough, it acquires the conformation that allows the formation of a perfect cholesteric helix. When the temperature is decreased, the mobility of the main chains decreases, which significantly retards the packing of mesogenic side groups. The result is the appearance of imperfections in the helix structure distorting the pitch of the helix. The consequence is the broadening of the light reflection peak, observed at lower temperature. This phenomenon is more liable to occur for a copolymer of AC-5 than for AM-5.

The experimental results presented demonstrate that the optical properties of cholesteric copolymers are determined mainly by the nature of a nematogenic monomer. This conclusion is confirmed by the data of Finkelmann and Rehage[5], who examined the optical properties of cholesteric polysiloxanes. These authors revealed that at equal compositions of copolymers the wavelength of selective reflection depends on the length of the spacer bindi nematogenic units.

When investigating the optical properties of the cho lesteric mesophase induced in nemato-cholesteric blends, the concept of the helical twisting power (htp) of the chiral additive is introduced:

$$htp = \frac{dp^{-1}}{dX_{ch}} \quad at \ X << 1$$

where X_{ch} is the molar fraction of the chiral additive. Taking into account that $\lambda_R=nP$, and assuming n to be constant, one may write

$$htp=n\frac{d\lambda_R^{-1}}{dX_{ch}}$$

i.e., htp is determined from the slope of λ_R^{-1} versus X_{ch} dependence.

For low molecular liquid crystals, htp is usually within the $0.5-1 \times 10^{-2}$ nm^{-1} range. Approximate evaluation of htp for cholesterol-containing monomers has given the values of 0.5×10^{-2} nm^{-1} in ChA-5/AM-5 copolymers. This gives reason to assert that the mechanism of cholesteric mesophase formation is the same as in blends of low molecular nematics and cholesterics.

The results of the present work demonstrate that the cholesteric mesophase of polymers having mesogenic side groups is, in many aspects, similar to that of low molecular cholesterics. The presence of the polymeric chain results, however, in the appearance of some specific features in the behavior of such systems. The possibility to quench the cholesteric structure by vitrification of polymers is worth emphasizing explicitly. The properties of polymeric cholesterics discussed above are far from covering the array of specific features particularly intrinsic to polymeric liquid crystals. The essential effect of molecular parameters on their structure and optical properties constitute one of these features. This problem will be treated in subsequent publications.

REFERENCES

1. N.A. Platé, V.P. Shibaev, Grebneobraznye Polimery i Zhidkiye Kristally, Khimiya, Moscow (1980).
2. V.P. Shibaev, N.A. Plate, Termotropnye Zhidkokristallicheskiye Polimery - Sovremennoye Sostoyaniye i Perspektivy, ZhVChO im. D.I. Mendeleyeva, 28, 165 (1983).
3. H. Finkelmann, G. Rehage, Makromol. Chem. Rapid. Commun., 3, 859 (1982).
4. H. Finkelmann, J. Koldehoff, H. Ringsdorf, Angew. Chem., 90, 992 (1978).
5. H. Finkelmann, G. Rehage, Makromol. Chem. Rapid Commun., 1, 733 (1980).

6. Ya. S. Freidzon, A.V. Kharitonov, S.G. Kostromin, The Fourth International Liquid Crystal Conference of Socialist Countries, Abstracts, Tbilisi, 2, 219 (1981).

7. A.M. Mousa, Ya. S. Freidzon, V.P. Shibaev, N.A. Plat Polymer Bull., 6, 485 (1982).

8. V.P. Shibaev, Ya. S. Freidzon, N.A. Platé, Vysoko-mol. Soyed., A-20, 82 (1978).

9. Ya. S. Freidzon, A.V. Kharitonov, V.P. Shibaev, N.A. Platé, Mol Cryst. Liq. Cryst., 88, 87 (1982).

10. Ya. S. Freidzon, V.P. Shibaev, N.N. Kustova, N.A. Platé, Vysokomol. Soyed., A-22, 1083 (1980).

11. Ya. S. Freidzon, A.V. Kharitonov, V.P. Shibaev, N.A. Platé, "Condition of Formation of Liquid Crys talline State in Cholesterol Containing Polymers", in Advances in Liquid Crystal Research and Applica tions, L. Bata, Ed., Pergamon Press, Oxford, Akade mai Kiada, Budapest, 2, 899 (1980).

PROPERTIES OF SOME LIQUID CRYSTALLINE SIDE CHAIN POLYMERS IN THE ELECTRIC FIELD

W. Haase and H. Pranoto

Institut für Physikalische Chemie
Technische Hochschule Darmstadt, Petersenstr.20
D-6100 Darmstadt

ABSTRACT

The switching properties of liquid crystalline side chain polymers with either siloxane or acrylic groups as main chain are described, especially in relation to the possibility of dual-frequency addressing of fields effects. Also the optical birefringence data are reported.

For one polysiloxane compound the dielectric properties as a function of temperature and frequency are reported. Different relaxation processes are described and an interpretation of the molecular dynamics is given.

INTRODUCTION

In principle liquid crystalline side chain polymers contain the same mesogenic groups as the low molecular liquid crystals. Therefore the comparison of their properties, especially in the electric field, is a point of interest. During the last few years, many attempts have been made to develop new substances [1,2,3]. Nevertheless, up to this time, little about their physical properties is known.

In this work we report on the switching properties in the electric field and on the optical birefringence of the substances as well as on the dielectric properties of one compound.

EXPERIMENTAL

Substances Investigated

The liquid crystalline side chain polymers investigated have
either a polysiloxane [4] (compounds SiCl and SiCN) or a polyacrylic[5]
(compound AcCN) polymer chain. These are coupled via flexible spacers with the mesogenic p-substituted benzoic acid phenylester
groups. SiCl and SiCN are copolymers with p-methoxy components in
excess.

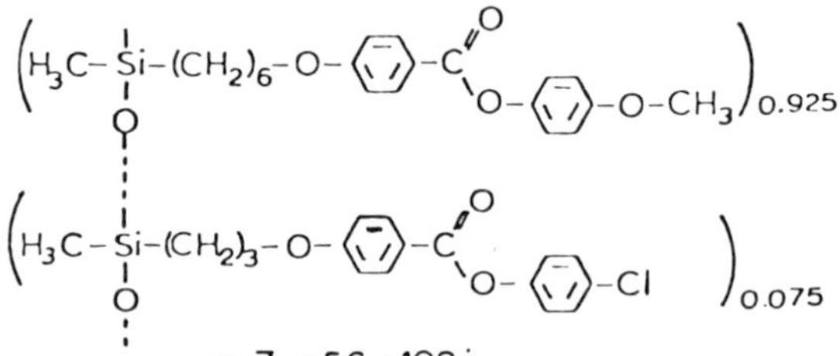

g 7 s56 n108 i

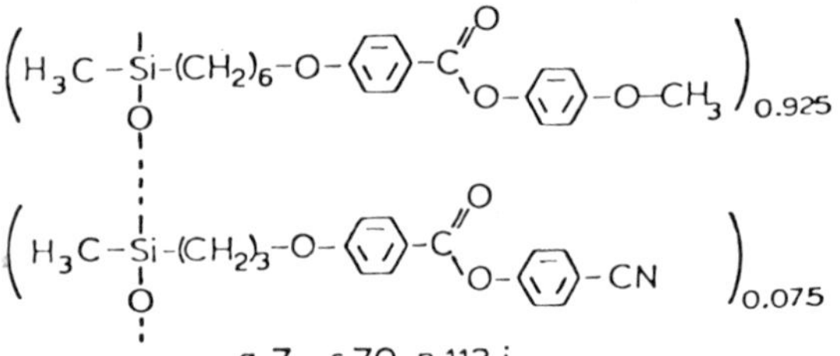

SiCl

g 7 s70 n 113 i

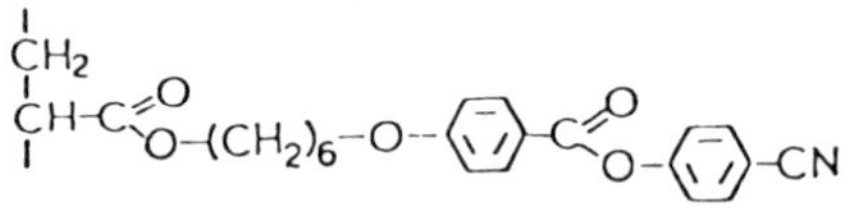

SiCN

g 36 n 133 i

AcCN

The polymerization degrees of SiCl and SiCN are approx. 95 [6],
of AcCN approximately 50 [5]. SiCl and SiCN were kindly prepared by
Dr. Finkelmann, AcCN was prepared in accordance to the procedure

314

described elsewhere [5]. All of the three substances investigated contain a strong polar chlorine or cyano group in p-position. SiCl and SiCN show smectic phases, whereas AcCN forms only a nematic liquid crystalline phase. It should be noted that the broadness of the liquid crystalline range is $\backsim$ 100 K for each of the substances.

Switching Properties

The cell preparation follows the method described elsewhere[7,8]. The glass plates were coated with a thin polyimide film to obtain a uniform orientation. The higher viscous polymer AcCN was put into the cell by capillary forces as well. The filling of these cells needs a time period of approximately 40 h at a temperature about 5 K above the clearing point. The display quality of AcCN is not uniform but good for the following experiments in most areas. The display quality of SiCl and SiCN is uniform over $\backsim$ 0.5 cm^2 proved by microscopic observations. The cell thickness in the experiments is 8 to 20 μm.

The experimental set-up used for the Freedericksz cell as well as for the TN-cell is described elsewhere [7,8].

Birefringence Measurements

The $n_{\parallel}$ and $n_{\perp}$ values as functions of temperature were determined with the aid of a Leitz-Jelley-microrefractometer, using a sodium lamp as light source. The birefringence was also determined by a conoscopic method using a polarizing microscope. The temperature in both experiments was adjusted by a Mettler hot stage FP-52 with the control unit FP-5.

For measurements with the microrefractometer the sample was put on the prism surface in the isotropic phase. Then the system was cooled down to nematic phase with a gradient of 12 K/h. For AcCN a quite good orientation was only obtained if a magnetic field of 1.2 T was applied for 12 h at a temperature of 5 K below the clearing point. Afterwards the system was cooled down with a temperature gradient of approximately 3 K/h.

Dielectric Measurements

For the dielectric measurements a General Radio equipment (1615 A bridge, 1310 oszillator, 1232 A detector) in the frequency range from 0.1 to 100 kHz was used. The cell for the measurements was constructed of two glass plates coated with gold in a certain array and fixed by springs. The spacers were made of a polyimide foil that had a thickness of 75 μm. For calibration we used toluolene. For the orientation of the molecules we applied the following procedure: The cell was heated to a temperature about 5 K belwo the clearing point being in a magnetic field of 1.2 T for a period of

12 h. Then the cell was cooled down to the temperature in question with a cooling rate of 3 K/h. This procedure is important to get a good orientation which is necessary especially for measurements of $\varepsilon_\parallel$. The system was thermostated; the constance of temperature is better than 0.1 K. The temperature was determined by a Ni-Cr/Ni thermocouple.

The measurements below 30°C can be done without a magnetic field as the preferred orientation did not change during the time of measurement.

RESULTS AND DISCUSSION

Switching Properties

For polymeric liquid crystals the formulas valid for low molecular liquid crystals [9] were applied:

$$\frac{1}{t_r} = \frac{1}{t_d^{\;o}} \left[\left(\frac{U}{U_o}\right)^2 - 1 \right] \tag{1}$$

$$\frac{1}{t_d} = \frac{1}{t_d^{\;o}} \left[\left(\frac{U}{U_o^+}\right)^2 + 1 \right] \tag{2}$$

t_r is the rise time and t_d the decay time as usually defined for TN-cells. $t_d^{\;o}$ is the passive decay time depending on the elastic constants k_{ii} and on the viscosity parameter η_i at a fixed cell thickness d:

$$\frac{1}{t_d^{\;o}} = \left(\frac{\pi}{d}\right)^2 \cdot \frac{k_{ii}}{\eta_i} \tag{3}$$

The threshold voltages U_o (for $\Delta\varepsilon > 0$) and U_o^+ (for $\Delta\varepsilon < 0$) depend on k_{ii} as well as on the dielectric anisotropy $\Delta\varepsilon$:

$$U_o = \pi \left(\frac{k_{ii}}{\varepsilon_o \Delta\varepsilon}\right)^{1/2} \tag{4}$$

$$U_o^+ = \pi \left(\frac{k_{ii}}{\varepsilon_o |\Delta\varepsilon|}\right)^{1/2} \tag{5}$$

In Table 1 some experimental data for the three compounds investigated are presented. Each time a TN-cell with dual-frequency addressing was used.

316

Table 1. Time constants and related values for the compounds
investigated using a TN-cell
$f = 0.2$ kHz $< f_o <$ 200 kHz

$T(^\circ C)$	$T_{ni}-T(K)$	$U_o(V)$	$U_o^+(V)$	$t_d^\circ(s)$	$t_d(s)$	$t_r(s)$
SiCl,		$d = 8.6$ μm,		$T_{ni} = 108^\circ C$		
87	21	17.9	15.9	19.5	6.2	9.8
90	18	12.8	10.4	16.1	5.5	3.9
96	12	10.0	12.6	8.9	4.1	1.0
112	6	7.4	11.9	5.7	2.7	0.4
115	3	6.0	13.6	4.9	2.3	0.2
SiCN,		$d = 18.4$ μm,		$T_{ni} = 113^\circ C$		
74	39	5.4	3.3	815	80.6	32.7
80	33	4.0	3.3	381	24.5	7.7
90	23	3.7	3.5	97.5	6.9	1.7
100	13	4.9	6.5	33.1	5.3	0.9
116	7	5.4	12.5	20.7	8.3	0.7
AcCN,		$d = 18.4$ μm,		$T_{ni} = 133^\circ C$		
85	48	1.1	3.4	3700	176	4.6
95	38	1.4	4.5	900	148	2.0
105	28	1.1	4.1	550	65.7	0.7
120	13	1.1	–	202	–	0.3
130	3	0.9	–	99	–	0.1

The ranges of the nematic phases accessible for the measurements of SiCl and AcCN were limited because of their substance
parameters. Owing to the low values of $\Delta\varepsilon$ of SiCl at lower temperatures the threshold voltage U_o increases rapidly to high values.
In the case of AcCN the high viscosity at low temperatures restricts
the measurement.

Table 1 shows that all three samples are switchable under the
influence of an electric field. Altering the frequency of the
electric field from 0.2 kHz to 200 kHz, the orientation changes
from homeotropic to homogenous and vice versa.

For a cell thickness of 10 μ m the rise time is in the range of
a few seconds. The smallest value is about 100 ms at a temperature
just below the clearing point. This was found for AcCN because of

its small threshold voltage. On the other hand the passive decay times $t_d{}^o$ are very long. For AcCN $t_d{}^o$ at low temperatures is even a thousand times greater than the rise time. The active decay times, however, are remarkably shorter. For SiCl and SiCN the active decay times are in the range of the rise times. The voltages used are 15 V for the active turn-off and 30 V for the turn-on process depending on experimental conditions.

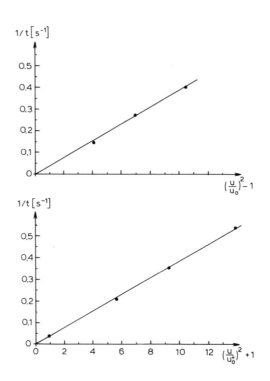

Fig.1. Reciprocal rise time vs $\left(\dfrac{U}{U_o}\right)^2 - 1$ at 80°C for SiCN (top) and reciprocal active decay time vs $\left(\dfrac{U}{U_o^+}\right)^2 + 1$ at the same temperature (bottom).

In Figure 1 the reciprocal time is plotted versus $\left(\dfrac{U}{U_o}\right)^2 - 1$ and $\left(\dfrac{U}{U_o^+}\right)^2 + 1$. The linearity of the plot demonstrates the validity of the relations given in equations (1) to (5) for polymeric liquid crystals. This was also confirmed by experiments with Fréedericksz cells [7,8]. For SiCl the elastic constant $k_{11} = 7.3 \cdot 10^{-12}$ N was calculated from the experimental value of $(\Delta\varepsilon)_{static} = 0.17$.

318

Based on this calculation we estimated a viscosity coefficient η_1 $1.5 \cdot 10^4$ cP at T = 99°C.

Information about the dynamics of liquid crystals in a planar cell after applying an electric field can be obtained from the measured phase difference $\frac{1}{2}\delta$. These experiments were described elsewhere [7]. Using the relation [10]

$$I = I_p \cdot \sin^2 (2\phi) \sin^2 (\tfrac{1}{2}\delta) \tag{6}$$

with I the registered intensity, I_p the intensity of the transmitted polarized light and ϕ the angle between the optic axis and the vector of the electric field. For the phase difference follows

$$\frac{1}{2}\delta = \frac{\pi}{\lambda} \int_0^d (n_{(z)} - n_\perp) \, dz \tag{7}$$

where z is the axis normal to the plane. With φ as angle between the optic axis and the glass plates of the planar cell [11] for $n_{(z)}$ follows

$$n_{(z)} = \frac{n_{||} \, n_\perp}{(n_{||}^2 \sin^2\varphi + n_\perp^2 \cos^2\varphi)^{1/2}} \tag{8}$$

This leads to

$$\frac{1}{2}\delta = \frac{\pi d}{\lambda} \frac{n_{||} \, n_\perp}{(n_{||}^2 \sin^2\overline{\varphi} + n_\perp^2 \cos^2\overline{\varphi})^{1/2}} \tag{9}$$

$\overline{\varphi}$ is defined by

$$\overline{\varphi} = \arccos (<\cos^2\varphi>_d)^{1/2} \tag{10}$$

and can be received by a small number of iteration steps.

The $\overline{\varphi}$ data are presented in Figure 2 as a function of the rise time. It is shown that a linear relation is valid for small $\overline{\varphi}$ data. This is equivalent with the validity of the relation [9]

$$\overline{\varphi} = \varphi_o \exp (t_r/\tau) \tag{11}$$

Here φ_o and τ are constants.

Formula (11) is invalid for higher values of $\overline{\varphi}$. This can also be seen in Figure 2. For this range a relation

$$(\overline{\varphi}_{max} - \overline{\varphi}) = \overline{\varphi}_{max} \exp(-t_r/\tau) \tag{12}$$

was found showing a linear relation between $\ln(\overline{\varphi}_{max} - \overline{\varphi})$ and t_r.

This is analogous to the pulse-response function of a dipole in the electric field [12]. This will also be reported later [13].

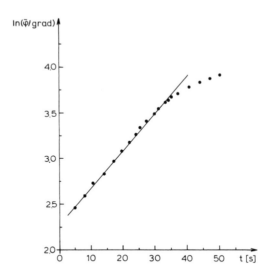

Fig. 2. ln ($\overline{\varphi}$) vs the turn-on time at 74°C for SiCN

Optical Birefringence

In Figure 3 the $n_{\parallel}$ and $n_{\perp}$ data of SiCl and SiCN as function of temperature are presented. We observed that the transition from the smectic to the nematic state is continuous if the sample is heated slowly with about 12 K/h. If the heating rate is higher (10 K/min) the $n_{\parallel}$ curve shows a discontinuity at the phase transition smectic → nematic. A satisfactory interpretation cannot be given at present time, but this behaviour is obviously dependent on phase transition kinetics.

It is known that for liquid crystalline polymers transition temperatures depend on their polymerization degree. The temperature ranges for the phase transitions are generally broader than for low molecular liquid crystals.

The smectic phases of SiCl and SiCN were not described before. The X-ray patterns however prove the existence of smectic A phases.

In Figure 4 the $n_{\parallel}$ and $n_{\perp}$ data for AcCN as function of temperature are given for the nematic and the glassy state. In the last one, the $n_{\parallel}$ and $n_{\perp}$ values are independent of temperature. Applying the $n_{\parallel}$ and $n_{\perp}$ data for the nematic and the glassy state, the order parameter for AcCN as a function of temperature has been determined

320

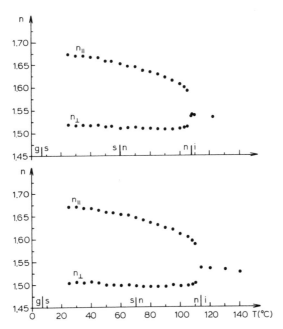

Fig. 3. Temperature dependence of $n_{||}$ and $n_{\perp}$ for SiCl(top) and SiCN (bottom).

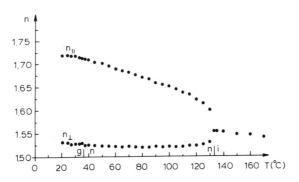

Fig. 4. Temperature dependence of $n_{||}$ and $n_{\perp}$ for AcCN.

based on the model of Vuks [14] using the extrapolation procedure described by Haller [15]. Similar calculations were done by Finkelmann [16] for liquid crystalline side chain polysiloxanes.

The formula related to the Vuks-model is

$$\alpha_i = \frac{3}{N}\left(\frac{n_i^2 - 1}{\bar{n}^2 + 2}\right) \tag{13}$$

$N = \frac{\rho}{M} \cdot N_A$ is the particle density.

Using the relation for the order parameter S

$$S = \frac{\alpha_{\parallel} - \alpha_{\perp}}{\alpha_l - \alpha_t} \tag{14}$$

it follows that

$$S \cdot \frac{\Delta\alpha}{\bar{\alpha}} = \frac{n_{\parallel}^2 - n_{\perp}^2}{\bar{n}^2 - 1} \tag{15}$$

with $\bar{\alpha} = \frac{1}{3}(\alpha_{\parallel} + 2\alpha_{\perp})$, $\bar{n} = \frac{1}{3}(n_{\parallel} + 2n_{\perp})$ and $\Delta\alpha = \alpha_l - \alpha_t$

The Haller extrapolation uses the expression

$$\ln\left(\frac{\Delta\alpha}{\bar{\alpha}} \cdot S\right) = A + C\left(1 - \frac{T}{T_{ni}}\right). \tag{16}$$

$\ln\left(\frac{\Delta\alpha}{\bar{\alpha}} \cdot S\right)$ is plotted versus $\left(1 - \frac{T}{T_{ni}}\right)$ assuming $S = 1$ for $0\,^{\circ}K$.

With $\frac{\Delta\alpha}{\bar{\alpha}} \cdot S = \frac{\Delta\alpha}{\bar{\alpha}} = k^{-1}$ for $T = 0\,^{\circ}K$ we got

$$S = k \cdot \frac{n_{\parallel}^2 - n_{\perp}^2}{\bar{n}^2 - 1} \tag{17}$$

The result for S as a function of T is presented in Figure 5. S is found to be ~ 0.8 at the transition point $n \to g$.

Dielectric Properties of SiCl

In Figure 6 the dielectric constants [17] at 1 kHz, $\varepsilon_{\parallel}$ and $\varepsilon_{\perp}$, for SiCl as functions of temperature are presented. The relaxation of the rotation of the mesogenic group around the main chain occurs at $62\,^{\circ}C$. This relaxation process is a single relaxation [18] as the plot in the Cole-Cole diagram shown in Figure 7 forms a half circle. The relaxation of the rotation around the mesogenic long axis is found at $20\,^{\circ}C$. This relaxation process is not of a single type, as shown in the Cole-Cole plot in Figure 8. In this region there are various relaxation processes around the polymeric main chain.

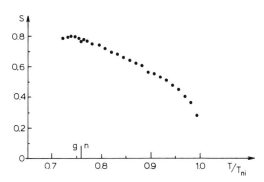

Fig. 5. Calculated order parameter as function of a reduced
temperature T/T_{ni} for AcCN.

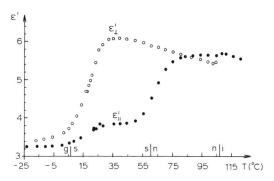

Fig. 6. Real part of the dielectric constants – $\varepsilon'_{\parallel}$ and $\varepsilon'_{\perp}$ –
for SiCl as a function of temperature at 1 kHz.

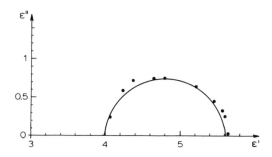

Fig. 7. Cole-Cole plot for the parallel dielectric constant
– ε' real part and ε'' imaginary part – for SiCl at 80.7°C.

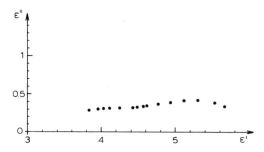

Fig. 8. Cole-Cole plot for the perpendicular dielectric constant –
ε' real part and ε'' imaginary part – for SiCl at 21.1°C.

At 84°C the dielectric anisotropy changes from positive to negative values. This is demonstrated in Figure 9, showing a magnified section of Figure 6. We interpret this sign changing by use of the Maier-Meier [17] theory, an extended Onsager theory. The dielectric anisotropy depends on the anisotropy of polarizability $\Delta\alpha$ and the value and angular position of the permanent electric dipole moment μ

$$\Delta\varepsilon = \frac{N\,h\,F}{\varepsilon_o} \left[\Delta\alpha - F\frac{\mu^2}{2kT} (1-3\cos^2 \beta) \right] S \qquad (18)$$

h is the cavity factor and F the Onsager reaction field. ß is the angle between the molecular dipole moment and the optical axis. S is the order parameter. In case of ß being smaller than 54.7° the dipole part is positive. Otherwise the dipole part compensates the

324

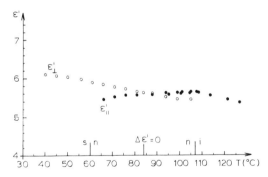

Fig. 9. Quasistatic real part of dielectric constant – $\varepsilon'_{\parallel}$ and $\varepsilon'_{\perp}$ – as a function of temperature for SiCl.

induced part of the polarization. $\Delta\varepsilon$ may even reach negative values. In the case of a temperature at which $\Delta\varepsilon = 0$, the orientational and induced parts are of the same magnitude compensating each other. We interpret our observation of the sign changing of $\Delta\varepsilon$ as caused by a changing of the orientational part of dielectric anisotropy as a function of temperature. This is confirmed by the continuous in-crease of $\Delta\varepsilon$ as a function of temperature in the region in question ($65^{\circ}C$ to T_{ni}), in accordance with eq. 18.

Concerning the frequency dependence of the real part of dielec-tric constants two cases can be distinguished, shown in Figure 10. At low temperatures $\Delta\varepsilon$ is negative in the whole frequency region whereas at temperatures above $84^{\circ}C$ there is a change of sign.

For low molecular liquid crystals the temperature dependence of f_o – the frequency of dielectric isotropy derived from dielectric measurements – can be described by an Arrhenius equation. This can also be done for the relaxation process leading to

$$\ln f_i = -\frac{E_a}{RT} + const ; \quad i = o, R \quad (19)$$

with E_a as activation energy.

As described in the literature [9,19] the relaxation frequency f_R for low molecular liquid crystals for the relaxation around the short axis is in the same order of magnitude as the frequency f_o of the dielectric isotropy at the same temperature. Also their temperature dependences are similar. In Figure 11 we present the temperature dependences of f_o and f_R for SiCl. It can be seen that

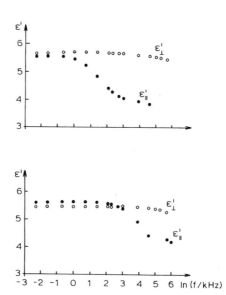

Fig. 10. Frequency dependence of the real part of the dielectric con-
stant - $\varepsilon'_{\parallel}$ and $\varepsilon'_{\perp}$ - for SiCl at 99.2°C(top) and 75.6°C(bott

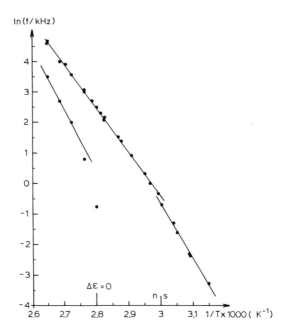

Fig. 11. Frequency of relaxation f_R as a function of temperature in
the nematic and smectic phases (upper lines) for SiCl oriented parallel to electric field and temperature dependence of
the dielectric isotropy f_o (lower lines).

326

$\ln f_o \propto \frac{1}{T}$ is not linear in the vicinity of the temperature (T = 84°C) where $\Delta \varepsilon$ becomes zero at 1 kHz. The curve approaches a parallel to the $\ln f_o$ -axis at T = 84°C. From the straight part of the line we calculated the activation energy, mentioned in Table 2.

Table 2. Molecular processes and their activation energies

Symbole	Interpretation	Activation energy (kJ/mol)
f_o	dielectric isotropy	188
f_R	relaxation for the rotation of the short axis of the nematic group around the main chain:	
	in the nematic phase	118
	in the smectic phase	154
f_R	relaxation primary of the main chain in the smectic phase	412
f_R	relaxation of the ester groups in the glassy state	48

In Figure 11 the temperature dependence of the relaxation around the short axis in the plot $\ln f_R \propto \frac{1}{T}$ follows a straight line in the nematic state as well as in the smectic state [19,20]. Their activation energies are slightly different (Table 2). In Figure 12 the temperature dependence of $\ln f_R$ in the smectic and also in the glassy state is presented. The described relaxation process at room temperature is interpreted as relaxation of the main chain. This process has a high activation energy (Tab.2). Perpendicular dipole components of the mesogenic side chain also contribute to this process (see Fig.6). We assume that at least three different processes contribute to this relaxation process as can be interpreted from Figure 8. Also below the glass temperature a relaxation process takes place. This we interpret as a relaxation of the ester group. This relaxation process needs the lowest activation energy (Tab.2). It should be mentioned that also SiCN shows similar dielectric properties [21].

327

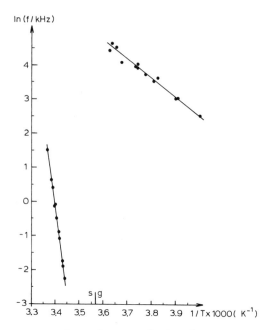

Fig. 12. Temperature dependence of the frequency of relaxation
oriented parallel and perpendicular to electric field
at low temperatures.

ACKNOWLEDGEMENT

We thank the Deutsche Forschungsgemeinschaft for financial
support. SiCl and SiCN have been kindly prepared by Dr. Finkelmann
and some samples of AcCN by Prof. Ringsdorf.

REFERENCES

1. H.Finkelmann, J.Koldehoff, and H. Ringsdorf, Angew.Chem.,Int. Ed. Engl. 17 : 12 , 935 (1978).
2. V.P.Shibaev, N.A.Platé and Ya.S.Freidzon, J.Polym.Sci.,Polym. Chem. Ed. 17 : 1655 (1979).
3. H.Finkelmann, and G.Rehage, Makromol. Chem., Rapid Commun. 1 : 31 (1980).
4. H. Finkelmann, U. Kiechle, and G. Rehage, Mol. Cryst.Liq. Cryst. 94 : 343 (1983).
5. M.Portugall, H. Ringsdorf, and R. Zentel, Makromol.Chem. 183 : 2311 (1982).
6. H. Finkelmann and U. Kiechle, personal communication.
7. H. Pranoto and W. Haase, Mol. Cryst. Liq. Cryst. 98 : 299 (1983).
8. W. Haase and H. Pranoto, Colloid and Polymer Sci., in press.
9. G. Baur, A. Stieb, and G. Meier, Appl. Phys. 6 : 309 (1975).
10. R. A. Soref and M.J. Rafuse, J. Appl. Phys. 43 : 2029 (1972).
11. C. Maze and D. Johnson, Mol.Cryst.Liq. Cryst. 33 : 213 (1976).
12. C. J. F. Böttcher and P. Bordewijk, in : " Theory of Electric Polarization ", 2nd Ed., Vol. II, Elsevier Sci. Publ. Comp. Amsterdam, p. 38, (1978).
13. H. Pranoto and W. Haase, to be published.
14. M. F. Vuks, Optics and Spectroscop., 20 : 361 (1966).
15. I. Haller, H. A. Huggins, H.R. Lilienthal and T.R. McGuire, J. Phys. Chem. 77 : 950 (1973).
16. H. Finkelmann, H. Benthack,and G.Rehage, J.Chimie Physique, 80 : 163 (1983).
17. W. Maier and G. Meier, Z. Naturforschg., 16a : 262 (1961).
18. H.Kresse and R.V. Talrose, Makromol.Chem., Rapid Commun., 2 : 369 (1981).
19. W. H. de Jeu and Th. W. Lathowers, Mol.Cryst.Liq.Cryst., 26 : 225 (1974).
20. H. Ringsdorf, G. Strobl, and R.Zentel, Freiburger Arbeitstagung Flüssigkristalle, 31.3. – 2.4. 1982; R. Zentel, Dissertation, Mainz (1983).
21. H. Pranoto, W. Haase, H. Finkelmann, and U. Kiechle, to be published.

SOME ELECTRO-OPTICAL PHENOMENA IN COMB-LIKE LIQUID

CRYSTALLINE POLYMERIC AZOMETHINES

R.V. Talroze, V.P. Shibaev, V.V. Sinitzyn,
and N.A. Platé

Department of Polymer Chemistry
Moscow State University
Moscow, USSR

INTRODUCTION

The ability to orient in an electric field, a well-known property of liquid crystals, is the basis of their numerous technical applications. This is why interest is expressed in electro-optical phenomena in liquid crystalline (LC) polymers. Comb-like LC polymers with nitrile groups in mesogenic moieties can be oriented in an alternating electric field[1,2,3]. In most cases the ability to quench this orientation in polymers below the glass transition temperature has been proven. The similarity in structure and properties of LC polymers and low molecular mass liquid crystals lead to the study of the electro-optical effects in polymers.

This paper deals with the study of the mechanism of orientation process, the guest-host effect and the electrohydrodynamic (EHD) instabilities in LC comb-like polymers with azomethine and cyandiphenyl mesogenic groups. In the last part of the paper the cooperative structure transition in LC nematic polymers is discussed.

EXPERIMENTAL

Synthesis of monomers and polymers obtained by radical polymerization of corresponding monomers in benzene solutions and phase behavior are described elsewhere[3,4,5]. The main characteristics of the polymers studied are given in Table 1.

331

Table 1. Main Characteristics of LC CN-containing Comb-like Polymers.

N	R	m	$T_g, °C$	$T_{Cl}, °C$	Mesophase Structure Type
$-(CH_2-CR)_{\overline{n}}$ $\quad\quad\mid$ $COO-(CH_2)_m-O-C_6H_4-CH=N-C_6H_4-CN$					
1a	H	6	20	158	Nematic
1b	H	11	10	169	Smectic
2a	CH_3	6	35	125	Smectic
2b	CH_3	11	25	155	Smectic
2c	copolymer of 1a and 2b (20 mole %)		20	132	Smectic
$-(CH_2-CR)_{\overline{n}}$ $\quad\quad\mid$ $COO-(CH_2)_m-O-C_6H_4-C_6H_4-CN$					
3a	H	2	2	112	Nematic
3b	H	5	40	120	Cybotactic Nemati
3c	H	11	25	127	Smectic
4a	CH_3	5	60	124	Smectic
4b	copolymer of 3b and 4a (20 mole %)		40	102	

The electro-optical measurements were carried out using an electro-optical cell described by Talroze, et al. which was mounted on a heating stage of a polarizing mi-croscope, MIN-8. A photo-element was used in conjunction with a recorder to monitor the transmitted light intensity The film thickness was about 12 μm, electric field inten-sity was 10^5 V/cm. The pleochroic dye for studying the "guest host" effect had an absorption band at 505 nm.

$$O_2N-C_6H_4-N=N-C_6H_4-N\begin{array}{l}\diagup C_2H_5\\ \diagdown C_2H_4OH\end{array}$$

RESULTS AND DISCUSSION

Mechanism of the Orientation Process

CN-containing LC polymers of comb-like structure can
be oriented in an alternating electric field with its side
chains along the field direction in spite of the chemical
bonding of mesogenic groups to the main chain. This means
that either the side groups behave independently of the
backbone chains or the conformational changes of macromolecules are influenced by the reorientation of the mesogenic
side groups and polymer chains follow these changes.

The orientation process in a LC polymer is a result
of its dielectric interaction with an electric field. The
duration of this process depends on the polymerization degree, P_W; the greater the polymerization degree, the longer the duration (Table 2).

A detailed analysis of the orientation kinetics has
shown that the curves of the change in optical transparency (Fig. 1) are related to the field induced relaxation
process and can be described by the equation:

$$1 - \Theta = e^{-k\tau^n} \qquad (1)$$

where Θ is the degree of completeness of the process,
τ the time, and K and n constants (n=1 in a wide
temperature range does not depend on the polymerization
degree; the case of n>1 will be discussed later). The
constant k which can be regarded as a rate constant of the
orientation process depends exponentially on the temperature (Fig. 2). The calculated values of the effective
activation energy are given in Table 2. The activation
energy does not depend on the degree of polymerization of
nematic polymers having identical mesogenic groups (Table 2).

The decrease of the orientation rate with the decrease
of temperature seems to be related to the increase of the
viscosity of the LC melt. The absence of the initial homogeneous orientation of the polymeric liquid crystal in
the electro-optical cell implies that the average viscosity
of the LC polymer is the important factor, instead of the
rotational viscosity.

The E_A values given in Table 2 exceed the activation
energy of the rotational viscosity of CN-containing low
molecular nematics, which is equal to 36-52 kJ/mole[6]. At
the same time, they are close to the E_A values of processes

Table 2. Activation energy (E_A) and orientation time
($\tau 1/2$) of polymers with the different polymerization degree ($\overline{P}w$).

N	Mesophase Structure Type	$[\eta]$ dl/g	$\overline{P}_w$	$\tau 1/2$ sec.	E_A kJ/mole
1a	Nematic	0.02	5-10	4.0	–
		0.11	250	18.0	92
		0.12	350	25.0	100
		0.20	1200	510.0	100
1b	Smectic	0.10	–	10.0	85
2c	Smectic	–	–	120	160
3b	Cybotactic Nematic	0.06	70	0.54	180
		0.08	140	0.8	172
		0.14	550	3.0	172
		0.24	2000	14.4	–
4b	Smectic	0.10	–	4.0	228

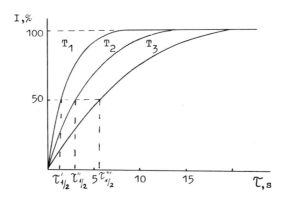

Fig. 1. Kinetics of an orientation process of a LC polymer in an electric field (temperature
$T_1 > T_2 > T_3$).

related to segmental movements of macromolecules of acrylic polymers such as the viscous flow of a nematic polymer melt. One can suppose that the segments of macromolecules are involved in the orientational movement. The longer duration of orientation with the increase of the polymerization degree means that the position of the macromolecule as a whole is also changed.

The activation energy of the acrylic comb-like polymers of smectic structure is about the same as that in nematic systems (Table 2). This is not typical for low molecular smectics which are oriented with more difficulty than nematics. The incorporation of methacrylic units increases E_A values up to 160 kJ/mole for azomethine (polymer 2c) and 228 kJ/mole for cyandiphenyl (polymer 4b) derivatives. These data show that the activation energy depends not so much on the mesophase but primarily on the chemical structure and the mobility of polymer backbones taking part in the orientation process.

The orientation of methacrylic homopolymers (polymers 2a,2b,4a) in the LC phase is severely hindered and the complete orientation of these polymers is not achieved. The slow "response" and high activation barrier in methacrylic derivatives are apparently caused by hindered mobility of the polymer backbone as a result of a change in conformation at the attachment sites of mesogenic side groups to the methacrylic chains.

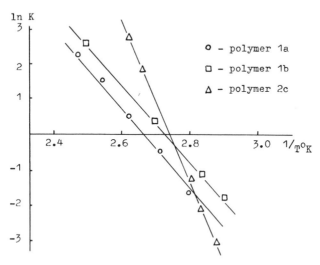

Fig. 2.   Logarithm of the rate constant of the orientation process versus the reciprocal temperature ($^{\circ}$K).

335

These data confirm the decisive role of chain mobility in the orientation process of comb-like polymers
in an electric field. However, the process itself, as in
low molecular liquid crystals, is caused by the orienting
tendency of mesogenic groups in the side chains which
forms the LC phase when the dielectric constant of anisotropy has a positive value (i.e., when $\Delta\varepsilon > 0$).

Guest-Host Effect

Some pleochroic dyes were shown to display a strong
contrast in electro-optical effects in LC polymers with
mesogenic side groups of structure identical to the structure of low molecular mass dye molecules[6]. The incorporation of 1-2% by weight of dye molecules has been shown
not to change the mesophase structure. An electric field
induces an alignment of a guest dye in a host polymer
which results in a partial loss of color of the polymer
film. Such optical behavior is explained by the orientation of dye molecules parallel to the direction of incident light and by an increase of transmitted light intensity.

According to Pelz[8], the determination of the orientational order parameter of dissolved dye molecules is
possible by comparing absorption measurements in the nematic phase with that in the orientation (homeotropic) and
the melt (isotropic) phases. The order parameter can be
calculated according to the equation:

$$S = 1 - \frac{E_h\, c_i\, n_o}{E_i\, c_h\, n_i} \qquad (2)$$

E_h, E_i – extinctions of the homeotropic sample and isotro-
          pic melt, respectively;
c_h, c_i – concentration of the dissolved substance in the
          nematic and isotropic phases, respectively;
n_i – refractive index of the isotropic melt;
n_o – refractive index of the ordinary ray.

The calculations have been carried out with the assumption of slight difference between the sample densities
in the two phases in the vicinity of the nematic $\leftrightarrow$ isotropic phase transition and of close values of n_o and
n_i ($n_o/n_i \sim 1$).

Figure 3 shows the temperature dependence of the
orientational order parameter of dissolved dye in the
nematic matrix of polymer 1a.

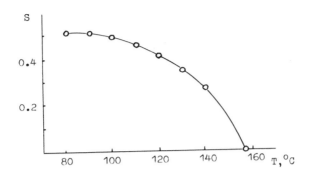

Fig. 3.   Orientational order parameter S as a function of
temperature for polymer 1a.

The similarity of the molecular structure of the dye
and of the mesogenic side groups suggests that this curve
corresponds to the temperature change of the orientational
order parameter S of the LC polymer. The limiting value
of S=0.5 is practically independent of the degree of polymerization ($P_w \sim 100$ and 1500). This value is in agreement
with the parameter S obtained for low molecular mass nematics and on the other hand, with the S value of dye
molecules which are chemically bonded to the polymer
matrix[9].

Electrohydrodynamic Instability

Application of an a.c. electric field to homeotropi-

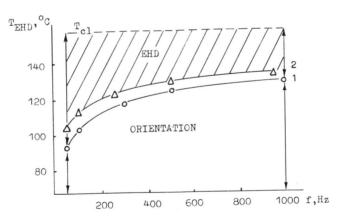

Fig. 4.   Dependence of the temperature of the appearance
of EHD instability (T_{EHD}) in homeotropically
oriented films of polymers 1a (1) and 1b (2) on
the electric field frequency.

cally oriented films of polymers la and lb induces intens
"boiling" of LC melts which results in a sharp distortion
of the optical transparency. This seems to be an EHD
process of dynamic scattering mode (DSM) type. Data show:
in Fig. 4 indicate that the appearance of DSM in polymeri
liquid crystals essentially depends on temperature.

Below T_{EHD} (Fig. 4) the formation of hemeotropically
oriented structures is observed; above the T_{EHD} process
the DSM takes place. This process is more pronounced at
higher temperatures, probably due to the presence of impurities which are more mobile at high temperatures.
With the increase of the electric field frequency the
lower limiting temperature T_{EHD} is displaced and the DSM
region becomes narrower (at higher frequency increased
ion mobility is needed for the flow of liquids). At lowe:
temperatures in the high frequency field the ions oscilla:
only, which does not disturb the homeotropic orientation.
This type of frequency dependence on T_{EHD} permits two-
frequency switching of optical characteristics of polymer
films. At the given temperature T (where $T_{EHD} < T < T_{EHD}^{f2}$)
one can change the regime of the homeotropic orientation
to an electrohydrodynamic by variation of the frequency
of the electric field (Fig. 5). By switching on the low
frequency field, the initial homeotropically oriented film
with high transparency starts to scatter the light due to

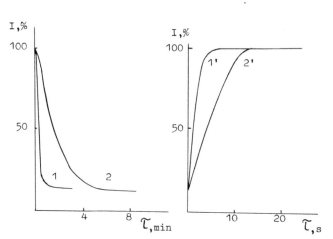

Fig. 5. Kinetics of the change in transparency of polymer
la (1) and lb (2) films at the discrete switching
of electric field frequency (U=250 V, d=12 μm,
T=125 °C); 400 H_Z (1), 4 KH_Z (1'), 200 H_Z (2),
and 1 KH_Z (2').

338

EHD instability. The transparency is 5-6 times decreased (Fig. 5). The time of appearance of DSM in polymers exceeds the time of the orientation process and depends on the field frequency.

As shown above, EHD instability manifests itself at high temperatures at which the viscosity of the LC melt is not too high. The influence of the molecular weight of the polymer on the temperature T_{EHD} seems also to be related to the viscosity. If in the polymer la with P_w of about 100-150 the EHD process takes place in a wide temperature range (Fig. 4), DSM in the polymer with $P_w \sim 1500$ in an electric field of f=50 Hz appears only a few degrees below T_{cl}.

The question arises: what mechanism is responsible for the EHD instability in these polymers? Such effects in liquid crystals are the result of the ionic conductivity. The free charge-carrier can appear either due to ionic impurities in the substance, or due to the injection or exclusion of electrons by neutral molecules on electrodes. The ionic conductivity of CN-containing Shiff's bases, for instance, depends on small quantities of p-aminobenzonitrile.

As long as the value of $\Delta\varepsilon$ is positive, the EHD instability in the CN-containing comb-like polymers cannot be described by the Carr-Helfrich mechanism. The process seems to have an isotropic mechanism which is related to the movement of impurities' ions in the gradient of electric forces due to the inhomogeneity of the field and the charges' distribution in the sample. Ions moving toward electrodes attract the molecules of the liquid crystal causing the turbulent flow of the liquid. The isotropic mechanism of EHD instability is also confirmed by the independence of the process from the mesophase structure of nitrile containing polymers.

The main features of the process are thus established by the study of EHD instability in these comb-like polymers. The strong temperature and molecular mass dependence of the process is a unique characteristic of the polymeric liquid crystals.

Cooperative Structure Transition in Oriented Polymers

As mentioned, the electric field induces the orientation of polymer side groups. The homeotropic structure with an alignment of these groups ($\Delta\varepsilon > 0$) normal to the electro-optical cell areas is thus formed. The possibil-

ity of quenching of such uniaxial orientation in the solid
polymer below the glass transition temperature is shown
for a large group of polymers (Table 1, polymers 1b, 2a,
2b, 3b, 3c). At the same time, the homeotropically oriented structure in polymers 1a and 3a has been shown not to
be quenched[10]. Upon being cooled, the polymer film oriented in an electric field undergoes a stepwise change of optical transmittance due to the appearance of birefringence.
As an example, the dependence of the optical transmittance,
measured with crossed polarizers, on temperature for the
polymer 1a is given in Fig. 6. The transparency of the
film is not decreased (curve 2) and this fact shows that
disordering on the macroscopic level does not take place.
The transition from state I to the state II is reversible
in an electric field (Fig. 6). This fact, as well as the
narrow transition temperature interval (1-3), suggests
that the change of optical properties with the temperature is related to cooperative structure transition. This
is confirmed by dielectrical measurements. A maximum is
present on the curve giving the tangent of the dielectric
loss angle as a function of temperature for samples oriented in an electric field of the measuring cell (Fig. 7).

Since the position of that maximum does not depend
on the frequency of the electric field, the transition
may be interpreted as structural and not relaxational.

The role of the external field is clear from the experimental data discussed below. There is no change of
structural and thermodynamic characteristics in unoriented
polymers in the vicinity of the transition temperature
T_{tr}. This means that the transition is induced by the
field. On the other hand the transition also occurs when
the electric field is switched off at a temperature 10-15
above T_{tr} and further cooling is carried out in the absenc

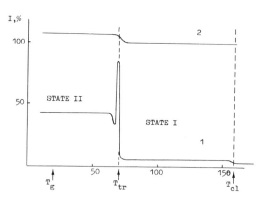

Fig. 6. Temperature dependence of the optical transmittance under crossed polarizers (1) and of the
transparency (2) for polymer 1a.

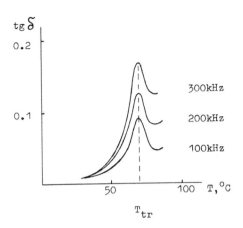

Fig. 7. Tangent of the dielectric loss angle versus the
temperature in the preoriented sample of polymer 1a.

of the field. Moreover, the transition temperature decreases with the increase of the field intensity (Fig. 8).
Such dependence means that the initial homeotropic structure is stabilized by the field and the transition itself
is caused by the internal thermodynamic properties
of the polymer which are responsible for the instability
of the homeotropically oriented state below T_{tr}.

Thus, the preliminary orientation created by the action of the external field is the main condition needed
for the transition. The chemical bonding of LC phase

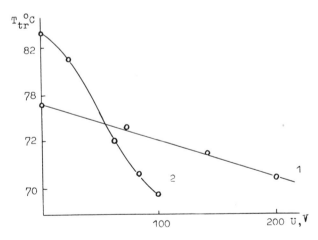

Fig. 8. Temperature of the structure transition in polymer 1a (1) and its copolymer with metylmethacry-
late (2) versus voltage.

341

forming mesogenic groups by polymer chains results in the appearance of this peculiar effect unknown for low molecular mass analogs. This transition may be caused by the conformational change of polymer backbones at the attachment sites of side chains.

At the same time, however, the transition is also influenced by the type of mesophase structure. In the polymers 1a and 3a of nematic structure, the transition process manifests itself very clearly, but the incorporation of 20 mole % long chain units such as

$$-CH_2-C(CH_3)-$$
$$COO-(CH_2)_{11}-O-C_6H_4CH=N-C_6H_4-CN$$

in the polymer 1a induces smectic alignment of polymer 1a units in an oriented state and prevents the transition The homeotropic structure can be quenched in this copolymer and other homopolymers having the smectic structure (Table 1). The formation of smectic regions in an oriented polymer apparently results in the increase of the "side" interactions of mesogenic groups. The position of mesogenic groups in smectic layers is stabilized and the mobility of polymer backbones between the layers is restricted which results in the fact that the conformation of the macromolecule can be quenched.

It must be pointed out that the rebuilding of the homeotropic structure during cooling of the sample below the transition temperature induces the formation of peculiar optical texture. Under crossed polarizers a dark cross with concentric colored rings appears (Fig. 9). By means of optical microscopy and X-ray methods, the

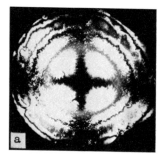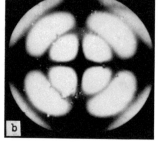

Fig. 9. Optical texture of polymer 1a (a) and its copolymer with MMA (b) under crossed polarizers below the transition temperature T_{tr}.

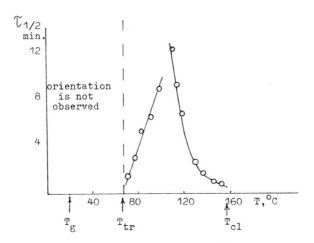

Fig. 10. Temperature dependence of the orientation time
for polymer 1a.

the colored rings are shown to be caused by a regular increase of the side chain inclination to the normal to the
film area. In the middle of the film the homeotropic
orientation of side groups is maintained and at the boun-
dary the inclination in the radial direction is the greatest.

It can be supposed that the centro-symmetric texture
may be caused by the formation of the inclined structure
at the phase transition. The phase transition is accompanied by a decrease in volume of the visco-elastic polymer sample with a constant film thicknes and, as a result,
by a contraction of the LC polymer.

The structural transition discussed is accompanied
by a change in the kinetics of the orientation process.
As it has been shown, at high temperatures an exponential
increase of the orientation time with decreasing temperature takes place. Within the temperature range immediately preceeding T_{tr}, a marked increase of the orientation
rate and a decrease in the orientation time with a minimum at T_{tr} are observed (Fig. 10). The form of the kinetic
curve is also changed and that leads to an increase of
the parameter n in the equation (1). At high temperatures
"n" is equal to 1, with decrease of temperature this value
goes up to 2. Below T_{tr}, the orientation of the polymer
along the field direction is not observed even in the
field of 2.5×10^5 V/cm.

We cannot give a full explanation of the kinetic

343

trends at present. One can only suppose that anomalous changes of the mechanism of the orientational process may be related to the pretransitional phenomena taking place just at the temperature T_{tr}.

The study of the kinetics of the orientational process of LC comb-like polymers in an electric field shows the main peculiarities of the electro-optical behavior of LC polymers in comparison with the low molecular mass liquid crystals. The electro-optical effects in comb-like polymeric liquid crystals, namely guest-host effects in the presence of a dye of low molecular mass, EHD instability and the structure transition phenomenon can be use in making LC polymer films with adjustable optical characteristics.

REFERENCES

1. H. Finkelmann, D. Naegele, H. Ringsdorf, Makromol. Chem., 180, 803 (1979).
2. R.V. Talroze, S.G. Kostromin, V.P. Shibaev, N.A. Plate, H. Kresse, K. Sauer, D. Demus, Makromol. Chem. Rapid Commun., 2, 305 (1981).
3. R.V. Talroze, V.V. Sinitzyn, V.P. Shibaev, N.A. Plate "Influence of the Structure of Mesogenic Groups on th Structure and Dielectric Properties of the Comb-like Liquid Crystalline Azomethine Containing Polymers", in Advances in Liquid Crystal Research and Applications, L. Bata, Ed., Pergamon Press, Oxford-Akademiai Kiada, Budapest, 2, 915 (1980).
4. R.V. Talroze, V.V. Sinitzyn, V.P. Shibaev, N.A. Plate Mol. Cryst. Liq. Cryst., 80, 211 (1982).
5. V.P. Shibaev, S.G. Kostromin, N.A. Plate, Eur. Polym. J., 18, 651 (1982).
6. V.V. Beljaev, M.P. Grevenkin, E.I. Kovshev, G.G. Slae "Koeffitzient Obyomnoi i Vraschtatelnoi Vyaskosti Jidkokristallicheskikh Materialov", in Abstracts of Communications of the III^d Nauchnotechnicheskogo Seminara, Opticheskie Svoistva Jidkikh Kristallov i ikh Primenenie, M.G. Tomilin, Ed., Leningrad, 75 (1983).
7. V.P. Shibaev, V.G. Kulichikhin, S.G. Kostromin, N.V. Vasiljeva, L.P. Breverman, N.A. Plate, Dokl. AN SSR, 263, 152 (1982).
8. G. Pelzi, D. Vetters, D. Demus, Krystal und Technik, 14, 427 (1979).
9. H. Benthack, H. Finkelmann, G. Rehage, 27th International Symposium on Macromolecules, Strasbourg, 2, 961 (1981).
10. R.V. Talroze, V.V. Sinitizyn, V.P. Shibaev, N.A. Plat Polymer Bulletin, 6, 309 (1982).

THERMO-RECORDING ON THE LIQUID CRYSTALLINE POLYMERS

V.P. Shibaev, S.G. Kostromin, N.A. Platé,
S.A. Ivanov, V.Yu. Vetrov and I.A. Yakovlev

Departments of Chemistry and Physics, Moscow
State University, Moscow, USSR

In recent years the study of the dynamics of structural transformations in polymeric liquid crystals under an applied electric field has attracted increasing interest. It is now well-known that liquid crystalline (LC) polymers with mesogenic side groups are capable of orientation in an electric field in the same way as low molecular weight LC compounds[1,2]. The essential difference is that there exists the possibility of freezing the orientation induced by the electric field by cooling LC polymers below the glass transition temperature[1]. This special feature, which as absent in low molecular mass liquid crystals, permits the preparation of highly oriented polymeric films with the homeotropic orientation of mesogenic groups.

In this paper, one of the examples of the use of such oriented films as film matrix for information recording is examined. We used the principle of thermo-recording, which is well known for low molecular mass liquid crystals[3,4]. On the transparent film of the homeotropically oriented LC polymers (Fig. 1a), regions of local overheating are created with the aid of a laser beam. In these local regions the liquid crystal passes to an isotropic melt and the homeotropic orientation is destroyed. Instead of a transparent monodomain of homeotropic texture, a polydomain texture that scatters light is formed on cooling. In this way, information can be recorded on the transparent film. This information may be projected onto a screen but may be "wiped off" by applying an electric field or a temperature rise. For the thermo-recording to take place, it is necessary for an oriented state to be stable in the absence of an electric field for a rather long time (i.e., the orientation relaxation rate should be sufficiently low). It should be noted that low molecular weight nematics do not fulfill this requirement, as they are rapidly disoriented when the electric field is switched off. That is why in thermo-recording devices based on low molecular weight liquid crystals a "smectic-nematic" transition is most often made use of, as homeotropic texture is sufficiently stable in a smectic LC state.

Another situation is presented by LC polymers. As it has recently been shown[1], the homeotropic orientation from the nematic CN-containing polymers:

$$-CH_2-CH-$$
$$\underset{\underset{O}{\parallel}}{C}-O-(CH_2)_5-O-\bigcirc-\bigcirc-CN \qquad (1)$$

induced by an applied electric field remains unchanged a long time after switching off the field. In this case, the orientation relaxation rate is very low and decreases when the molecular weight of the polymer increases; this is explained by the high viscosity of the polymer sample. This result permits us to use the transition "nematic-isotropic melt" for thermo-recording of information on LC polymers.

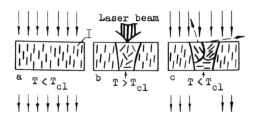

Fig. 1. Scheme of thermo-recording using the film of the homeotropically oriented nematic LC polymer. Only the mesogenic groups (I) are shown.

The polymer[5,6] was CN-containing polyacrylate (1), having $T_{cl}=106°C$ with weight average degree of polymerization $\overline{P}_w=150$[7]. The scheme of the device is given in Fig. 2. A sample of the polymer was put in an electro-optical glass cell with transparent electrodes, separated by 40 μm thick holder. The cell was maintained in a special thermostate (±0.1 °C). The helium-neon laser with power 10 mW was used as a source of light. The laser beam passed through the attenuator and focused on the LC polymer surface. The diameter of the laser beam focus was 70 μm. The laser beam power passed through the polymeric sample which was fixed by a photoresistor.

To determine the optimum recording conditions, the dependence on temperature of light intensity passed through the sample for varying light power with or without electric field was studied (Fig. 3). The polydomain light scattering texture formed on cooling the sample from isotropic melt without electric field takes place at T_{cl} (Fig. 3, curve 1). At the same time, the power of transmitted light through the sample is equal to 10^{-3} part of power of light transmitted through the isotropic melt. Being cooled in the electric field (U=100 V, f=500 Hz; curve 2) the polymeric sample becomes slightly turbid at T_{cl} (formation of LC domains) and a transparent homeotropic texture is formed. In this case, the shape of the laser beam transmitted through the sample does not change (Fig. 4a).

An increase of the laser beam power caused local heating of the sample with $\Delta T \sim 0.6°C$: the minimum on

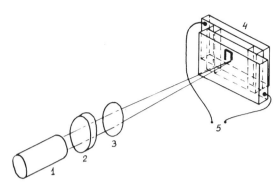

Fig. 2. Schematic drawing of the measuring arrangement for thermo-recording: 1-laser, 2-attenuator, 3-focused lens, 4-electro-optical cell, 5-a.c. voltage generator.

347

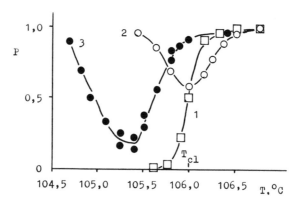

Fig. 3. Temperature dependence of laser beam power passed through sample for different average densities of laser beam power (W) on the surface of a LC polymer: 1- U=0, W=15 W/cm^2; 2- U=100 V, W=15 W/cm^2; 3- U=100 V, W=130 W/cm^2. Laser beam power is given in relative units: the power at T=106.5 °C is accepted for 1.

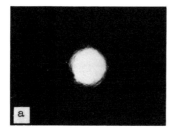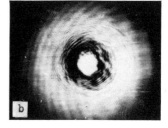

Fig. 4. The profile of a laser beam passed through a sample of homeotropically oriented LC polymer under an applied electric field (U=100 V). Average density of laser beam power on the surface of polymer was equal to 15 W/cm^2 (a) and 130 W/cm^2 (b).

curve three (Fig. 3), corresponding to transition from isotropic melt to LC phase, was observed at T=105.4°C.

In addition to local heating of the sample, nonlinear refraction took place (i.e., the beam diverged and its shape was disturbed (Fig. 4b).

After two to three seconds under the action of laser beam with the same power, a turbid spot with non-oriented texture (Fig. 3, curve 1) appeared. The diameter of this spot by two or three times exceeded the diameter of the laser beam. On subsequent illumination of polymeric film by an unfocused laser beam, the contrasting dark spots were projected onto a screen corresponding to zones of focused beam action. Fig. 5a shows the letters recorded on a polymeric film. The recorded symbols are completely "wiped off" in two to three seconds (Fig. 5b) by an alternating electric field (U=100 V, f=500 Hz).

The laser-recorded information can be kept for a long time if the sample is cooled below the glass transition temperature. From this viewpoint, LC polymers differ beneficially from low molecular mass liquid crystals. The effect described demonstrates the possibilities of regulating structural and optical properties of LC polymeric materials and using LC polymers as the film matrix for information recording.

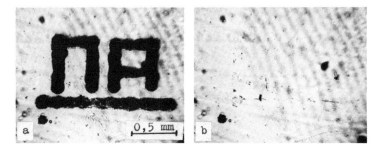

Fig. 5. a - transparent film with laser thermo-recorded letters; b- this film after application of an electric field.

LITERATURE

1.   R.V. Talroze, S.G. Kostromin, V.P. Shibaev, N.A. Pla
     H. Kresse, K. Sauer, D. Demus, Makromol. Chem. Rap
     Commun., 2, 305 (1981).
2.   H. Ringsdorf, R. Zentel, Makromol. Chem., 183, 1245
     (1982).
3.   H. Melchior, F. Kahn, D. Maydan, D.B. Fraser, Appl.
     Phys. Lett., 21, 392 (1972).
4.   F. Kahn, Appl. Phys. Lett., 22, 111)1973).
5.   S.G. Kostromin, V.P. Shibaev, N.A. Plate, "Zhidkokri
     tallicheskie Poliakrilaty i Polimetakrilaty, Soder
     hashie Ziandiphenilnye Gruppy i Obladayushiye Spo-
     sobnostyu k Orientazii v Elektricheskom Polye",
     Authors' Certificate USSR, N 887574 (1981).
6.   V.P. Shibaev, S.G. Kostromin, N.A. Plate, Eur. Polym
     J., 18, 651 (1982).
7.   S.G. Kostromin, R.V. Talroze, V.P. Shibaev, N.A. Pla
     Makromol. Chem., Rapid Comm., 3, 803 (1982).

ELECTRO-OPTIC EFFECTS IN LIQUID CRYSTAL SIDE CHAIN POLYMERS

H. J. Coles* and R. Simon

Liquid Crystal Group, Schuster Laboratory, Physics
Department, University of Manchester
Manchester M13 9PL, United Kingdom

INTRODUCTION

The recent work of Krigbaum[1] and Blumstein[2] on main chain
polymer liquid crystals and that of Finkelmann[3] and Ringsdorf[4] on
side chain systems has generated considerable interest in the potential of these materials for use in electro-optic devices. Indeed
over the last three years or so papers have started appearing repor-
ting on electro-optic effects in a variety of polymer liquid crystal
systems[5,6]. The main thrust of this initial work has been to examine
nematic and cholesteric polymer liquid crystals presumably in the
hope of producing electro-optic effects and displays analogous to
those observed with monomeric mesophases. However, as observed by
Ringsdorf and Zentel[6] and Finkelmann et al[5], for such polymer materials, the operating parameters are not as convenient as for monomeric systems, i.e. the threshold voltages are higher, the response
times are two to ten times slower and the operating temperatures are
not usually convenient. Whilst the promise of these new polymer
materials would seem to be in combining the polymer specific with
the monomeric specific liquid crystal properties the resultant high
viscosity associated with the polymer chain would seem to be a major
drawback to these systems. The obvious conclusion is therefore that
such polymers, in their pure state, will always have inferior performance to equivalent monomers. However as Finkelmann pointed out[7]
the operating voltages for side chain systems are not impractical and
it is possible to use the mesophase to glass transition (at T_g) to
store information written electro-optically in the nematic or
cholesteric polymer phases. This storage property is fundamental
to these polymer liquid crystal materials and appears to be their
most important feature. For such a storage device to be useful T_g
must be above ambient temperatures. This implies very slow response

times (minutes to hours) if the material is heated just above T_g due to the inherent high viscosity of the polymer or very high operating temperatures ($\sim200^\circ$C) if response times of less than a second are to be achieved in these nematic or cholesteric polymer liquid crystals. These features would not seem to be desirable for practical storage devices requiring contrast between written and unwritten regions.

Our approach reported recently[8-10] has been somewhat different. For a number of years we have also believed that the main potential for polymer liquid crystals would be in optical storage devices. The problem, however, was how to overcome the difficulties posed by the high polymer viscosity without destroying the storage property. Our novel approach has been to examine smectogenic side chain polymer liquid crystals, with T_g's well below ambient temperatures, when subject to electric or magnetic fields. As is well known for monomeric liquid crystals, the smectic phase is inherently a bi-stable storage medium when suitably addressed by thermal and electrical fields[11]. We have carried over some of the ideas from monomeric smectic storage devices and applied them to the polymer systems. The polymers used herein were synthesised at Hull[12] and are based on

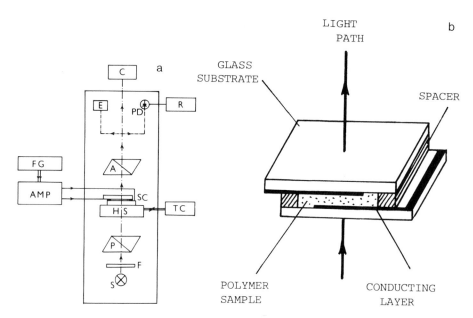

Figure 1 (a) Schematic of the optical arrangement where FG = function generator, AMP = high voltage amplifer, TC = temperature controller, R = fast chart recorder, C = camera, E = eyepiece, PD = photodiode detector, SC = sample cell, HS = hot stage, S = lamp source, F = optical filter and P, A = polariser and analyser, respectively. (b) Enlarged view of sample cell construction.

the highly successful monomeric cyanobiphenyls. Preliminary results that concentrate on a variety of electro-optic storage effects including guest-host and laser addressing on a variety of polymers are being published elsewhere. In this paper we will concentrate on the basic electro-optic storage effect in polysiloxane side chain polymers. Data will be presented for both a homopolymer and a copolymer system, and the main features or advantages of such materials will be emphasised.

EXPERIMENTAL

Apparatus

A schematic of the electro-optic apparatus is given in figure 1. The basic apparatus was an Olympus BH-2 transmission polarising microscope adapted to give simultaneous photodiode detection, photographic facilities and also direct sample observation. The sample cell shown enlarged in figure 1(b) was heated in a thermostatically controlled Lin-Kam 600 hot stage system. This hot stage could be adjusted to any temperature in a range -20° to $+600^{\circ}C$ with a stability of $0.1^{\circ}C$. Fields were applied across the sample using the amplified output from a low voltage function generator. The amplifier which produced voltages up to 400V (rms) over the frequency range of interest (0 to 100kHz) was an Electro-Optic Developments model LA10A. Voltages were switched across the cell using a suitable reed relay (not shown), and the photodiode output was amplified and recorded using a fast response chart recorder.

The sample cells were constructed from glass slides coated using conducting In/SnO_2 and etched to give a 2mm square electrode configuration. The glass slides were then spaced, at typically ~40μm, using a non conducting epoxy glue. Before assembly the cells were thoroughly degreased and then washed in isopropyl alcohol. They were then baked to remove residual solvents and filled using capillary action at temperatures ~10$^{\circ}C$ above the polymer liquid crystal's clearing temperature (T_c). On cooling back to ambient temperatures the texture was then uniform and air bubble free.

Materials

The two polymers under study and their transition temperatures are given in figure 2. The homopolymer, denoted PG253, was an alkyloxycyanobiphenyl side chain system and the copolymer (PG296) was a similar alkyloxycyanobiphenyl but also included a methylated benzoic ester side chain moiety grafted in equal percentage on to the fully saturated polysiloxane backbone. For both polymers the degree of functionality was 50, and the materials were prepared for us by Professor Gray's group at Hull University. One significant

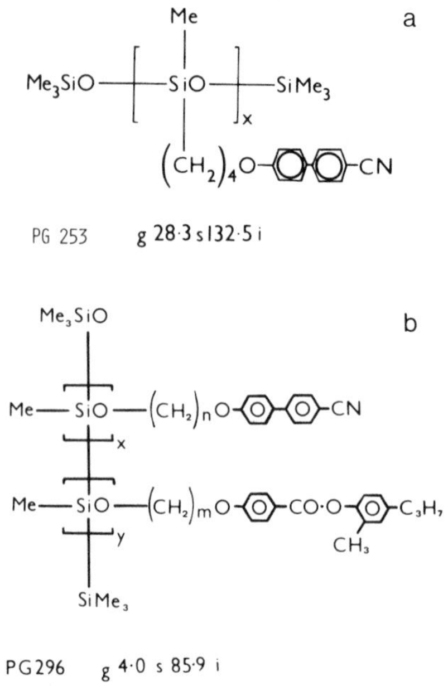

Me$_3$SiO—[—SiO—]$_x$—SiMe$_3$

Me (on the Si)

(CH$_2$)$_4$O—⬡⬡—CN

a

PG 253   g 28·3 s 132·5 i

Me$_3$SiO

Me—SiO—(CH$_2$)$_n$O—⬡—⬡—CN
     |$_x$

Me—SiO—(CH$_2$)$_m$O—⬡—CO·O—⬡—C$_3$H$_7$
     |$_y$ CH$_3$

SiMe$_3$

b

PG 296   g 4·0 s 85·9 i

Figure 2 Phase transitions and structure of (a) PG253 and (b) PG296.
g = glass transition, s = smectic phase and i = isotropic phase (from
DSC ref 12).

feature of both polymers is their high positive dielectric anisotropy
($\Delta\varepsilon$).

RESULTS

Before either monomeric or polymeric nematic and cholesteric
liquid crystals can be studied in electro-optic devices it is necessary to use some surface alignment agent to produce the desired
optical texture prior to field switching. This observation is also
true for monomeric smectic materials addressed using high or low
frequency electric fields. However in the case of smectic polymers
this condition is not evident. The existence and trajectory of the
polymer main chain appears to be sufficient to give a highly scattering non field aligned texture without the use of surface alignment
techniques. The removal of this prerequisite is an important simpli-
fication for this type of device and in the work reported below no
surface alignment techniques were used.

Homopolymer PG253

Thermo-optic analysis is carried out in the absence of aligning

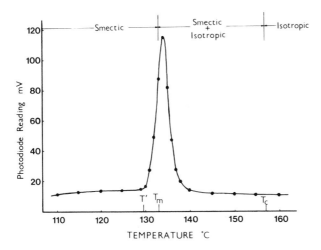

Figure 3   Thermo-optic analysis of PG253.

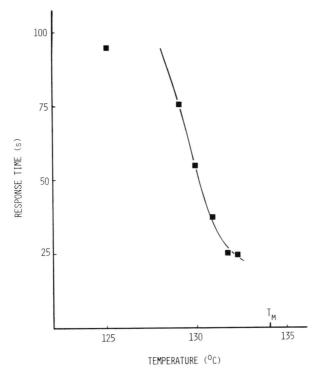

Figure 4   Optical response time for PG253, as a function of temperature, to an applied sinusoidal ac field of 207V rms, 4kHz.   Sample
thickness = 45+2μm.

fields across a sample and it consists of measuring the transmission of the sample, held between crossed polars, as a function of temperature. For the isotropic phase of a material the transmission will therefore be a minimum and the 'texture' will appear black. The thermo-optic analysis for PG253 is given in figure 3. At low temperatures (below 130°C) the material is in the smectic phase and the transmission of the turbid immobile scattering texture is very low. At ∿130°C the optical texture becomes mobile and the transmission increases by almost an order of magnitude. Near to T_m, the maximum in the DSC thermogram, the transmission starts to decrease. This corresponds to small dark isotropic regions appearing in the texture. Further increases in temperature lead to an increase in the dark 'isotropic' regions until the final traces of the smectic A focal conic texture disapppear at T_c, the so-called clearing temperature. The wide biphasic region between T_m and T_c appears to be a function of the polymer polydispersity (see later).

Electro-optic response times were measured by heating the material from ambient to the required temperature, applying a pulsed ac field and measuring the response time for the light

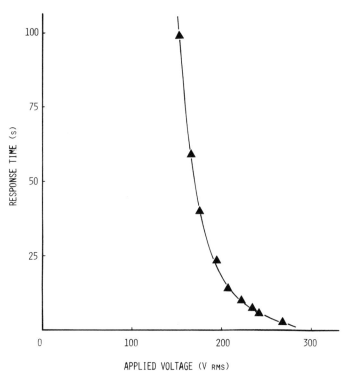

Figure 5 Optical response times for PG253 as a function of applied voltage (f = 2.5kHz sinewave). T = 129.7°C and thickness = 45 $\pm$ 2μm.

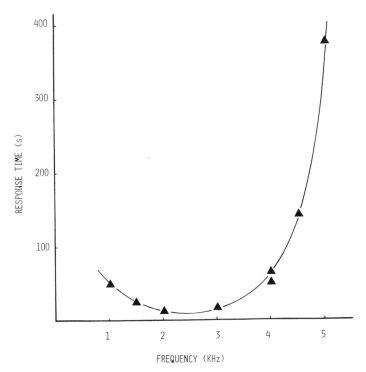

<u>Figure 6</u> Optical response time for PG253 as a function of frequency (V rms = 207 sinewave). T = 129.7°C and thickness = 45 ± 2μm.

transmission to drop to 50% of its initial value. [Note the optically clear isotropic texture will give a transmission minima using crossed polars]. The electro-optic response times are given for various temperatures, voltages and applied field frequencies in figures 4, 5 and 6 respectively. The significant observations are:

(i) despite the fact that the polymers are smectogenic in nature the timescales are of the order of seconds in the mobile or biphasic regions,

(ii) below T' in figure 3 the response times increase by an order of magnitude over a few degrees,

(iii) the response times show a distinct threshold effect decreasing rapidly with the increasing applied rms voltage, figure 5, for a fixed temperature and field frequency, and

(iv) the response time shows a minima in the 2 to 3kHz region, figure 6 for a constant temperature and fixed rms voltage.

<u>Copolymer PG296</u>

As in the case of PG253 the thermo-optic analysis, figure 7, was carried out without the use of aligning fields (surface or

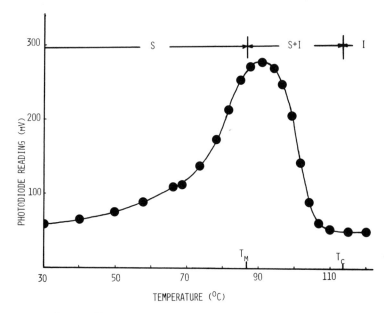

Figure 7 Thermo-optic analysis for PG296.

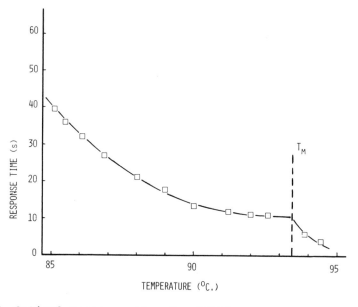

Figure 8 Optical response time for PG296 as a function of temperature. V rms = 300 sinewave at f = 2kHz. Thickness = 45 ± 2μm.

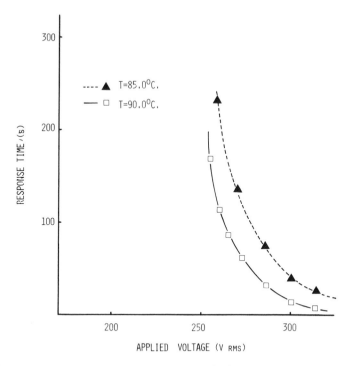

Figure 9 Optical response time for PG296 as a function of applied
voltage (f = 2kHz sinewave). Temperatures as indicated and thickness
= 45 ± 2μm.

electric). At low temperatures the scattering texture is immobile
(between 30° and 80°C) and the transmission through crossed polars
is low. For this polymer, however, the texture starts to become
mobile at around 80° and the transmission again increases markedly
(increasing by a factor of ∿5). At T_m the texture starts to disappear and black optically isotropic regions start appearing.
Gradually the transmission decreases until at T_c all smectic A like
texture disappears. This copolymer has a similar polydispersity to
the homopolymer and the thermo-optic trace again demonstrates a wide
biphasic region.

The electro-optic response curves under varying conditions of
temperature, voltage and frequency are given in figures 8, 9 and 10
respectively. The general trends observed for the homopolymer (PG253)
are repeated for the copolymer (PG296). The response time is of the
order of seconds near to T_m, it decreases rapidly with increasing
voltage, there is a distinct threshold voltage for ac fields and
there is an optimum frequency for the applied field. Application
of an ac square wave, rather than a sine wave, also alters the

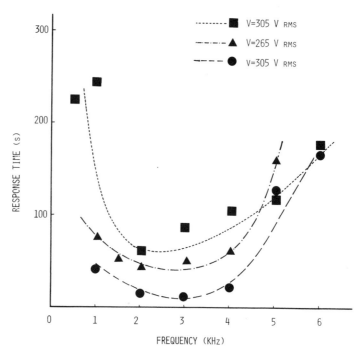

Figure 10 Optical response time for PG296 as a function of frequency and waveform (■ = square wave, ▲ and ● = sinewaves). T = 89.9°C and thickness = 45 ± 2μm.

response times. Under equivalent rms voltages the square wave increases the response time by a factor of ∿4 at the optimum frequency, figure 10. This interesting observation must reflect the frequency power spectrum of the square wave as opposed to the pure single frequency spectrum of the sine wave. For PG296 the threshold voltage also increases to ∿250V rms as compared with ∿150V rms for PG253. This is a reflection of the reduced $\Delta \varepsilon$ for PG296 due to the replacement of 50% of the cyanobiphenyl side groups by the methylated benzoic esters.

The optical microscopy of these polymer liquid crystals identifies both polymers as having smectic A phases only. However unlike monomer materials the evolution of a particular well defined texture may take several hours. As is shown in plates 1(a-c) the smectic fan texture can be seen to grow gradually over 15 hours on annealing just below T_m. All of the photomicrographs are for PG296 which exhibited similar optical textures as PG253. An important observation is that the evolution of the fan texture leads to a lower light scattering texture than for a non annealed sample in the absence of applied fields. Therefore for high contrast it is important not to anneal the polymer. Application of a low voltage dc electric field,

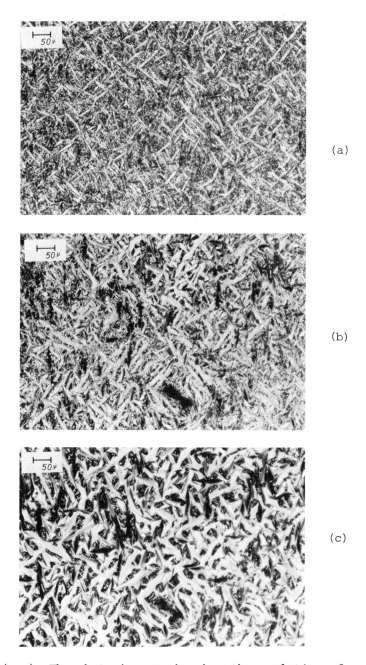

Plates 1(a-c) The photomicrographs show the evolution of a smectic
fan texture on annealing the copolymer (PG296) for (a) 6 hours,
(b) 12 hours and (c) 15 hours at a temperature of 84.3°C. Textures
observed using crossed polarisers.

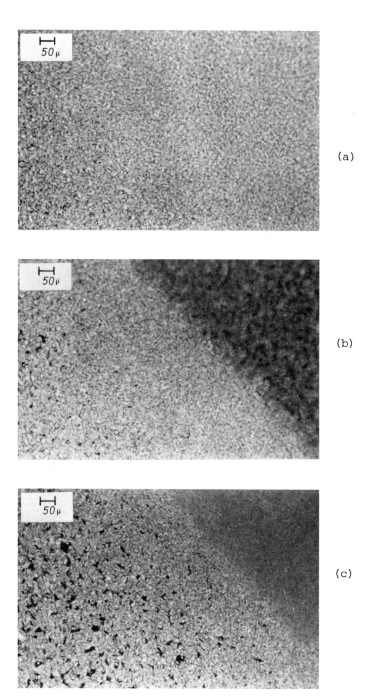

(a)

(b)

(c)

Plates 2(a-c) Photomicrographs showing the response of PG296 to a dc electric field. Texture with (a) no field applied, (b) 5 minutes after the application of a 100V dc voltage and (c) the texture 30 minutes after removal of the field. The electrode area is in the top right hand corner of the plates and T = 83°C. Crossed polarisers.

362

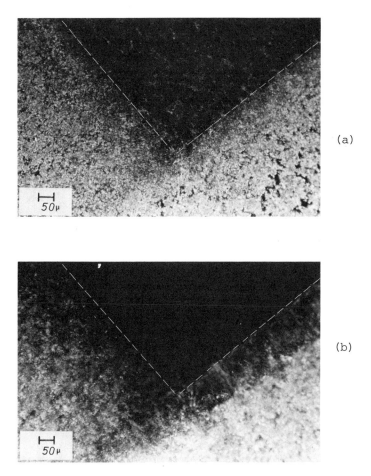

(a)

50μ

(b)

50μ

Plates 3(a and b)   Pseudohomeotropic alignment induced by (a) 300V dc
and (b) 350V dc in PG296.   The electrode area is shown by the dashed
lines, T = 86°C and textures observed through crossed polarisers.

plates 2(a-c),leads to a field induced scattering texture different
to that in the field free region.   This field induced scattering
texture demonstrates a storage effect as the texture still persists
some 30 minutes after removal of the field with the sample maintained
at the same high temperature, just below T_m.   On continued annealing
of the sample (for hours or days) this texture would gradually relax
away.   Application of higher dc voltages, plates 3(a and b) induce
a pseudo-homeotropic alignment.   With polarisers removed the texture
has a slightly grey appearance but with much lower optical density
than the field free region.

Heating the sample into the biphasic region gives the typical texture shown in plate 4(a). The dark areas correspond to the melted optically isotropic polymer. Application of a dc field induces a turbulence that smears out the biphasic texture, plate 4(b). This turbulent texture may then be stored on cooling the sample back to ambient temperatures, plate 4(c). Annealing the samples well into the biphasic region, plate 5, gives typical smectic bâtonnets and focal conics as well as the optically isotropic regions. These textures are stable at constant temperature. Application of an ac field at frequencies above 1kHz induces a true homeotropic texture. Under crossed polars this texture appears black and on cooling with no polarisers the field aligned region appears optically clear, plate 6(a). Verification that the alignment is homeotropic comes from the conoscopic observations plates 6(b and c). Use of a quarter wave plate, plate 6(c) establishes that the texture is uni-axial and positive, i.e. the optically anisotropic mesogenic segments are aligned perpendicular to the glass substrates, i.e. along the electric field direction. A marked feature of this texture is its clarity even at very wide viewing angles. Further this clear texture is stored in the smectic A phase above T_g and no deterioration of the optical quality has been observed in eighteen months. The contrast ratio between the clear and scattering regions may be as high as $10^3:1$ to $10^4:1$ as compared with $10:1$ for a monomeric smectic A material measured under identical conditions.

CONCLUSIONS

An electro-optic storage effect has been demonstrated in smecto-genic side chain polymer liquid crystals. The storage effect is observed above T_g and would appear to be a result of the high visco-sity and bistable nature of these systems. The optical response times at temperatures of $\sim 60^\circ C$ above ambient may be of the order of seconds or less. A number of factors that influence the response time have been identified:

(i)   Temperature. This influences the response time in two ways. In the first the viscosity is reduced for large values of $T-T_g$ and in the second the temperature determines whether the material is in the mobile or biphasic regions.

(ii)   Voltage. As well as exhibiting a threshold effect these materials show faster response times for higher voltages.

(iii) Field Frequency. The frequency of the ac field shows a distinct optimum. Below this optimum (i.e. low frequency ac or dc fields) the induced texture is of the scattering type whilst above this optimum the textures appear optically clear or homeotropic.

(iv)   Waveform. Pure sine waves give faster response times than harmonic containing square waves.

(v)   Thickness. Although not specifically described herein we have observed that thinner cells lead to both faster response times and lower threshold voltages.

364

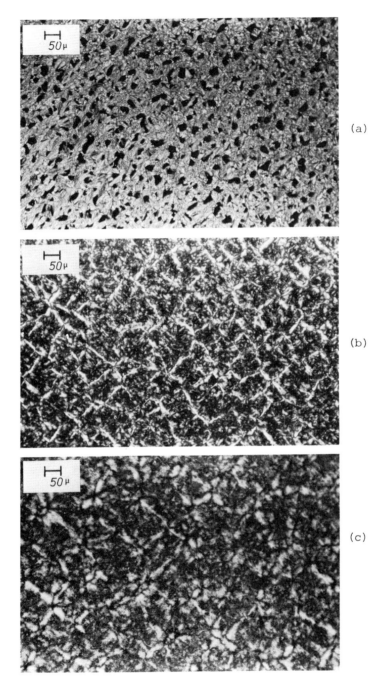

Plates 4(a-c)   Photomicrographs of PG296 in (a) the biphasic region
and (b) dc field induced turbulence both at 90ºC.   (c) field induced
texture of (b) stored at room temperature.   Crossed polarisers.

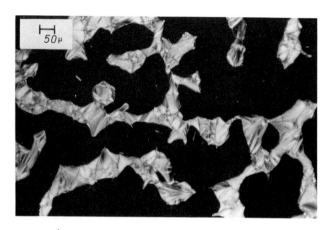

Plate 5 Smectic bâtonnets and focal conics induced in PG296 on annealing in the biphasic region at T = 99.1°C. Crossed polars.

    (vi) Polydispersity. The polymers studied herein are fairly polydisperse (i.e. Mw/Mn ∿ 2) and this is reflected in the width of the mobile/biphasic region. As the response time varies with temperature in this region then the polydispersity must also influence this response time.
    (vii) Dielectric Anisotropy. The polymer with the highest $\Delta\varepsilon$ shows the lower threshold voltage and therefore for equal fields the faster response times.

    The above variables appear to determine the response time for the polymers studied herein. Judicious combination and variation of these factors have allowed us to achieve response times on the 100ms time scale whilst at the same time maintaining the optical storage effect. It is evident that further changes in the polymer properties i.e. length of the polysiloxane backbone, flexible spacer length, degree of side chain substitution, and changes in the type of side chain, i.e. optical and dielectric properties, chemical structure etc, will lead to significant improvements in and control over their electro-optic switching properties. These parameters when combined with those summarised above lead to the exciting possibility of new fast response electro-optic storage devices based on single materials engineered for a specific application.

ACKNOWLEDGMENTS

    The authors are very grateful to Professor G W Gray, Dr D Lacey and Dr P Gemmell for kind provision of the polymer samples and DSC data. One of us (HJC) thanks the SERC for the provision of the research grants under which this work was carried out and RS was employed.

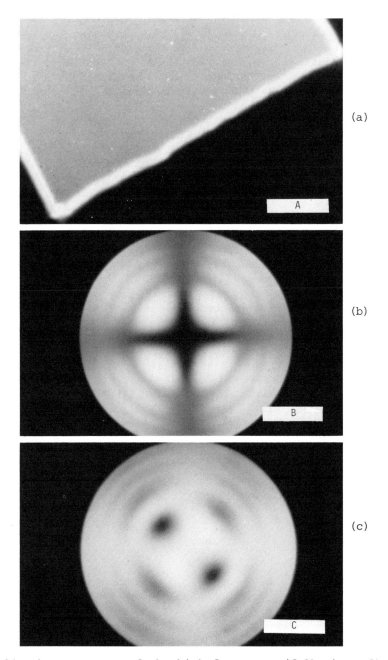

Plates 6(a-c) A-storage of the high frequency (f>1kHz) ac field
induced texture on cooling PG296 to room temperature. The light
region depicts the electrode area (with no polarisers). B and C.
Conoscopic observation of the electrode area under (b) crossed
polars and (c) including a λ/4 waveplate.

367

1. W R Krigbaum and H J Lader. Electric field-induced flow insta-
    bilities in low molecular weight and polymeric nematics. <u>Mol.
    Cryst.Liq.Cryst</u>. 62: 87 (1980).
2. A Blumstein, S Vilasagar, S Ponrathnam, S B Clough and
    R B Blumstein. Nematic and Cholesteric Thermotropic Polyesters
    with Azoxybenzene Mesogenic Units and Flexible Spacers in the
    Main Chain. <u>J.Poly.Sci: Poly.Phys</u>. 20: 877 (1982).
3. H Finkelmann, H Ringsdorf and J H Wendorff. Model Consideration
    and examples of Enantiotropic Liquid Crystalline Polymers.
    <u>Makromol.Chem</u>. 179: 273 (1978).
4. H Ringsdorf and A Schneller. Synthesis, Structure and Proper-
    ties of Liquid Crystalline Polymers. <u>Brit.Pol.J</u>. 13: 43 (1981).
5. H Finkelmann, U Kiechle and G Rehage. Liquid Crystalline Side
    Chain Polymers in the Electric Field. <u>Mol.Cryst.Liq.Cryst</u>. 94:
    343 (1983).
6. H Ringsdorf and R Zentel. Liquid Crystalline Side Chain Poly-
    mers and their Behaviour in the Electric Field. <u>Makromol.Chem</u>.
    183, 1245 (1982).
7. H Finkelmann, Liquid Crystalline Side Chain Polymers. <u>Phil.
    Trans.R.Soc.Lond</u>. A309: 105 (1983).
8. R Simon and H J Coles. Investigations of Smectic Polysiloxanes
    1 - Electric Field Induced Turbulence. <u>Mol.Cryst.Liq.Cryst</u>.
    102 (Letts): 43 (1984).
9. H J Coles and R Simon. Investigations of Smectic Polysiloxanes
    2 - Field Induced Director Reorientation. <u>Mol.Cryst.Liq.Cryst</u>.
    in press (1984).
10. H J Coles and R Simon. Electro-Optic Effects in a Smectogenic
    Polysiloxane Side Chain Liquid Crystal Polymer, in "Recent
    Advances in Liquid Crystalline Polymers", L L Chapoy (ed),
    Applied Science Publishers Ltd (Elsevier) London (1984).
11. E P Raynes, Electro-Optic and Thermo-Optic Effects in Liquid
    Crystals. <u>Phil.Trans.R.Soc.Lond</u>. A309: 167 (1983).
12. G W Gray, P A Gemmell and D Lacey, private communication.

PHASE SEPARATION OF POLYMERIC LIQUID CRYSTALS BASED ON CELLULOSE

Derek G. Gray

Pulp and Paper Research Institute of Canada, and
Department of Chemistry, McGill University, Montreal
Quebec H3A 2A7

The papers presented in this symposium give some indication
of the wide variety of polymers which are now known to form liquid
crystalline phases. Polymeric liquid crystals are usually classified according to the mesophase structure e.g., nematic, cholesteric, smectic A, etc.). However, these classes are quite broad.
For example, the cholesteric lyotropic phases formed by synthetic
polypeptides in suitable solvents differ markedly from the cholesteric thermotropic phases formed from silicone polymers with cholesteryl ester side chains. In particular, the driving forces
behind the formation of the mesophases are quite different for
these two examples, being essentially due to chain stiffness in
the first case and to anisotropic dispersion force interactions in
the second case. It may therefore be useful to classify polymeric
liquid crystals according to the polymer chain structure.

Two distinct classes of polymeric chain structures may be
envisaged. In one class of mesophases, the polymers contain mesogenic groups, either in the main chain or in side chains, linked
by flexible spacers. The mesogenic groups may resemble conventional low molecular weight liquid crystals or may be rigid chain
segments. Examples of segmented chain mesophases include the
side chain cholesteryl derivatives of methacrylates and polysiloxanes and main chain aromatic polyesters containing flexible aliphatic chain segments. Polymers with mesogenic groups in the main
chain or side chains are conveniently classified as segmented
chain mesophases. The second class of mesophase is derived from
polymers with uniformly stiff or semiflexible backbones which are
chemically homogeneous on a scale very much shorter than their
contour or persistence lengths. This class may be called uniform
chain mesophases. Examples of the latter class are the synthetic

369

polypeptides, aromatic polyamides and cellulose derivatives.

Many cellulose derivatives form lyotropic liquid crystals in suitable solvents and several thermotropic cellulose derivatives have been reported (1-3). Cellulosic liquid crystalline systems reported prior to early 1982 have been tabulated (1). Since then, some new substituted cellulosic derivatives which form thermotropic cholesteric phases have been prepared (4), and much effort has been devoted to investigating the previously-reported systems. Anisotropic solutions of cellulose acetate and triacetate in trifluoroacetic acid have attracted the attention of several groups. Chiroptical properties (5,6), refractive index (7), phase boundaries (8), nuclear magnetic resonance spectra (9,10) and differential scanning calorimetry (11,12) have been reported for this system. However, trifluoroacetic acid causes degradation of cellulosic polymers; this calls into question some of the physical measurements on these mesophases, because time is required for the mesophase solutions to achieve their equilibrium order. Mixtures of trifluoroacetic acid with chlorinated solvents have been employed to minimize this problem (13), and anisotropic solutions of cellulose acetate and triacetate in other solvents have been examined (14,15). The mesophase formed by (hydroxypropyl)cellulose (HPC) in water (16) is stable and easy to handle, and has thus attracted further attention (10,11,17-19), as has the thermotropic mesophase of HPC (20). Detailed studies of mesophase formation and chain rigidity for HPC in dimethyl acetamide (21) and for the benzoic acid ester of HPC in acetone and benzene (22) have been published. Anisotropic solutions of methylol cellulose in dimethyl sulfoxide (23) and of cellulose in dimethyl acetamide/ LiCl (24) were reported. Cellulose tricarbanilate in methyl ethyl ketone forms a liquid crystalline solution (25) with optical properties which are quite distinct from those of previously reported cholesteric cellulosic mesophases (26).

The first useful and tractable theories for the formation of uniform chain polymeric liquid crystals have been developed by Flory and coworkers (27), based on a lattice consideration of the asymmetry of molecular shape. A theory for hard non-interacting rods (28,29) predicts that rods with axis ratios greater than 6.4 will form a liquid crystalline phase in the absence of diluent. The presence of a non-interacting diluent increases the axis ratio required to form coexisting isotropic and nematic phases (28,29); the effect of changing isotropic interactions between rod segments and diluent is small for good solvents (28). However, if there is a substantial anisotropy of polarizability of the rod segments, then this anisotropic segment – segment interaction (characterized by an energy T) stabilizes the mesophase for axis ratios less than 6.4 (30). The effect of a diluent is predicted to reduce this energy of interaction to T v_2, where v_2 is the volume fraction of the rod (31). In the classical Maier-Saupe theory for low molar mass liquid crystals, the asymmetry of molecular shape is ignored and only the asymmetric energy of interaction is considered (27).

Most liquid crystalline polymers are not rod-like, but possess some chain flexibility. The lattice theories may be applied to the separation of a liquid crystalline phase from concentrated solutions of semiflexible polymers. Flory (32) showed that the critical concentration for chains composed of a set of freely jointed rods depends almost completely on the axis ratio of the rods and not on the length of the chain. Thus if the semiflexible polymer may reasonably be modelled as a set of freely jointed rods of known length and diameter, then the critical concentration may be estimated. A value for the rod length is required; for long chains obeying random flight statistics the Kuhn segment length has been suggested (33). However, the conformation of semiflexible or stiff chain polymers is often better modelled by the worm-like chain (34) which for contour lengths much shorter than its persistence length approaches a rigid rod, and for contour length much longer than its persistence length follows random flight statistics. The worm-like chain also approximates more detailed rotational isomeric state models (35). The stiffness of a worm-like chain may be characterized by a persistence length, q, or by an equivalent Kuhn segment length, $k_w = 2q$. The value of k_w, from dilute solution measurements on fractions of the stiff chain polymer in a given solvent (36) may then be taken as the rod length in Flory's treatment of the phase separation of freely jointed rods, and the predicted values for the critical concentration may be compared to experimental values measured in the same solvent. However, the worm-like chain model (where chain flexibility is distributed uniformly along the chain) is very different from the freely jointed rod model (where chain flexibility occurs only at discrete points on the chain); a phase separation theory based on a worm-like chain model rather than the freely jointed rod model would seem more realistic for semiflexible chains. Ronca and Yoon (37) developed the theory for rod-like particles (29) to cover the case where the polymer chain is represented by a worm-like chain, modified to incorporate a maximum permissible curvature. Extensions of the modified worm-like chain model to consider the effects of chain length, diluent and isotropic and anisotropic interactions will be most useful.

An alternative approach to phase separation for rigid and freely jointed rods (38) and for worm-like chains (39,40) has been proposed by Khokhlov and coworkers. Their work is based not on lattice models but on Onsager's theory for the phase separation of rigid rods. These theories are valid only for very low concentrations, so their applicability to the phase separation of cellulose based mesophases may be questionable.

Recently, ten Bosch, Sixou and coworkers have extended theories for low molar mass nematic liquid crystals to cover aspects of the phase separation of semiflexible polymers (41-43). The polymer is taken to be an elastic worm-like chain with orientation dependent Van der Waals interactions. These theories have been applied to the phase separation of (hydroxypropyl)cellulose – dimethyl acetamide mesophases (17).

How well do the current theories predict the phase separation of cellulosic and other uniform chain polymer mesophases? The experimental observations by many workers may be summarized as follows. Most cellulosic mesophases form at critical volume fractions of polymer ranging from 0.3 to 0.5 for high molecular weight samples at room temperature. The critical volume fractions decrease with increasing molecular weight to approach these asymptotic values. The critical volume fractions for a given polymer and solvent increase with temperature; with high boiling solvents, the critical volume fraction approaches unity at around 190° ± 30° C, depending on the derivative. For a given polymer, the critical volume fraction is dependent on the solvent; in general, solutions in highly polar acidic solvents such as trifluoroacetic acid form mesophases at lower critical concentrations than in simple organic solvents such as acetone. Furthermore, whereas heavily substituted cellulosic derivatives with long flexible side chains form mesophases easily in a wide range of solvents, polymers such as cellulose trinitrate, triacetate or tricarbanilate seem to form mesophases with some specific solvents but not with others.

These observations are generally in line with the view that chain stiffness is the major factor in mesophase formation. The critical volume fractions are roughly in accord with values predicted by the freely jointed chain model with Kuhn segment lengths from dilute solution measurements. A molecular weight dependence for the critical polymer concentration would of course be expected when the chain contour length approached the order of magnitude of the Kuhn segment length. The increase in critical concentration with temperature is in line with the well known decrease in chain dimensions with temperature of cellulosic polymers (44). The effect of solvent on critical concentration is generally mirrored by the effect of solvent on chain stiffness, as indicated qualitatively by viscosity measurements (45).

To test the theories quantitatively, careful measurements on well-characterized fractions of polymer are required. The contour length, persistence length and chain diameter may be found from dilute solution measurements on fractions in a given solvent. The critical concentrations for mesophase formation in the same solvent may be calculated from the contour length, persistence length and chain diameter by means of the theories for freely jointed or worm-like chains mentioned above. Results have been reported for the benzoic acid ester of (hydroxypropyl)cellulose in acetone and benzene at 25° C (22), for (hydroxypropyl)cellulose in dimethyl acetamide at temperatures from 20° C to 130° C and for (acetoxy-propyl)cellulose in dialkyl phthalate from 25° C to 170° C (46). In the first of these references (22) the measured critical concentration for high molecular weight polymer lay between the high values predicted by the lattice theory and the low values predicted by the Onsager theories for freely jointed rods. (The experimental values also fell between the Onsager predictions for freely jointed and for worm-like chains.) Conio et al. noted (21) that the lattice theories predicted a critical concentration larger than that observed experimentally. They also concluded that the major factor in the temperature dependence of the critical concentration was the change in the chain stiffness, but that anisotropic interactions made some small contribution (21). These conclusions also apply to the (acetoxypropyl)cellulose - dibutyl phthalate system (46). In both of these systems, the composition and molecular weight of the anisotropic and isotropic phases in the two-phase region showed that some fractionation had occurred, but further work is required to define and interpret behavior in this part of the phase diagram.

The phase separation of cellulosic mesophases is generally in accord with the following speculative and qualitative ideas. The cellulosic chain in dilute solution is pictured as being made up of relatively stiff helical regions with regular conformations around the anhydroglucose links, interspersed with regions of one or more anhydroglucose links with alternative 'flexible' conformations. The distribution of the flexible conformations along the chain is statistical, rather than fixed by chain chemistry as in the case of segmented chain mesophases. The transition between regular helical and alternative flexible conformations must be a function of solvent and temperature, but for most cellulosics there is no evidence for the extreme cooperativity observed for the helix - coil transition of polypeptide chains. The isotropic - liquid crystalline phase transition is governed predominantly by the chain stiffness, which according to this picture is a function of the average length of the hypothetical helical chain segments in the isotropic phase. Anisotropic chain - chain interactions may play a role for very short chains, at elevated temperatures, or for polymers with very polar or polarizable substituents rigidly attached to the cellulose chain. The helical chain segments, oriented at random in the isotropic phase, become oriented about a preferred direction in the mesophase, so that individual chains tend to straighten out and the whole structure takes up the characteristic supermolecular helicoidal arrangement of the cholesteric state. The helical chain configuration may resemble that exhibited by many cellulose derivatives in the crystalline state, but in the liquid crystalline state there is no interchain lateral order and the orientational order is imperfect. The tendency of cellulosics to crystallize directly without displaying the liquid crystalline state may be inhibited by the presence of long, flexible and somewhat non-uniform side chains, or by the use of solvents with strong specific affinity for the cellulose derivative. It must be emphasized that the picture for mesophase formation in this paragraph is speculative and suffers from weaknesses, not the least being the absence of any convincing evidence for helical cellulosic chain conformations in dilute solutions or in the mesophase!

ACKNOWLEDGEMENTS

The support of the Natural Sciences and Engineering Council of Canada is gratefully acknowledged. I thank Drs. Cifferi, Sixou, Yoon and Zugenmaier for preprints of their work.

REFERENCES

1. D. G. Gray, J. Appl. Polym. Sci., Appl. Polym. Symposium 37:179 (1983).
2. R. D. Gilbert and P. A. Patton, Prog. Polym. Sci. 9:115 (1983).
3. S. P. Papkov, Yu. Belousov, and V. G. Kulichikhin, Khim Volokna 3:8 (1983); Chem. Abs. 99:55204u.
4. S. N. Bhadani and D. G. Gray, Mol. Cryst. Liq. Cryst. 99:29 (1983).
5. G. H. Meenten and P. Navard, Polymer 23:1727 (1982); ibid 24:815 (1983).
6. P. Sixou, J. Lematre, A. ten Bosch, J.-M. Gilli, and S. Dayan, Mol. Cryst. Liq. Cryst. 91:277 (1983).
7. G. H. Meenten and P. Navard, Polymer 23:483 (1982).
8. S. Dayan, P. Maissa, M. J. Vellutini, and P. Sixou, J. Polym. Sci., Polym. Letters Ed. 20:33 (1982).
9. D. L. Patel and R. D. Gilbert, J. Polym. Sci., Polym. Phys. Ed. 20:1019 (1982).
10. S. Dayan, F. Fried, J.-M. Gilli, and P. Sixou, J. Appl. Polym. Sci., Appl. Polym. Symposium 37:193 (1983).
11. P. Navard, J. M. Haudin, S. Dayan, and P. Sixou, J. Appl. Polym. Sci., Appl. Polym. Symposium 37:211 (1983).
12. P. Navard and J. M. Haudin, Polymer Preprints 24(2):267 (1983).
13. D. L. Patel and R. D. Gilbert, J. Polym. Sci., Polym. Phys. Ed. 21:1079 (1983).
14. S. M. Aharoni, J. Macromol. Sci. Phys. B21:287 (1982).
15. B. Yu. Yunusov, O. A. Khanchich, A. T. Serkov, and M. T. Primkulov, Vysokomol. Soedin., Ser. B 25:395 (1983); Chem. Abs. 99:89783x.

16. R. S. Werbowyj and D. G. Gray, <u>Mol. Cryst. Liq. Cryst. (Letters)</u> 34:97 (1976).
17. M. J. Seurin, A. ten Bosch, and P. Sixou, <u>Polym. Bull. (Berlin)</u> 9:450 (1983).
18. K. Shimamura, <u>Makromol. Chem. Rapid Commun.</u> 4:107 (1983).
19. T. Asada, <u>in</u>: "Polymer Liquid Crystals," A. Cifferi, W.R. Krigbaum, and R.B. Meyer, eds., Academic Press, p. 247 (1982).
20. S. Suto, J. L. White, and J. F. Fellers, <u>Rheol. Acta</u> 21:62 (1982).
21. G. Conio, E. Bianchi, A. Cifferi, A. Tealdi, and M. A. Aden, <u>Macromolecules</u> 16:1264 (1983).
22. S. N. Bhadani, S.-L. Tseng, and D. G. Gray, <u>Makromol. Chem.</u> 184:1727 (1983).
23. P. A. Patton and R. D. Gilbert, <u>J. Polym. Sci., Polym. Phys. Ed.</u>, 21:515 (1983).
24. C. L. McCormick, P. A. Callais, and B. H. Hutchinson, Jr., <u>Polymer Preprints</u> 24(2):271 (1983).
25. P. Zugenmaier and U. Vogt, <u>Makromol. Chem. Rapid Commun.</u> 4:759 (1983).
26. R. S. Werbowyj and D. G. Gray, <u>Macromolecules</u>, in press (1984).
27. P. J. Flory, <u>in</u>: "Polymer Liquid Crystals," A. Cifferi, W.R. Krigbaum, and R.B. Meyer, eds., Academic Press, p. 103 (1982).
28. P. J. Flory, <u>Proc. Roy. Soc. London, Ser. A</u> 234:73 (1956).
29. P. J. Flory and G. Ronca, <u>Mol. Cryst. Liq. Cryst.</u> 54:289 (1979).
30. P. J. Flory and G. Ronca, <u>Mol. Cryst. Liq. Cryst.</u> 54:311 (1979).
31. M. Warner and P. J. Flory, <u>J. Chem. Phys.</u> 73:6327 (1980).
32. P. J. Flory, <u>Macromolecules</u> 11:1119 (1978).
33. R. S. Werbowyj and D. G. Gray, <u>Macromolecules</u> 13:69 (1980).
34. O. Kratky and G. Porod, <u>Rec. Trav. Chim. Pays-Bas</u> 68:1106 (1949).
35. M. Mansfield, <u>Macromolecules</u> 14:1822 (1981).
36. H. Yamakawa and M. Fujii, <u>Macromolecules</u> 7:128 (1974).
37. G. Ronca and D. Y. Yoon, <u>J. Chem. Phys.</u> 76:3295 (1982); 80:925 (1984).
38. A. Yu. Grosberg and A. R. Khokhlov, <u>Adv. Polym. Sci.</u> 41:53 (1981).
39. A. R. Khokhlov and A. N. Semenov, <u>Physica</u> 108A:546 (1981).
40. A. R. Khokhlov and A. N. Semenov, <u>Physica</u> 112A:605 (1982).
41. A. ten Bosch, P. Maissa, and P. Sixou, <u>J. de Phys.</u> 44:105 (1983).
42. A. ten Bosch, P. Maissa, and P. Sixou, <u>J. Chem. Phys.</u> 79:3462 (1983).
43. A. ten Bosch, P. Maissa, and P. Sixou, <u>Polymer Preprints</u> 24(2):246 (1983).

44. P.J. Flory, O. K. Spurr, Jr., and D. K. Carpenter, _J. Polym. Sci._ 27:231 (1958).
45. S. M. Aharoni, _J. Macromol. Sci.-Phys._ B21:105 (1982).
46. G. V. Laivins, Ph.D. Thesis, McGill University, Montreal, to be submitted (1984).

LIQUID CRYSTALLINE POLYMER SOLUTIONS AND MIXTURES

M.J. Seurin, J.M. Gilli, F. Fried,
A. Ten Bosch, and P. Sixou

Laboratoire de Physique de la Matière Condensée
Parc Valrose, 06034 Nice Cedex, France

Polymer solutions and blends have been extensively studied, motivation being both the fundamental and technological interest[1,2]. When one component is a liquid crystal polymer, synergy could produce exceptional properties due to the anisotropy of the mesomorphic polymer. It is important to determine the conditions for the existence of either a uniform, mesomorphic phase or an incompatible biphasic system. In such studies, the rigid or semi-rigid character of the chains and the anisotropic interaction between the chains play a role. The presence of mesomorphic macromolecules is "labeled" by their ability to align under certain conditions, a fact which can be exploited in the experimental determination of phase diagrams.

Several other specific interactions (isotropic, formation of hydrogen bonds...) can also be of importance. In order to better understand the different mechanisms involved we have studied systems of increasing complexity :

a) Thermotropic polymers

b) Mesomorphic polymers in an isotropic or anisotropic solvent.

c) Mixture of a mesomorphic and an non-mesomorphic polymer.

d) Mixture of a mesomorphic and an non-mesomorphic polymer in an isotropic solvent. Polymers investigated were thermotropic or lyotropic cholesterics (cellulosic derivatives [3,4]) or nematics (aromatic polyesters such as DDA-9[5] or polyalkylisocyanates [6].)

Comparison of experimental results with the theoretical

predictions is made. We focus also on the role of the molecular weight, a well-known, fundamental parameter in the non-mesomorphic polymer mixtures.

The mesomorphic caracter of the mixture can be observed in several ways. The simplest is microscopy between crossed polarizers. If the solution is anisotropic, a characteristic texture appears, essentially determined by the surface alignment. Alignment can be modified (but with less success) using the same methods developed for the small molecule liquid crystals. Roughly, the textures are similar to these observed in small molecule liquid crystals. However some specific caracteristics have been observed (dependance on the molecular weight, rigidity...). An example of polygonal field texture (well-known in the case of small molecule cholesterics[7]) is shown in fig. 1.

Fig. 1. Polygonal field texture of a cholesteric solution of hydroxypropyl cellulose (HPC) in acetic acid.

The alternatively black and white retardation lines are connected to the cholesteric pitch or more precisely to the anisotropy of polarizability of the macromolecules and to their orientation relative to light propagation[8]. Figure 2 illustrates how in the case of the same polymer, these finger print patterns are modified by dilution of the cholesteric polymer in a simple solvent. A quantitative measurement of the disappearance of the texture at the order-disorder transition temperature can be obtained by recording the light

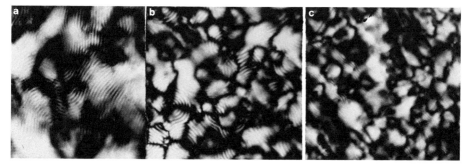

Fig. 2.   Fingerprint patterns with visible retardation lines
          in HPC solution in acetic acid at various polymer con-
          centration a) c = 32,6%   , b) c = 45% , c) c = 49,5%.

intensity (with a photomultiplier) when the sample is heated or
cooled.
An example of recorded curve is shown in fig. 3a. Fig. 3b corresponds to the case of a small molecule liquid crystal (PAA). The intensity transmitted by the sample varies abruptly at the transition
temperature ; the observed difference of transition temperature during cooling and heating is small. Contrary observations are made
in a polymer (DDA9)[x] where there is a relatively large biphasic
area. To describe the transition, we trace only the temperature at
which the polymer becomes completely isotropic (shown by an arrow).
The variations of this temperature can be compared with pseudo-transition temperature as defined by Brochard[9] and calculated[10], al
though deviations are expected for large biphasic separation.

EXPERIMENTAL RESULTS

Thermotropic polymer

        Transition temperatures obtained with well-fractionated poly-
mers show that there is an increase of these temperatures with molecular weight for the low molecular weight and a saturation at
high molecular weights. An example is shown fig. 4 in the case of
polynonylisocyanate. The same behaviour is observed for other semi-
rigid polymers (benzoic acid ester of hydroxypropyl cellulose) or
mesomorphic polymers of alternating rigid mesogenic part and flexible spacers[5] or comb-like polymers[11], and was predicted in the worm-
like chain model[12].

_____

[x] We are very grateful to professors A. and R. Blumstein for
     the kind gift of the sample.

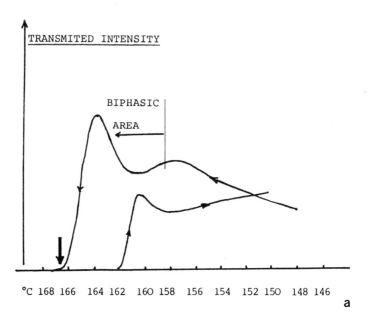

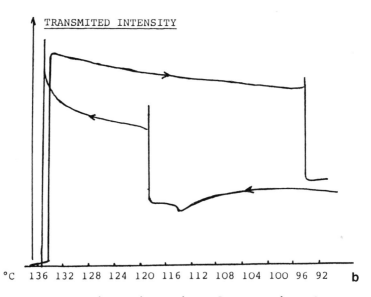

Fig. 3. Transmitted intensity of a nematic polymer
DDA [9] (fig. 3 a) and a small molecule liquid
crystal (3b) (PAA) as a function of temperature.

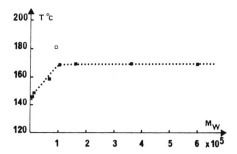

Fig. 4. Nematic-Isotropic transition temperature of polynonylisocyanate as a function of molecular weight.

Lyotropic polymer in isotropic solvent

Two different cases occur depending on the presence of strong
interactions with solvent.

In the case of relatively inert solvent of quasi-spherical molecules, the transition temperature in the mixture is given simply
by a lowering of the pure polymer transition temperature as the polymer concentration decreases. The concentration law of the transition temperature seems to depend on the variation of the persistence
length with temperature[13]. In general a persistence length varies
as $1/T$ and square weight concentration dependence of the transition
temperature will give a fairly good fit. An example of observed
a) and predicted b) curves for different molecular weights is shown
figure 5.

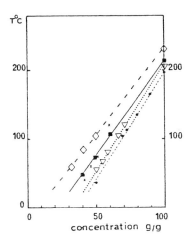

Fig. 5a
Experimental phase
diagram for hydroxypropyl
cellulose in dimethyaceta-
mide for different molecular
weights M_w

◇ M_w = 1000000

■ M_w = 300000

▽ M_w = 100000

. M_w = 60000

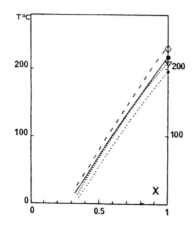

Fig. 5 b.
Theoretical phase diagrams
for hydroxypropyl cellulose
in dimethylacetamide for different molecular weights.

-- M_W = 1000000

— M_W = 300000

... M_W = 100000

.. .. M_W = 60000

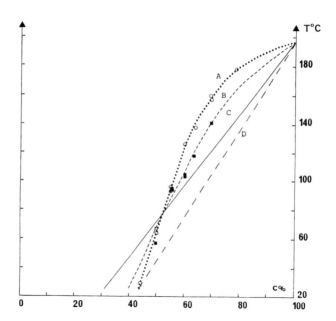

Fig. 6. Phase diagrams of solutions of : A) HPC/Ethylene glycol
B) HPC/diethylene glycol. C) HPC/Polyethylene glycol
M_W = 300   D) HPC/dimethylacetamide.

When the polymer-solvent interactions are strong (as in a hydrogen bonding solvent) the phase diagram shows distinct curvatures
as illustrated in fig. 6.

Lyotropic polymer in an anisotropic solvent

If the polymer and the solvent are appropriately choosen there is complete miscibility. An example is the case of the mixture of DDA-9 in PAA. Miscibility is promoted by steric similarity and the mesogenic rigid part of the polymer is identical to the rigid part of the small molecule liquid crystal. The transition temperature of the pure polymer is greater than that of the small molecule liquid crystal and the mixture shows the predicted[10] continuous intermediate temperature between the two pure components (fig. 7).

Mixture of a mesomorphic and a non-mesomorphic polymer

An increase of the rigidity of the chains of one component of the mixture increases the tendency to segregation, depending on the relative molecular weights of the two components and on the importance of specific interactions (Flory) (fig. 8). On increasing the molecular weight of one component of the mixture, incompatibility and segregation can occur[14].

Mixture of a mesomorphic and a non mesomorphic polymer in a solvent

In addition to the previous effects of molecular weight, specific interactions between polymers and strong interactions (sometimes preferential) of one or the other polymer with the solvent can occur. Some very interesting effects can be obtained.

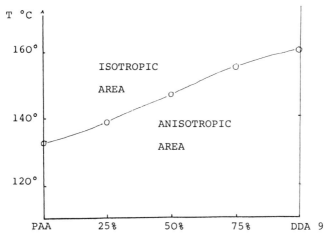

Fig. 7 . Isotropic-Anisotropic transition in a mixture of PAA/DDA9.

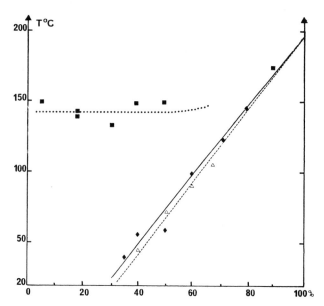

Fig. 8. Phase diagrams of HPC in polyethylene glycol (PEG)
♦ HPC 60000/PEG 300   △ HPC 60000 / PEG 600
■ HPC 60000/PEG 1000

1) If in a mixture of a lyotropic polymer in a solvent, we replace the solvent partially by a flexible polymer with preferential affinity for the solvent, segregation leads to enclosures of concentrated semirigid polymers. The transition temperature is consequently relatively high (fig. 9).

2) In the case of existence of a "gel" (as in solution of hyroxypropyl cellulose in water) in addition to the previous effects, we find a modification of the gel formation when a part of solvent is replaced by a flexible polymer. (fig. 10).

IN CONCLUSION

Mesomorphic phases can be obtained in various cases (thermotropic, lyotropic in a simple solvent or in a liquid crystal solvent, mixtures of polymers). As predicted, the effect of the molecular weight on phase diagrams and on segregation phenomena is observed. The experimental phase diagrams can be relatively well described by the theory in simple cases. Specific interactions should be included in future work as well as a more detailed examination of biphasic separation[10]. Further experimental studies on the biphasic regions are in progress.

This paper has focused on phase diagrams. Viscosity and light scattering measurements in dilute systems have also been used

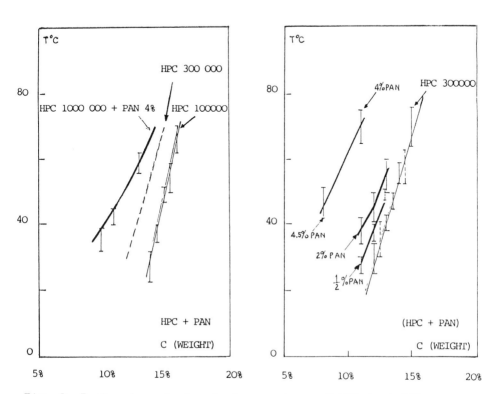

Fig. 9. Isotropic-anisotropic temperature of HPC in trifluoroacetic acid (TFA) and of hydroxypropylcellulose + polyacrylo-
nitrile (HPC + PAN) in TFA.

to determine chain flexibility[15,16]. In concentrated solutions,
systematic studies of the NMR spectrum[17], viscosity[18],DSC[19], cholesteric pitch[20] for semi-rigid polymers and the effects of electric and magnetic fields[21] give further insight in these systems.

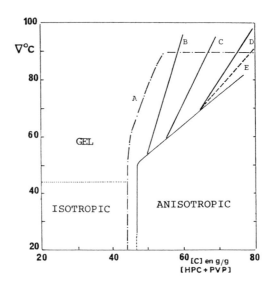

Fig. 10. Phase diagram of HPC/H$_2$O and modifications due to
polyvinyl (p) pyrrolidone (PVP) A) HPC 60000/PVP 10000
(c) PV P = 10%   B) HPC 60000 / PVP 10000 (c) HPC = 50%
C) HPC 60000 / PV P10000   (c) HPC = 55%, D) HPC 60000/
PVP 10000
(C) HPC = 65%   E) HPC 60000 / PVP 40000   (C) HPC = 65%

REFERENCES

1. D.R. Paul, S. Newman, Polymer blends, vol. 1 and 2,
   Acad. Press, (1978).
2. O. Olabisi, L.M. Robeson, M.T. Shaw, Polymer-Polymer
   Miscibility Acad. Press, 1979.
3. S. Dayan, P. Maïssa, M.J. Vellutini, P. Sixou, J. Polym.
   Sci, 20, 33, (1982).
4. R.R. Werbowy , D.G. Gray, Macromolecules 13, 69, (1980).
5. A. Blumstein, S. Vilasagar, S. Ponratham, S.B. Clough,
   R.B. Blumstein, J. Polym. Sci., 20, 877, (1982).
6. S.M. Aharoni, Macromolecules, 12, 537, 1979.
7. Y. Bouligand, J. Phys. (France), 33, 715, (1972).
8. C. Robinson, J.C. Ward, R.B. Beevers, Discuss Faraday Soc.,
   25, 29, (1958).
9. F. Brochard, C.R. Acad. Sci., B 299, 229, (1979).
10. A. Ten Bosch, P. Maïssa, P. Sixou, J. Chem. Phys., 79,
    3462, (1983).
11. H. Finkelmann, B. Luhmann, G. Rehage, H. Stevens, Liquid
    Crystals and Ordered Fluids, Vol. 4, in press.
12. A. Ten Bosch, P. Maïssa, P.Sixou,J.Phys.Lett.44,105,(1983).

13. A. Ten Bosch, P. Maïssa, P. Sixou, Companion paper
14. M.J. Seurin, A. Ten Bosch, J.M. Gilli, P. Sixou, Polymer (in press).
15. S. Dayan, P. Maïssa, M.J. Vellutini, P. Sixou, Polymer, 23,800 (1982).
16. F. Fried, G. Searby, M.J. Seurin, S. Dayan, P. Sixou, Polymer 23, 1755, (1982).
17. S. Dayan, F. Fried, J.M. Gilli, P. Sixou, J. Appl. Polym. Sci. (in press).
18. P. Navard, M. Haudin, S. Dayan, P. Sixou, J. Appl. Polym. Sci. (in press).
19. S. Dayan, J.M. Gilli, P. Sixou, J. Appl. Polym. Sci., 28, 1527, (1983).
20. F. Fried, J.M. Gilli, P. Sixou, Mol. Cryst. Liq. Cryst., (in press).
21. J.M. Gilli, J.F. Pinton, P. Sixou, A. Blumstein, R. Blumstein, submitted to Mol. Cryst. Liq. Cryst.

THERMAL BEHAVIOR OF MESOMORPHIC CELLULOSE DERIVATIVES

Patrick Navard and Jean-Marc Haudin

Centre de Mise en Forme des Matériaux
ERA CNRS n° 837
Ecole Nationale Supérieure des Mines de Paris
Sophia Antipolis O6565 VALBONNE FRANCE

INTRODUCTION

As a part of our study of the thermal properties of mesomorphic cellulose derivatives [1-3],this paper is devoted to two sub-
jects. The first one is a first step for analyzing the mesomorphic-
isotropic transition, in order to separate the first and second
order contributions. The second part is a DSC study of hydroxypropylcellulose, both in the bulk and in solution.

EXPERIMENTAL

p-Azoxyanisole (PAA) was supplied by Fluka. A smectic and
cholesteric sample for thermometry, labelled TM75A,was used as received from BDH. Hydroxypropylcellulose (HPC), M_w=60,000, from
Hercules, was dehydrated at 100°C prior to use.Cellulose triacetate
(CTA) supplied by Fluka was mixed with trifluoroacetic acid (TFA)
at room temperature for several hours and studied in the following
three days. Measurements were performed with a Perkin Elmer DSC 2
calorimeter.

PHASE TRANSITIONS

The nature of the nematic-isotropic transition of low molar
mass liquid crystals is described by two main theories. The Maier
and Saupe (MS) approach [4] uses a mean field theory with a quadrupolar interaction between adjacent molecules. It gives a first order
transition, usually in agreement with experiments. Nevertheless, it

389

was supposed that this first order transition could be due to the mean field approximation, other methods giving a second order transition [5] . An extension of this theory coupled with the use of the Flory lattice theory [6] gives a first order transition with a heat capacity diverging near the transition temperature.

Another approach was used by de Gennes [7-8] . This macroscopic theory is based on the Landau theory of phase transition, and its findings give a reasonable qualitative agreement with experiments[9]. This theory predicts a weak first order phase transition with a strong second order component, i.e. a diverging heat capacity. Recently doubts were cast on the quantitative validity of the theory [10]

The problem is more difficult when polymers are considered. If they are supposed to be rigid rods, without "soft" interactions, the Flory lattice theory [11] provides a model in agreement with experiments. If the macromolecular chain is semi-flexible, the situation is less clear. Several theories are currently developed [12-15], and their main features are described in this volume. They also describe the transition as first order, but with a strong second order component.

Experimentally, the careful measurement of the heat capacity of low molar mass liquid crystals shows this "diverging" behavior at the nematic-isotropic transition, and also at the smectic-nematic transition [16]. For polymers, there is a limitation in the measurement of heat capacities, since the transition takes place in a relatively large temperature range, and thus, there is at each temperature a mixture of heat capacity and heat of transition. In order to solve this problem, we have built a tool for being able to make the difference between the first and the second order component of a DSC transition peak. The basic idea is that these two contributions of the total recorded heat power will behave in a different manner when the heating rate or the mass of sample are changed.

Theoretical Background

In this paper, a thermodynamic phase transition is studied using Differential Scanning Calorimetry (DSC). This phase transition, which will be described according to the current thermodynamic theories as a first order or a second order one, is recorded on the DSC trace as an anomalous change in the differential power ΔP, different from the normal ΔP variation only due to the heat capacity of the material. This variation, sharp or smooth, will be called the "transition peak". We define the height h of the peak as the distance between the heat capacity trace, or baseline, and the maximum ΔP during the course of the phase transition. In the case of a pure second order phase transition, this height is the difference between the heat capacities before and after the transition. In the case of a pure first order phase transition, it is simply the maximum height of the peak above the baseline.

A number N is defined as $N = h'/h$, h' being the height of the transition peak when the mass m or the heating rate $\dot{T}_p$ are multiplied by two. The theoretical values of N will be determined in the case of an isothermal first order phase transition, a second order phase transition and a non-isothermal first order phase transition (case of an impure material). Only results will be given here, the detailed theory being described elsewhere [17].

For a first order phase transition :

$$N = \frac{2[-1 + (1 + \dfrac{\Delta h}{mC'_p{}^2 R_o \dot{T}_p})^{\frac{1}{2}}]}{-1 + (1 + \dfrac{2 \Delta h}{mC'_p{}^2 R_o \dot{T}_p})^{\frac{1}{2}}} \qquad (1)$$

Eq.1 gives the value of N if the mass m and the heating rate $\dot{T}_p$ are small. Δh is the transition enthalpy per unit mass, C'_p is the specific heat capacity of the component (supposed to be a constant) and R_o is a thermal resistance. It can be shown that $1 < N < \sqrt{2}$.

It is practically easy to multiply $\dot{T}_p$ by two. It is more difficult to do so for m, and in the case of two masses m_1 and m_2, leading respectively to two peak heights h_1 and h_2, N is given by:

$$N = \exp \left(\frac{0.69 \ln (h_2/h_1)}{\ln (m_2/m_1)} \right) \qquad (2)$$

For a second order transition, as a DSC curve is the recording of the differential power ΔP versus time t, it is possible to write :

$$\Delta P = \frac{d \Delta W}{dt} = \frac{d \Delta W}{dT_p} \times \frac{dT_p}{dt} \qquad (3)$$

where T_p is the programmed temperature. This equation is valid only when no first order phase transition occurs.

$$\frac{d \Delta W}{dT_p} = mC'_p \quad \text{and} \quad \frac{dT_p}{dt} = \dot{T}_p \qquad (4)$$

$$\Delta P = mC'_p \dot{T}_p \qquad (5)$$

When m or $\dot{T}_p$ are multiplied by two, ΔP is also multiplied by two. A second order transition is recorded by DSC as a change of heat capacity, without involving a transition energy. Equation 5 is thus valid for describing such a transition. In this case, $h=m\dot{T}_p(C'_{p2}-C'_{p1})$ where C'_{p1} and C'_{p2} are the specific heats before and after the transition. As a consequence, $N=2$.

The case of an impure material is more difficult to treat. When the material is impure, the first order phase transition is generally not isothermal. The presence of impurities lowers the melting point and broadens the melting peak. If the impurity is soluble at any quantity in the material, one can describe the melting curve using the Van't Hoff equation. It is thus possible to show that $N=2$ in this case.

The measurement of N for a transition can lead to several interesting conclusions. The one which is of interest here is the occurrence of an anomalous second order component in the DSC peak of a mesomorphic-isotropic transition. The first applications that we made are described in the next section.

Results

p-Azoxyanisole - Fig.1 shows a DSC trace of p-azoxyanisole : the first peak at 117°C is the crystal-nematic transition peak. Its N value is 1.4 (measured with $\dot{T}_p=2.5$, 5 and 10°C/min). The second peak at 135°C is the nematic-isotropic transition peak. Its N value is 1.3 (same heating rates). These two N values are typically the ones of first order phase transitions. The heat capacity of PAA is small compared to the height of the peak. Even if an anomalous second order phenomenon occurs, increasing the heat capacity jump under the nematic-isotropic peak by 100% or 200%, it cannot shift N towards two in a detectable way.

TM75 A - Fig. 2 shows a DSC trace for this material. The two peaks at 41°C and 53°C correspond respectively to the smectic-cholesteric and cholesteric-isotropic transitions. The N values were measured for low heating rates (0.62, 1.25 and 2.5 °C/min) since the peaks are close together, and found to be 1.9 for the first peak and 1.8 for the second one. TM75 A is a favorable case since its heat capacity is large compared to the height of the peaks. Even by taking into account the variation of heat capacity between two phases under the peak, it is impossible to find a N value so close to two. The only way is to suppose that a second order component plays an important role.

Cellulose triacetate-trifluoroacetic acid cholesteric solutions - This kind of lyotropic polymer liquid crystals undergoes a mesomorphic-isotropic phase transition upon heating. The peak is well defined but very small [2]. The determination of N for this

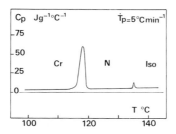

Fig.1. Experimental DSC trace of p-azoxyanisole

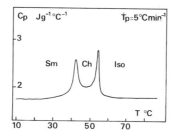

Fig. 2. Experimental DSC trace of TM75 A

transition gives 1.85 (concentration of polymer : 26 % by weight). N is very close to 2 and this can be explained if it is the melting of an impure solution, or the end of a dissolution process or if there is a strong second order component in the transition. The materials used are pure, and the second hypothesis is not valid [2]. So, this high value of N probably reflects that a liquid crystal-isotropic transition is only weakly first order. This result is close to the case of TM75A.

These first results show that the use of number N can be a powerful tool, showing very easily and quickly if a transition can be treated as a first order one. If so, a transition enthalpy can be measured. The diverging character of the heat capacity has been studied for some compounds by adiabatic calorimetry, and the above theory will be applied to them soon. Another field of application is the study of the transitions of thermotropic polymers.

HYDROXYPROPYLCELLULOSE

Hydroxypropylcellulose/water

As is true for many cellulose derivatives, hydroxypropylcellulose is able to form an ordered structure in solution when the
polymer concentration is above a certain critical concentration.
HPC/water solutions were studied over a wide range of concentrations, and the ordered structure which appears above a concentration C** of about 40 % by weight was found to be cholesteric [18-20]
as is the case for all liquid crystalline cellulose derivative
solutions studied until now. Upon heating, isotropic and cholesteric solutions undergo a phase transition at about 40°C. This new
phase is turbid, and it is called "white suspension" or "white
gel "[19]. The critical concentration is not sensitive to temperature
or to molecular weight. HPC also forms cholesteric solutions when
dissolved in organic solvents. Tsusui and Tanaka [21] show that the
phase diagram for the latter is very different from HPC/water, but
very similar to that of the cellulose acetates/trifluoroacetic
acid [1]. The main difference is that an HPC/water liquid crystalline
solution is transformed into a turbid state (gel or suspension)
upon heating, whereas an HPC/organic solvent liquid crystalline solution is transformed into an isotropic solution. In the phase diagram given by Werbowyj and Gray [19] this turbid state is termed
"white suspension" when C<C** and "white gel" when C>C**,C** being
the critical concentration.

Figure 3 shows a collection of several DSC traces of four HPC/
water solutions. The transition toward the turbid state, as seen by
visual appearance, optical microscopy, or light scattering, is easily detectable by DSC. From the traces, one can see that there is an
important difference between an isotropic and a cholesteric solution. The transition from the isotropic to the turbid state is recorded as a rather sharp endothermic peak. When the concentration
increases, the surface of the endothermic peak decreases. But
another phenomenon arises : a decrease in heat capacity occurs
along a large temperature range. This can be related to measurements
of turbidity made by Werbowyj and Gray [19]. They found that the
change in turbidity as a function of temperature has a different
behavior when the solutions are below or above C**. Below C**, there
is a sharp change in turbidity at the transition temperature T_t.
Above C**, there is a small change in turbidity at T_t, followed by
a slow turbidity increase with temperature. The DSC measurements
show the same behavior.

Without forming any hypothesis about the nature of the turbid
phase, it can be assumed that there is probably no difference between its nature at low and high concentrations since for a 60 %
solution the characteristic peak of low-concentration solutions is
present. Thus, the differences found in the traces between low and

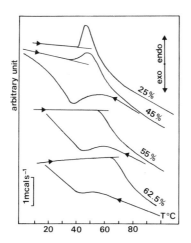

Fig.3. DSC traces of HPC(M_w=60,000)/water solutions. Concentrations
are given above each curve. Arrows indicate direction of
temperature scan. Scanning rate : 10°C/min.

high concentrations are probably due to the breaking of inter - or
intra- molecular bonds. The enthalpy being relatively high, this
transition is probably governed by an intermolecular phenomenon :
the energetic balance between water-HPC bonds, which is energetically favorable in the isotropic state; and HPC-HPC bonds, favorable
in the turbid state. The difference in energy between these two
sets of bonds is recorded as an endothermal peak. The sharp decrease
in enthalpy between 45 and 60 % cannot be explained by a simple
concentration effect. It could probably be related to the cholesteric structure of the solution which is stable because of strong
interactions between HPC chains.

From the determination of T_t, it is possible to plot a phase
diagram (Fig.4). T_t is taken as the temperature where the thermal
transition begins on the DSC trace. The phase diagram can be compared with the dashed line in Fig.4, which is the diagram proposed
by Werbowyj and Gray [19], plotted from the visual appearance of the
solutions. Upon cooling, there is an exothermic peak and/or a heat
capacity transition, depending on the concentration. This peak is
connected with the reverse transition from the turbid state to the
isotropic or cholesteric state, as seen both by visual appearance
and by turbidity measurements [19]. There is a difference of 2 to
5°C between T_t and the reverse transition temperature in the same
way as for the supercooling of numerous materials. When a solution
is cycled between the turbid and the isotropic or cholesteric states, there is no change in the temperatures of transition, the
shape of the curves, or the enthalpy.

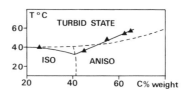

Fig.4. HPC/water phase diagram.Triangles indicate experimental DSC
data. Dashed line is the diagram given in Ref. 19

Hydroxypropylcellulose (bulk)

The starting point of this work is to understand the DSC curve
of HPC. Upon heating, HPC undergoes a transition between 160°C and
205°C, recorded as an endotherm in Fig.5, as previously seen by
SUTO et al [22].There is considerable evidence that in this temperature range, HPC is a liquid crystal, most probably a cholesteric
one. As noticed by these authors, this broad endotherm should dis-
guise two endotherms. These are separated upon cooling, as seen in
Fig.5. Such a material is called a monotropic liquid crystal, when
two transitions are so close that they cannot be separated upon
heating. Upon cooling, supercooling allows the two transitions to
occur at different temperatures. Such a phenomenon is usual for
small molecule liquid crystals, and was sometimes seen in polymers
[23]-[25].The transition enthalpies are the same (3.4 J/g) if deduced
from the heating or cooling curves. From the cooling curve, it is
possible to separate the two contributions, the peak at lower temperature being the liquid crystal-crystal transition, with $T_m=180°C$
and $\Delta H_m=2.1$ J/g, and the other peak being the liquid-liquid crystal transition, with $T_i=188°C$ and $\Delta H_i=1.3$ J/g. This is the usual
behavior of a thermotropic polymer.

The problem consists in having an idea about the meaning of
these two figures ΔH_m and ΔH_i. Their values are given per gram of
polymer. It is well known that the polymer is not fully crystalline
and so, the transition enthalpy for a perfect crystal(ΔH_{ms} and
ΔH_{is}) is far from these values. $\Delta H_{ms}+\Delta H_{is}$ was found to be 27 J/g
by Samuels [26]. In our case, two possibilites can be imagined in
order to measure ΔH_{ms} and ΔH_{is} : either the liquid crystalline phase of the sample above T_m is coming from the crystalline phase existing below T_m(case a) or it exists below T_m a mixture of crystalline and liquid crystalline phases (case b). The two possibilities
have a different influence on the specific ΔH_{is}. We shall see which
possibility could be preferred in the light of three physical experiments : thermal analysis, birefringence and spectrophotometry.

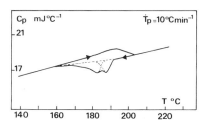

Fig.5. Experimental DSC trace of HPC (M_w=60,000)

Since the specific enthalpy for the crystal-liquid transition is known, it is possible to deduce that our sample had a crystallinity of 12.4% (similar to what Samuels obtained).
In case a :

$$\Delta H_{ms}^a = 16.8 \text{ J/g and } \Delta H_{is}^a = 10.2 \text{ J/g. So } \Delta H_{is}^a/(\Delta H_{ms}^a + \Delta H_{is}^a) = 37.8\%$$

In case b :

$$\Delta H_{ms}^b = 16.8 \text{ J/g and } \Delta H_{is}^b = 1.3 \text{ J/g. So } \Delta H_{is}^b/(\Delta H_{ms}^b + \Delta H_{is}^b) = 7.2\%$$

These calculations are performed with the hypothesis that the theoretical value of 27 J/g was calculated from a sample having the same kind of transition as ours (whether it is a or b).

Results, both theoretical and experimental, show that the contribution of the liquid crystal-liquid enthalpy over the crystal-liquid enthalpy is small, less than 15%-20%. If HPC follows this rule, case b should be preferred.
Birefringence and spectrophotometry suggest [27] that at room temperature, the "amorphous" phase is a low order cholesteric phase, and so case b is preferred. This leads to the conclusion that the specific cholesteric-isotropic transition enthalpy is 1.3 J/g for HPC (M_w=60,000).

REFERENCES

1. P.Navard, J.M.Haudin, S.Dayan and P.Sixou, J. Polym.Sci.,Polym Lett.Ed.,19 : 379 (1981)
2. P.Navard, J.M.Haudin, S.Dayan and P.Sixou, J.Appl.Polym.Sci., Appl. Polym. Symp.,37 : 211 (1983)
3. P.Navard, J.M.Haudin and S.M.Aharoni, J.Polym.Sci.,Polym. Lett. Ed.,21 : 271 (1982)
4. W.Maier and A.Z. Saupe, Naturf.,A13 : 564 (1958);A14:882(1959) A15:287 (1960)
5. G.W.Smith, in : "Liquid Crystal and Plastic Crystals,Vol.IV", G.W.Gray and P.A. Winsor eds., Ellis Horwood, Chichester, England (1974),p.189

6. P.J.Flory and G.Ronca, Mol.Cryst.Liq. Cryst.,54:311 (1979)
7. P.G.de Gennes, Phys.Lett., 30A:454 (1969)
8. P.G.de Gennes, Mol. Cryst. Liq.Cryst.,12:193 (1971)
9. P.G.de Gennes, "The Physics of Liquid Crystals",Clarendon Press,
    Oxford (1974)
10. L.Senbetu and C.W.Woo, Mol.Cryst.Liq.Cryst.,84:101 (1982)
11. P.J.Flory, Proc.Roy.Soc.London Ser.A, 234:73 (1956)
12. G.Ronca and D.Y.Yoon, J.Chem. Phys.,76:3295(1982)
13. G.Ronca and D.Y.Yoon, J.Chem. Phys.,to appear
14. A.Ten Bosch, P.Maissa and P.Sixou, J.Chem.Phys.,to appear
15. A.Ten Bosch, P.Maissa and P.Sixou, J.Physique, Lettres,44:
    L105 (1983)
16. H.Marynissen, J.Thoen and W.Van Dael, Mol.Cryst.Liq.Cryst.,97:
    149 (1983)
17. P. Navard and J.M.Haudin, submitted to J.Therm. Anal.
18. R.S.Werbowyj and D.G.Gray, Mol. Cryst. Liq. Cryst . 34:97 (1976)
19. R.S.Werbowyj and D.G.Gray, Macromolecules, 13:69 (1980)
20. Y.Onogi, J.L.White, and J.F. Fellers, J. Polym. Sci.Polym.
    Phys. Ed., 18:663 (1980)
21. T.Tsusui and R.Tanaka, Polym.J., 12:473 (1980)
22. S.Suto, J.L.White and J.F. Fellers, Rheol.Acta,21:62 (1982)
23. P.Iannelli,A.Roviello and A.Sirigu, Eur.Polym.J.,18:753(1982)
24. A.Blumstein and O.Thomas, Macromolecules,15:1264(1982)
25. A.Roviello and A.Sirigu, Makromol.Chem. 180:2543 (1979)
26. R.J.Samuels,J.Polym.Sci.,A2(7):1197(1969)
27. P.Navard and J.M.Haudin, submitted to Polymer

RHEO-OPTICAL STUDIES ON THE STRUCTURAL RE-FORMATION OF NEMATIC LIQUID CRYSTALS OF ROD-LIKE POLYMERS AFTER CESSATION OF STEADY FLOW

Tadahiro Asada

Department of Polymer Chemistry
Kyoto University
Kyoto, Japan

INTRODUCTION

Polymer liquid crystals generally form unusual textures, accompanied by complex and unstable superstructures. Such textures depend on a sample history, especially mechanical and thermal histories, and also on the surface nature of the vessels. We have described the structural characteristics of polymer liquid crystals under a shear force by means of rheo-optical methods.[1,2]

To what extent the structural features of a nematic liquid crystal remain unchanged when subjected to a shear force after the removal of the applied force is a problem of great theoretical and practical interest. It is important to know whether an oriented continuous phase achieved during steady flow remains unchanged or not, even after cessation of the steady flow. Another interesting question is whether spontaneous orientation of molecules continues to increase or stops after the removal of the applied force. The structure seems to change gradually with time after the cessation of steady flow, while the stress decreases instantaneously to zero. Such a structural reformation process has been studied by a rheo-optical method. Effort has been made to estimate the spontaneous orientational relaxation and the rearrangement due to wall effect separately.

EXPERIMENTAL METHODS

The rheo-optical properties were measured by means of the polarized light technique, which was introduced in detail elsewhere.[1] The apparatus is a combination of a cone and plate type rheometer equipped with a transparent cone and plate made of quartz with an optical system. A monochromatic laser light beam (wavelength = 6328Å) passes through a polarizer, quartz plate, sample, quartz cone, analyzer, and finally is detected by a photomultiplier tube. Thus, this apparatus enables us to measure the rheological properties and transmission of polarized light through a sheared sample simultaneously. The diameter of the cone and plate was kept constant (8 cm), and the cone angle was 0.865°, unless otherwise described.

The shear stress and intensity of transmitted light (I_x, $I_{//}$, I_E) were measured simultaneously as functions of time, after cessation of steady flow at various shear rates ($\dot{\gamma} = 2 \times 10^{-1} \sim 1.1 \times 10^2$ s^{-1}) at 25°C. Here, I_x, $I_{//}$ and I_E are fractional transmitted light intensities with crossed polarizers (at $\psi = 45°$ in Fig. 1), parallel polarizers (at $\psi = 45°$), and crossed polarizers at the extinction position (at $\psi = 0°$), respectively. The geometry is shown in Fig. 1.

When the retardation Γ of uniaxially oriented phase undergoes continuous changes, I_x and $I_{//}$ vary in a quasi-periodic manner, as expressed by

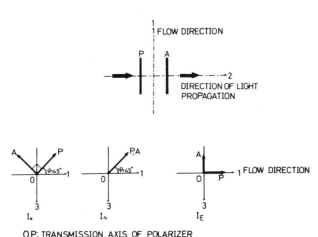

OP: TRANSMISSION AXIS OF POLARIZER

OA: TRANSMISSION AXIS OF ANALYZER

Fig. 1.   Geometry of the optical system. 1: flow direction, 2: direction of light propagation, and 3: radial direction of the cone and plate. OP is the transmission axis of the polarizer, and OA is the transmission axis of the analyzer.

$$I_X = K \sin^2(\pi\Gamma/\xi)$$

$$I_{//} = K[1 - \sin^2(\pi\Gamma/\xi)]$$
(1)

and shown schematically in Fig. 2. In Eq 1, ξ is the wavelength of the light, and K is the amplitude that is unity when there is no absorption and scattering. In such a simple case, these equations enable us to evaluate the birefringence from the data of I_X and $I_{//}$ (transmission method). The same concept can be applied to a liquid crystal system, as long as the system is optically inactive and behaves as a monodomain.

Materials

Equal amounts of poly(γ-benzyl-L-glutamate) and poly(γ-benzyl-D-glutamate) having weight-average molecular weight 1.5×10^5 were dissolved in m-cresol to give solutions of racemic poly-γ-benzyl glutamate (PBG). The behaviors of Nematic Liquid Crystals of the PBG solutions listed in Table 1 will be introduced.

GENERAL FEATURES OF RHEO-OPTICAL BEHAVIORS

After the cessation of steady shear flow, the optical quantities I_X, $I_{//}$ and I_E always changed with time, while the shear stress decreased instantaneously to zero. As an example, results obtained for L2 after the cessation of steady flow at the rate $\dot{\gamma}$ of 6.9 s^{-1} is shown in Fig. 3. As can be seen in this figure, transmitted light intensities I_X and $I_{//}$ show wavy changes, as

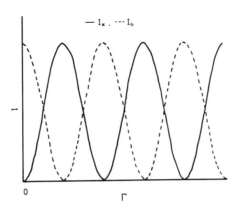

Fig. 2. Schematic representation of the variation of transmitted light intensities I_X and $I_{//}$ with retardation Γ.

Table 1. Sample Cordes

$\overline{M}_w$ Conc.	1.5×10^5	2.1×10^5	2.6×10^5
20 wt%	L2	M2	H2
30	L3	—	—
40	L4	—	—

expected from Eq 1, suggesting that the retardation changes with time. With the aid of a quarter wave plate, it was ascertained that the retardation Γ decreases with time.

The amplitude of wavy curves of I_x and $I_{//}$ decreases at first and then increases again. On the other hand, I_E changes with time as if it enveloped the wavy curves of I_x and $I_{//}$, and takes a maximum at a time t_M. Such a change in I_E raised a doubt whether the orientation axis remained unchanged. In order to make this clear, the variation of transmitted light intensity I with time was measured by the use of crossed polarizers at various ψ, angle between the flow direction and polarizer axis, after the cessation of steady flow. It was thus ascertained that the variation of I_E with time is not caused by the change in the direction of the main orientation axis.

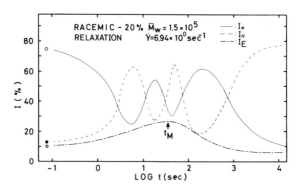

Fig. 3. The variation of transmitted light intensities, I_x, $I_{//}$ and I_E with time after cessation of steady flow for L2. t_M represents the time at which I_E takes its maximum.

The wavy changes of I_x and $I_{//}$ shown in Fig. 3 were converted to the variation of retardation Γ with time by the use of Eq 1. As an example, results obtained for the sample L2 after cessation of steady flow at various rates are shown in Fig. 4. The ordinate of this figure is the relative retardation Γ/Γ_0, where Γ_0 and Γ are retardations during steady flow and after cessation of the flow, respectively. As is evident from the figure, the higher the applied shear rate, the earlier the retardation decreases.

DISCUSSION

A Model for Interpreting the Optical Data

During steady shear flow at higher rates, a liquid crystalline system is considered to flow as an oriented continuous phase, which is shown schematically in Fig. 5(top). When the flow is ceased, molecular orientation relaxes, while liquid crystalline structure is re-formed under the influence of the wall and disclination points. Such structural changes must be reflected in optical data, as shown above. To interpret the optical data for a nematic polymer liquid crystal, a model for structural re-formation including relaxation of molecular orientation will be proposed.

Figure 6 shows the variation of relative retardation with time for the sample L2 at various thicknesses ranging from 200 to 1763 μm. As is evident from this figure, the thicker the sample is, the faster the retardation decreases, especially at long times. Curves for thicker samples seem to have the shape of a Maxwellian

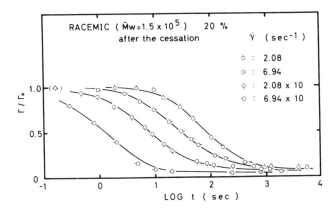

Fig. 4. The variation of relative retardation Γ/Γ_0 with time for L2 after cessation of steady flow.

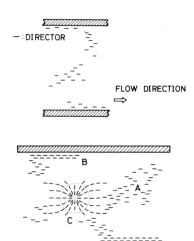

Fig. 5.   Schematic representation of liquid crystalline structure
of nematic polymer liquid crystal. Solid rods represent
directors.
[Top] Oriented continuous phase during steady shear flow.
[Bottom] Re-formation of structure after cessation of
steady flow.
A: The portion free from effects of both the wall and
   disclination points.
B: The portion strongly affected by the wall effect.
C: The portion affected by disclination points.

relaxation curve.   The calculated Maxwellian relaxation curve is
shown on the same figure by a solid line, for comparison.   The

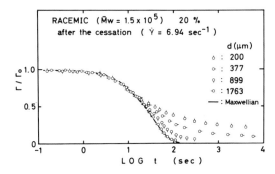

Fig. 6.   The variation of relative retardation with time for L2 at
various thicknesses d ranging from 200 to 1763 μm.
Solid-line curve is a calculated Maxwellian relaxation
curve.

optical relaxation behavior for the thinner sample deviates very much from the Maxwellian relaxation curve. Such deviation seems to arise from the wall effects, which promote the molecular reorientation against orientational relaxation and delay the decreasing of the total optical retardation. This idea has been taken into account in our present model.

For simplicity, the total volume of a sample is divided into three portions, depending on the effects of the wall and disclination points, as is shown schematically in Fig. 5 (bottom). The portion which is free from effects of both the wall and the disclination points is designated as A-phase. The portion where molecules are strongly affected by the wall and disclination points are designated as B- and C-phase, respectively.

The total time-dependent birefringence Δ_t of the system can be considered to be the sum of contributions from each phase, Δ_A, Δ_B, and Δ_C. Then, we obtain

$$\Delta_t = \Delta_A + \Delta_B + \Delta_C \tag{2}$$

$$\Delta_A = L \cdot S(t) \cdot X_A(t) \cdot f_A(t) \tag{3}$$

$$\Delta_B = L \cdot S(t) \cdot X_B(t) \cdot f_B(t) \tag{4}$$

$$\Delta_C = L \cdot S(t) \cdot X_C(t) \cdot f_C(t) \tag{5}$$

where the subscripts A, B, and C denote the three phases. $X(t)$ is the volume fraction of each phase as a function of time, and $f(t)$ is the orientation function of the director in each phase as a function of time. $S(t)$ is the order parameter as a function of time, and L is a constant inherent in a molecule manifesting the anisotropy of the polarizability. Then, $L \cdot S(t)$ corresponds to the birefringence. Since the relaxation of the order parameter can be considered to be completed in a very short time, we assume that the order parameter $S(t)$ is constant.

$$S(t) = S_o \tag{6}$$

Then, Eq 2 becomes

$$\Delta_t = L \cdot S_o [X_A(t) \cdot f_A(t) + X_B(t) \cdot f_B(t) + X_C(t) \cdot f_C(t)] \tag{7}$$

And,

$$\frac{\Gamma(t)}{\Gamma(0)} \equiv \frac{\Gamma}{\Gamma_o} = \frac{\Delta_t}{\Delta_o} = \frac{\Delta_A}{\Delta_o} + \frac{\Delta_B}{\Delta_o} + \frac{\Delta_C}{\Delta_o} \tag{8}$$

where

$$\Delta_o = L \cdot S_o \cdot f_A(0) \tag{9}$$

405

$$\frac{\Delta_A}{\Delta_o} = X_A(t) \cdot \frac{f_A(t)}{f_A(0)} \qquad (10)$$

$$\frac{\Delta_B}{\Delta_o} = X_B(t) \cdot \frac{f_B(t)}{f_A(0)} \qquad (11)$$

$$\frac{\Delta_C}{\Delta_o} = X_C(t) \cdot \frac{f_C(t)}{f_A(0)} \qquad (12)$$

At t = 0, the whole system is composed of A-phase. As soon as the steady flow is ceased, some parts of A-phase, denoted by $X(t) \equiv X_B(t) + X_C(t)$, transform into B- and C-phases.

In Fig. 8 an experimental result for the sample L2 of thickness 380 μm (curve a in Fig. 8) is compared with a Maxwellian relaxation curve (curve b in Fig. 8). Both curves deviate from each other at larger t. In the first approximation, the deviation is considered to have been brought by the wall effect, i.e. the appearance of B-phase. The B-phase will grow with time under the influence of wall effect to some extent [the volume fraction of which is denoted here by $X_B(\infty)$. This is schematically illustrated in Fig. 7. When the transformation into the B-phase is assumed to be analogous to one-demensional crystal-crystal transformation, the volume fraction of B-phase at t = t is given by

$$X_B(t) = X_B(\infty)\ (1 - e^{-k_B t}) \qquad (13)$$

where k_B is the rate constant for the transformation process from A- to B-phase. Some examples of calculated curves according to Eq 13 at different k_B values are shown in Fig. 8(bottom).

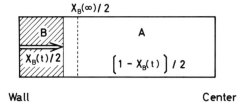

Wall   Center

Fig. 7. Schematic illustration of the transformation of a part of A-phase into B-phase. $X_B(t)$ is the volume fraction of B-phase.

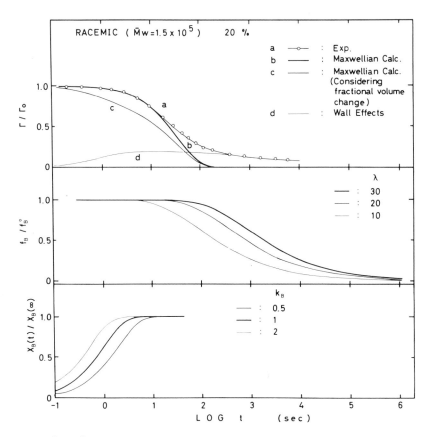

Fig. 8. [Top]. Experimental result for L2 of thickness 380 μm
(O) is compared with a Maxwellian relaxation curve.
Solid line (a): Combination result of curve c and d.
Thick solid line (b): Maxwellian relaxation curve.
Solid line (c): Maxwellian relaxation curve taking into
account the fractional volume change of A-phase. Solid
line (d): Contribution of wall effects.
[Middle]. Calculated curves for $f_B(t)$ according to Eq 17
at various λ values. $f_B^\circ = f_A(0)$.
[Bottom]. Calculated curves for the volume fraction of
B-phase against time at various k_B values in the Eq 13.
$X_B(t)$ is the volume fraction of B-phase.

The transformation into C-phase is assumed to be analogous to
three-dimensional crystal-crystal transformation to give

$$X_C(t) = X_C(\infty) \cdot (1 - e^{-k_C t^3}) \qquad (14)$$

where k_C is the rate constant for the transformation from A- to C-phase.

Considering Eqs 13 and 14, Eq 8 becomes,

$$\frac{\Gamma}{\Gamma_o} = [\{1 - X(\infty)\} + X_B(\infty)\, e^{-k_B t} + X_C(\infty)\, e^{-k_C t^3}]f_A(t)/f_A(0)$$

$$+ X_B(\infty)\{1 - e^{-k_B t}\}f_B(t)/f_A(0)$$

$$+ X_C(\infty)\{1 - e^{-k_C t^3}\}f_C(t)/f_A(0) \qquad (15)$$

where $X(\infty) = X_B(\infty) + X_C(\infty)$

Now, the time dependence of the orientation function for individual phase is assumed to be

$$f_A(t) = f_A(0)\, e^{-t/\tau}, \qquad (16)$$

$$f_B(t) = f_A(0)(1 - e^{-\lambda/\sqrt{t}}), \qquad (17)$$

and

$$f_C(t) = 0 \qquad (18)$$

where τ is a relaxation time of director's disorientation in A-phase, and λ is a rate parameter for orientational change in B-phase. The assumption employed for Eq 16 is based on the experimental evidence that the optical relaxation is Maxwellian, as already seen in Fig. 6, while that for Eq 17 is based on a consideration of the wall effect as well as experimental evidence.[2] The assumption for Eq 18 is considered to be reasonable, when the growth of C-phase is spherical.

Some examples of calculated curves for $f_B(t)$ according to Eq 17 at different λ values are compared in Fig. 8(middle). Since the ammount of B-phase, $X_B(t)$, is smaller at small t region and increases with time as shown in Fig. 8 (bottom), the contribution of B-phase to the optical retardation changes with time as shown by the curve d in Fig. 8 (top), which is a calculated curve for the case of $X_B(t)$ and $f_B(t)$ changing with time according to the thickest solid lines in Fig. 8 (bottom and middle). The combination of curve C and curve d gives curve a in Fig. 8 (top), which represents well the experimental curve (open circles), though the contribution of C-phase is neglected for simplicity.

Using Eqs 16, 17, and 18, Eq 15 becomes:

$$\frac{\Gamma}{\Gamma_o} = [(1 - X) + X_B \, e^{-k_B t} + X_C \, e^{-k_C t^3}] \, e^{-t/\tau}$$

$$+ X_B(1 - e^{-k_B t}) \, (1 - e^{-\lambda/\sqrt{t}}) \qquad (19)$$

where X, X_B and X_C denote $X(\infty)$, $X_B(\infty)$ and $X_C(\infty)$, respectively. This equation gives the time dependence of relative retardation $\frac{\Gamma}{\Gamma_o}$ for a nematic polymer liquid crystalline system after cessation of steady flow.

Analysis and Interpretation of the Optical Data

Equation 19 introduced above is useful for discussing a process of structural re-formation of a nematic polymeric liquid crystals. The analysis in terms of Eq 19 makes it possible to evaluate the orientational relaxation time τ in a force-free state of a nematic liquid crystal despite actual existence of wall and disclination effects. Figure 9 gives an example of computer-fitting for the experimental result shown in Fig. 4 ($\dot{\gamma}$ = 6.9 sec^{-1}). As is evident from this figure, the calculated values (solid line) coincide very well with the experimental ones (circles) over the entire region of time covered. The values of parameters are also sited on the same figure. The change in the optical relaxation curve will be discussed in terms of the change in the parameters, hereafter.

Similar calculations were carried out also for other cases to determine the six parameters in Eq 19. The results thus obtained

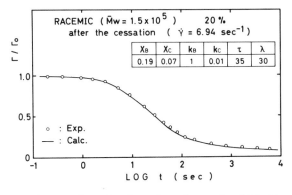

Fig. 9. A comparison of calculated values (solid line) of relative retardation with experimental ones (circles) for L2.

Table 2

%	σ (dynes/cm^2)	X_B	X_C	τ	λ
20	1.70×10^1	0.18	0.07	100	40
	5.69×10^1	0.19	0.07	35	30
	1.70×10^2	0.18	0.07	9	20
	5.69×10^2	0.18	0.07	1.5	3
30	2.29×10^1	0.20	0.10	100	100
	3.83×10^1	0.20	0.10	50	60
	7.63×10^1	0.21	0.10	20	50
	2.14×10^2	0.20	0.10	5	40
	3.48×10^2	0.25	0.10	3	40
	6.59×10^2	0.20	0.10	2	40
40	2.08×10^2	0.20	0.15	25	30
	3.37×10^2	0.15	0.16	5.5	6
	4.93×10^2	0.14	0.16	4.5	5
	9.36×10^2	0.16	0.15	2.4	3
	1.28×10^3	0.17	0.13	1.5	2
	2.81×10^3	0.20	0.13	1	2

for L2, L3 and L4 are listed in Table 2, and those for L2, M2, H2 in Table 3.

It can be seen from these tables that τ and λ decrease very rapidly with increasing shear rate or shearing stress of the steady flow given to the samples. From Table 2, it seems that X_C gets larger with increasing concentration. X_C for each sample seems to remain constant as long as the given steady flow is in the Newtonian flow region (Region II). In Fig. 10, the orientational relaxation time τ in A-phase is plotted logarithmically against shearing stress σ for L2, L3 and L4. As is evident from this figure, the variation of τ with concentration is rather slight.

The molecular weight dependence of the parameters for 20 % solution is seen in Table 3. X_C decreases with increasing molecular weight. X_C for each sample seems to be constant as long as

Table 3

$\bar{M}_w \times 10^{-4}$	σ (dynes/cm^2)	X_B	X_C	τ	λ
15	1.70×10^1	0.18	0.07	100	40
	5.69×10^1	0.19	0.07	35	30
	1.70×10^2	0.18	0.07	9	20
	5.69×10^2	0.18	0.07	1.5	3
21	1.15×10^1	0.16	0.05	500	100
	3.43×10^1	0.16	0.05	180	100
	6.86×10^1	0.17	0.05	80	100
	2.15×10^2	0.35	0.05	25	50
	5.41×10^2	0.30	0.04	4	50
	1.53×10^3	0.35	0.04	1	50
26	8.74×10^1	0.15	0.03	90	100
	1.62×10^2	0.15	0.03	50	90
	2.50×10^2	0.15	0.02	30	30
	4.23×10^2	0.20	0.02	7	10

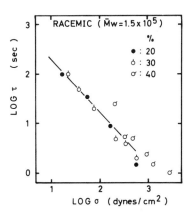

Fig. 10. Optical relaxation time τ plotted logarithmically a-
gainst shearing stress σ for L2, L3, and L4.

the given steady flow is in the Newtonian flow region. It becomes
smaller outside the flow region II. Such a result can be under-
standable assuming that disclination points are excluded outside
the system during the steady flow before cessation at the higher
shear rate. In Fig. 11, τ for L2, M2 and H2 are plotted logarithmically against shearing stress of the steady flow given to the
samples. The plots are represented by straight lines having the
same slope of -1, so long as the steady flow is in Newtonian flow
region. The higher the molecular weight, the larger the τ at
constant. τ obtained in the Newtonian flow region increases in
proportion to about $\overline{M}_w{}^3$, as is shown in Fig. 12.

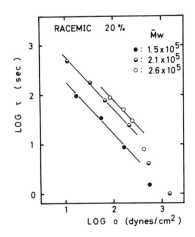

Fig. 11. Optical relaxation time τ plotted logarithmically against
shearing stress σ for L2, M2, and H2.

411

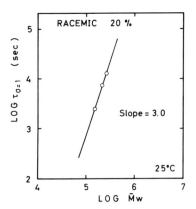

Fig. 12. Optical relaxation time τ plotted logarithmically against weight-average molecular weight.

It may be said that τ obtained here corresponds to the average orientational relaxation time of directors.

Flow curves for L2, M2 and H2 are shown logarithmically mically in Fig. 13. The shear-rate or shear-stress region where the plots are represented by straight lines having a slope of 1 is the so called Newtonian region (Region II). The Regions of the steady flow behavior may be sensitive to the structural re-formation process after the cessation of the steady flow as mentioned above.

When we fix one of four curves in Fig. 4 at a reference shear rate and shift other curves at arbitrary shear rates horizontally

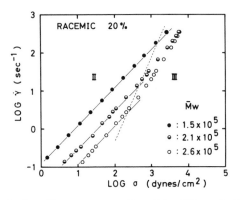

Fig. 13. Logarithmic flow curves for L2, M2 and H2 at 25°C. The dotted line divides Region II (Newtonian region) from Region III.

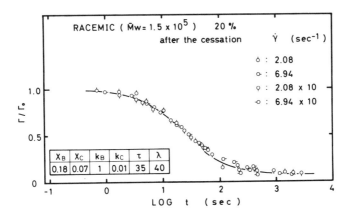

Fig. 14. Composite optical relaxation curve for L2.

along the abscissa, a composite curve shown in Fig. 14 can be obtained. A curve calculated according to Eq 19 is also shown in Fig.14 by the solid-line. Such a superposition can always be applied, so long as the shear rates at which relative retardation vs. log t curves are measured are in the Newtonian region. The fact that the optical relaxation curves can be superposed by the horizontal shift into a single composite curve implies that the structural re-formation including orientational relaxation for a sample measured at different shear rates occurs in the same manner so long as the given steady flow is in the Newtonian flow region.

REFERENCES

1. T. Asada, H. Muramatsu, R. Watanabe and S. Onogi, Macromolecules, 13, 876 (1980).
2. T. Asada, K. Toda and S. Onogi, Mol. Cryst. Liq. Cryst., 68, 231 (1981).
3. S. Onogi and T. Asada, "Rheology", pp. 127-147, Vol. 1, G. Astaria, G. Marrucci and L. Nicolais eds., Plenum Press, New York (1980).

THE DETERMINATION OF ELASTIC CONSTANTS IN

WEAKLY BIREFRINGENT LIQUID CRYSTALS

Donald B. Dupré and Jacob R. Fernandes

Department of Chemistry
University of Louisville
Louisville, KY   40292

ABSTRACT

A procedure is described that can measure optical phase retardations of birefrigent materials with a resolution 2×10^{-4} radians. The method relies on phase modulation with alternate right and left handed circularly polarized light. Phase sensitive detection is employed to reduce noise and thermal fluctuations in the optics and light source. The method is useful in Frederik's transition measurements to determine the elastic constants of weakly birefringent polymer liquid crystals with long equilibration times.

INTRODUCTION

The three elastic constants of a liquid crystal are important physical parameters which depend on the interaction between the molecules in the liquid crystalline state. While a large number of theoretical and experimental investigations[1] on the elastic constants are contained in the literature for thermotropic liquid crystals, very little is known about them in the case of lyotropic polymer liquid crystals such as those formed by poly-γ-benzyl-L-glutamate (PBLG) in various organic solvents. Some theoretical investigations have been carried out[2,3] but the experimental data is limited largely to measurements of the twist elastic constant[4] and a few recent measurements of the bend and splay constants[5,6].

In earlier publications[5,6] we have presented the results of our measurements of the bend and splay elastic constants of PBLG liquid crystals using the Frederik's magneto-optical distortion procedure. The central experimental necessity of this technique is the measurement of the change in the phase retardation of polarized light caused by deformation of an aligned sample in a slowly changing magnetic field. The application of this method to lyotropic liquid crystals is complicated by the very small birefringence of the material which makes the determination of the critical field difficult[5,7]. To overcome this problem, we have developed a sensitive method for the measurement of phase retardations. We will describe the optical procedure below and show how it applies and differs from the conventional procedure to monitor the Frederik's transition.

A number of techniques have been employed for phase retardation and birefringence measurements in liquid crystals. In general, a sinusoidal phase modulation is applied to the state of polarization of the light incident on the sample and the transmitted beam is analyzed[8]. The common devices used for the phase modulation are the Faraday effect polarization modulator[9], the photoelastic modulator[10] or a rotating polarizer[11]. The signal of interest often exists at the extrema of the modulation. Intermediate values of the modulation therefore complicate the detection scheme because the signal related to the birefringence is interposed on a large background signal which is subject to drift.

In the technique described in this article, background signals are optically balanced out and the phase retardation manifests itself as a difference signal. The procedure does not put stringent requirements on the electronics. We have been able to achieve a sensitivity of about 2×10^{-4} radians in the measurement of the phase retardation by this method.

PRINCIPLE OF THE METHOD

The optical setup is shown in Figure 1. Light from a 2 mW He-Ne laser, linearly polarized at 45° to the vertical is incident on a beam splitter, BS. The reflected beam is steered back to the forward path of the transmitted beam by the mirrors M1 and M2 and the rotating half circle mirror HM. The polarization of the two beams is rendered vertical and horizontal as shown using the two half-wave plates which we denote $(\lambda/2)_T$ and $(\lambda/2)_R$, respectively. Polarizers P_T and P_R are used to increase the degree of polarization in the two beams. The two beams with orthogonal polarization states are then incident on a quarter wave plate which is placed with its axis at 45° to the vertical. The light emergent from the $\lambda/4$ plate which is in turn incident on the sample, S, is therefore circularly polarized with opposite senses during the two halves of the mirror rotation cycle. An attenuator A_t is used to compensate for the imbalanced reflection from the beam splitter and equalize the intensity of the two beams. An analyzer, A, is placed after the sample with its axis at 45° to the vertical. The light intensity is detected by a photodiode, D, and measured in synchronous detection by a lock-in

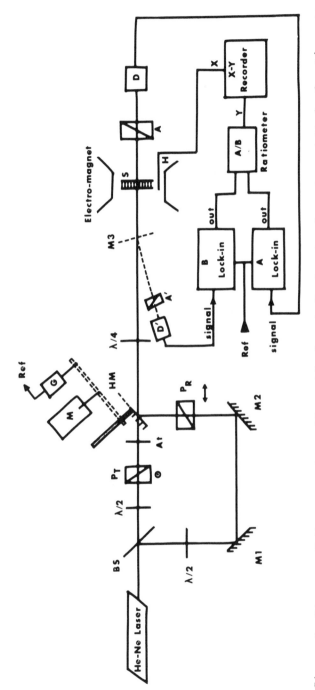

Figure 1 Diagram of the apparatus. The optical and electronic elements and their function are explained in the text.

amplifier A. The mirror is driven by a synchronous motor at 12.5 cps. A generator, G, is simultaneously driven by the motor and provides a reference signal for the lock-in detection.

When the sample does not introduce any phase retardation, as in the case of an isotropic material or an undistorted homeotropically aligned liquid crystal with the director along the optic axis, there is no change in the disposition of the electric vector in either of the beams and the detected signals during each half of the rotation cycle are equal. A DC signal is therefore presented to the lock-in amplifier which produces no output. In the presence of phase retardation, however, the two light beams emerging from the sample are elliptically polarized with the ellipses being of equal ellipticity but with their major axes at right angles to each other and at 45° to the uniaxial direction. The analyzer therefore passes the major axis of one ellipse during one half of the cycle and the minor axis of the other ellipse during the other half of the cycle. A difference signal is therefore present at the detector, D, with the period of the half-mirror rotation. This "ripple" in the signal is measured in synchronous amplification by the lock-in amplifier, whose output is proportional to the difference between the two alternating light levels from the left and right handed circularly polarized light making its way through the optical train.

Expressions for the intensities I_R and I_T passed by the analyser in the reflected and transmitted beams, respectively, are given by applying the matrix product $(\lambda/2)_R \times \underset{\sim}{P}_R \times (\lambda/4) \times \underset{\sim}{S} \times \underset{\sim}{A}$ and $(\lambda/2)_T \times \underset{\sim}{P}_T \times (\lambda/4) \times \underset{\sim}{S} \times \underset{\sim}{A}$ to the Stokes vector of the incident beam[12]. The matrix operation results in the following expressions for the intensities I_R and I_T in the reflected and transmitted beams:

$$I_R = \frac{I_0}{16} (3+\cos\Delta) (1 + \cos\Delta_1 \sin\delta) \qquad (1)$$

$$I_T = \frac{I_0}{16} (3+\cos\Delta)(1 - \cos\Delta_1 \sin\delta) \qquad (2)$$

where $I_0/2$ is in the incident intensity in the reflected beam, the effective incident intensity in the transmitted beam being adjusted to the same value by the attenuator A_t. Δ and Δ_1 are possible departures from perfect half-and quarter-wave retardation of the phase plates. The difference signal is therefore given by:

$$I = I_R - I_T = \frac{I_0}{8} (3+\cos\Delta)\cos\Delta_1\sin\delta \qquad (3)$$

If the half-wave and quarter-wave plates are perfect, $\Delta_1 = \Delta = 0$ and we obtain

$$I = \frac{I_0}{2} \sin \delta \qquad (4)$$

where δ is the phase retardation produced by the sample, S. The beam splitter has the additional effect of causing the reflected beam to become elliptically polarized. This effect is ignored in the above calculations as the linear polarization is restored by the polarizer P_R.

In practice, because of the finite thickness of the rotating mirror, there is a short time ($\sim 5\%$ of the rotation period) during which no light falls on the detector. This causes a finite signal to appear in the absence of birefringence. Fluctuations in the laser output cause the amplitude of this background signal to change giving rise to a drift in the baseline. This drift would be absent if there were no dark region as the signal in both beams would change equally and simply shift the DC output of the photodiode.

A partial compensation of this drift is obtained by reflecting a fraction of the beam at a small angle ($\sim 1°$) just before it enters the sample using a glass plate, M3. This signal is analyzed with another analyzer, A', placed at the same azimuth as A with respect to the direction of the beam. The signal at detector D' is identical to that at D except for the influence of the sample S. The output of the second lock-in amplifier B may thus be used to correct for optical expansion/contraction and laser thermal drift by taking the ratio A/B in the ratiometer. The output of the ratiometer is applied to the Y-input of an X-Y recorder whose X-axis is driven for our purposes by a calibrated Hall probe, H, placed next to the sample in the magnetic field.

RESULTS

In a test of the method, we have used a homeotropically aligned sample of the room temperature nematic liquid crystal 4-methoxy-benzylidene-4'-n-butylaniline (MBBA) as a source of tuneable birefringence. It is well known[13] that when a magnetic field is applied normal to the direction of alignment and the field increased, the liquid crystal stays in the underformed state until a critical field, H_c, is attained. Above H_c the sample continually deforms until it is completely aligned in the direction of the magnetic field). It is also well known that if the direction of alignment of the initial liquid crystal director deviates slightly from the normal to the magnetic field, the deformation starts below H_c. The deformation is small below the critical field, with most of the effect taking place above H_c. Such field induced deformations cause a change in the effective birefringence of the sample and therefore provides us with a source of tuneable retardation which is small below H_c and large above H_c.

Figure 2 is a comparison of the results obtained by our technique with the more conventional method[14] that uses crossed polarizers with the uniaxial direction of the sample being at 45° to the axes of the polarizers. The intensity transmitted in this case is given by

$$I = I_0\sin^2(\delta/2) \tag{5}$$

It is seen that the upper trace (A) of Figure 2 oscillates above and below the baseline, while the lower trace (B) does not have negative values. Maxima and minima of the upper and lower traces correspond to half integral and integral multiples of π in phase retardation, respectively. The experimental intensity profiles thus verify and illustrate the difference between the two optical arrangements that lead to equations (4) and (5) for the transmitted intensity.

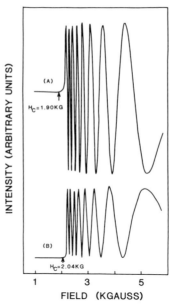

Figure 2   Intensity versus magnetic field traces for the Frederik's distortion of MBBA as monitored by (A) alternate senses of circularly polarized light as in Figure 1 and (B) the conventional method using linearly polarized light at a fixed angle with respect to the analyzer. Arrows indicate the onset of the distortion. (The intensity scales are identical for both traces and the sample was undisturbed between the two measurements).

The lack of coincidence of the critical field (arrows in Figure 2) for the onset of the distortion is due to the fact that the sample is not exactly normal to the magnetic field[15] and the greater sensitivity of method A. The method of alternate circularly polarized beam senses, used to obtain trace A, is capable of detecting a much smaller change in δ. Here the onset of the distortion of the liquid crystal is detected at about a 140 G lower field strength by method A. Method A detects a small change in δ well below the critical field which is not discernable by the conventional method B.

While the sensitivity of this procedure is ideally determined by the electronics, in practice temperature induced fluctuations in the optical elements reduces the sensitivity to $\sim 2 \times 10^{-4}$ radians. This compares favorably with the resolution obtained by other phase modulation techniques (9-11) which is in the range 2 μrad to 5 milliradians. The rotating half mirror has a cost advantage over the more commonly used phase modulation devices. The dual beam arrangement and the concomitant need for two lock-in amplifiers is required however to balance out thermal drift and laser fluctuations in experiments that must be run continuously over long periods. Very long field sweep times (as long as ~12 hrs) are necessary for an equilibrium Frederik's distortion trace in polymer liquid crystals which are composed of large molecules with long relaxation times.

ACKNOWLEDGEMENT

This work was supported in part by the National Science Foundation under grant DMR-8112975 and the National Institutes of Health under grant HL-24364.

REFERENCES

1. W. H. DeJeu, "Physical Properties of Liquid Crystalline Materials," Gordon and Breach, New York (1980), Chapter 6.
2. P. D. deGennes, Mol. Cryst. Liq. Cryst. (Letters), 34:177 (1977).
3. R. B. Meyer in "Polymer Liquid Crystals," A. Ciferri, W. R. Krigbaum and R. B. Meyer, Eds., Academic Press, New York (1982), Chapter 6.
4. D. B. DuPré in "Polymer Liquid Crystals," A. Ciferri, W. R. Krigbaum and R. B. Meyer, Eds, Academic Press, New York (1982), Chapter 7 and references therein.
5. J. R. Fernandes and D. B. DuPré, Mol. Cryst. Liq. Cryst. (Letters), 72:67 (1981).
6. J. R. Fernandes and D. B. DuPré, "Liquid Crystals and Ordered Fluids," Vol. 4, in press.

7. T. Haven, D. Armitage and A. Saupe, J. Chem. Phys., 75:352 (1981).
8. H. Janeschitz-Kriegl, Adv. Polymer Sci., 6:179 (1969).
9. M. R. Battaglia and G. L. D. Ritchie, Faraday Trans. II, 73:209 (1977).
10. G. Maret, M. V. Schickfus, A. Mayer and K. Dransfeld, Phys. Rev. Lett, 35:397 (1975).
11. H. M. Lim and J. T. Ho., Mol. Cryst. Liquid Cryst., 47:173 (1978).
12. A. Gerrard and J. M. Burch, "Introduction to Matrix Methods in Optics", John Wiley and Sons, New York (1975).
13. S. Chandrasekhar, "Liquid Crystals", Cambridge University Press, Cambridge (1977).
14. A. Saupe, Z. Naturforsch, 15a:815 (1960).
15. reference 13, p. 119.

MESOGENIC ORDER-DISORDER DISTRIBUTIONS

Siegfried Hoffmann

Department of Chemistry
Martin Luther University
DDR-4020 Halle/S., GDR

INTRODUCTION

Matter - at least within our scope of space and time - seems to have endeavored for a period of 10-20 billion years to gain a certain consciousness and understanding of itself (Fig. 1). Among the molecular species, screened by evolution in a Darwinian selection for suitable constituents of a dynamic reality-adaption pattern, amphiphilic molecules that did not only reflect hydrophilic-hydrophobic but also order-disorder distributions were rendered preferred survivors of the "grand" process.

(BIO)MESOGENS

A spectrum of quite differently shaped backbone structures (Fig. 2) that not only governed - depending on their one-, two- or three-dimensional preferred array and molecular specializations - aspects of information, function and compartmentation, but conquered and ruled the creative areas of their interdependences and interplays as well, proved especially useful.

As an artificial derivative and reflection of that awareness, the emerging pattern of man-made mesogens from its beginning towards the end of the last century has retraced the long developmental path of mesogenic life patterns. Starting with low-molecular-weight, rod-like thermotropics and polar-apolar ended lyotropics, which were enriched within their cooperative interactions, it enlarged

423

Fig. 1.   Mesogen pattern.   Adapted from ref. 3.

in our time not only with regard to the geometrical prerequisites of growing populations but also reached into
the fertile grounds of spreading macromolecular species
and simple biomesogen organizations (Figs. 3-5).   These
aspects are analogized by the ambiguous cover-painting
of Alexandre Blumstein's last Symposium-volume (Fig. 1).
It conveys much of what may have prompted Lehman[4], one
of the pioneers in the field of liquid crystals, to call
these lovely species "lebende Kristalle".   Anyone would
feel this gazing on the playful movements, the "living"
texture of liquid crystals - mysterious creations related
to us.

All that emerged and that might still develop from
these dynamic patterns of our creativity seems to reflect
aspects and characteristics of a grand evolutional pattern
that our artificial liquid crystals experienced developing
mesogen strategies.   So, perhaps, the creative pattern
that anticipated optimized mesogenic abilities, and the
derived artificial mesogenic pattern that developed in a

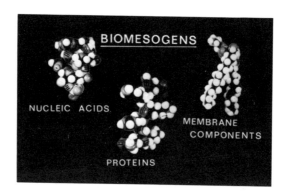

Fig. 2.   Biomesogenic backbone structures.

424

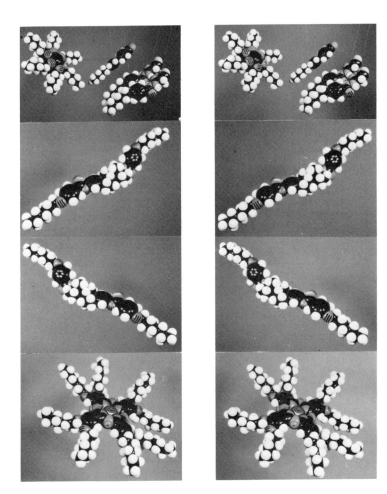

Fig. 3. Rod- and disc-like low molecular weight thermotropics (left to right and top to bottom): hexa-
n-alkanoyloxybenzenes,[5] cyanobiphenyls,[6] diacyl-
hydrazines,[7] enantiomeric estradiols,[8] metalloph-
thalocyanines.[9]

continuing process of rediscovery and "Nachempfindung",
might mutually stimulate each other and even improve their
respective facilities.

Even very simple and primitive liquid crystal moieties
seem to differ from non-mesogenic species in that they bear
some sort of rudimentary intelligence. At present, however,
we appear to be faced with difficulty in scientifically
treating this gleam of intelligent behavior. Our theories
and mathematics prefer rather abstract and etheric shadows

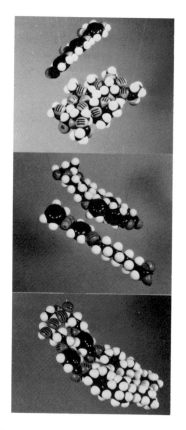

Fig. 4. Stiff, main and side chain high-molecular weight thermo- and lyotropics (left to right and top to bottom): aromatic polyamides,[10] cellulose triacetate,[11] azoxybenzene polymers with flexible main chain spacers,[12] polyacrylates with flexible side chain spacers,[13] polysiloxanes with a combination of flexible main and side chain parts.[14,15]

of real entities.

As we developed our liquid crystals in terms of thermo- and lyotropic characteristics, biomesogens both blurred and dialectically combined these extreme positions within their cooperative networks of interdependent complexity pattern and diverted our close-up view, originally aimed at the unifying constraints of large molecular ensembles, to the small boundaries of cooperatively acting domain organizations, utilizing a precisely tuned and refined instrumentation of domain-modulated phase transitions.[2,16,17] Though we are only beginning to understand life patterns,

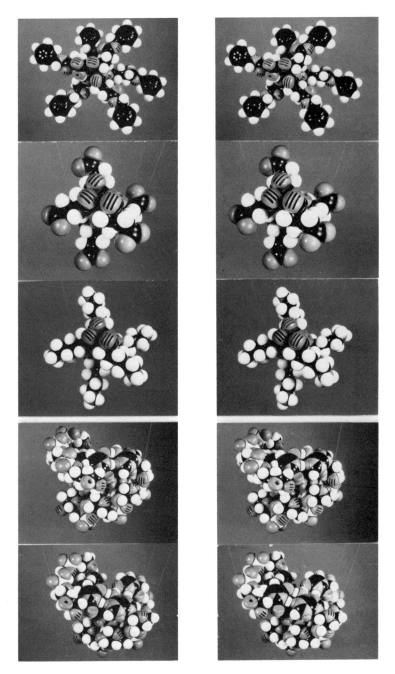

Fig. 5. Lyotropic polypeptides and polynucleotides (top
to bottom): poly(γ-benzyl-L-glutamate),[18] poly(L-
glutamic acid),[18] poly(L-lysine)[19] – double and
triple stranded RNAs.[20]

it seems their constituents enabled them to conquer the broad fruitful range between order and disorder by selecting mesogenic species that were provided by their special molecular design with optimizable free energy strategies, not only operating on a suitable imprinted affinity patter but also quite intelligently handling entropic order-disor der gradients.

ORDER-DISORDER PATTERNS

Proteins and nucleic acids experienced their first intimacy within the nucleation event. Later, they turned out to be the driving forces of the "grand game". It was representatives of the ordered patterns of these two part- ners in our life process that emerged first from our model building, X-ray assisted investigation as the two "hollow" structures: the protein α-helix and the DNA double helix. Since then, growing populations of more and more flexible and advanced helices and sheets (Figs. 6-9) have sprung from our X-ray diffractometer and computer output facili- ties.

Though there are quite obvious structural, and derive from them, enthalpic and entropic differences between the order expressions of the predominantly informational and functional components,[21] there is also a good deal of func tional overlap between them. Proteins - at least for the synthesis of rather primitive antibiotics - perform roles as informational molecules.[22] The rather archetypic RNAs, on the other hand, have recently been identified as cata- lysts.[23]

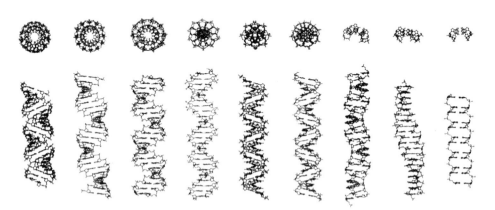

Fig. 6.   Nucleic acid polymorphs: Fibrous polynucleotides unwinding right- to left-handed helices. Olson- and Z-families omitted. Modified from ref. 24.

428

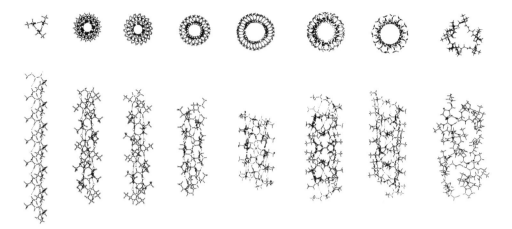

Fig. 7.   Polypeptide helical design between single and
          double stranded, right and left handed, parallel
          and antiparallel, helix-, spiral- and channel-
          expressions. Modified from refs. 25 and 26.

     Nucleic acids and proteins display both structural
and hierarchical complexity.   More complexity is displayed
by components of membranes.   Even a simple phospholipid
offers considerably higher degrees of structural and functional complexity than might originally have been needed
for purposes of pure barrier functions (Fig. 10).

     Order, however, a prerequisite for the storage of in-
formation and precise function, is actable, movable and
transferable (i.e. realizable only by mediating flexible
and dynamic elements).   Unfortunately, though our instrumentation with which to follow up these dynamic activities
is complex and rapidly enlarging, the problems themselves

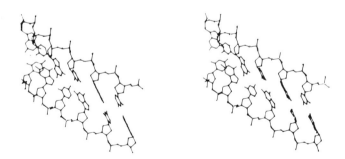

Fig. 8.   Olson-D/RNA[27] - example of a twisted polynucleo-
          tide sheet structure. Adapted from ref. 28.

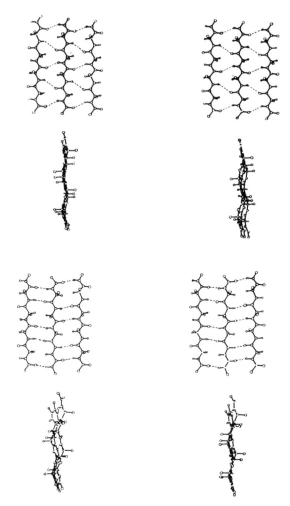

Fig. 9. Parallel and antiparallel polypeptide β-sheets that often display a right-handed twist. Adapted from ref. 29.

seem to offer even more complex difficulties. These might well be caused by some sort of a biological "Unschärferelation", that reflects limitations on an organizational level of closely interacting biomesogenic patterns,[22,30] which had been experienced earlier on the level of elementary particles.

Nevertheless, X-ray structural analysis in combination with dynamic assisting methods unraveled impressive pictures of both biopolymer-biopolymer and biopolymer-

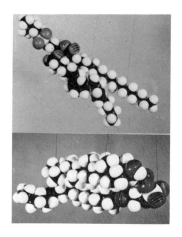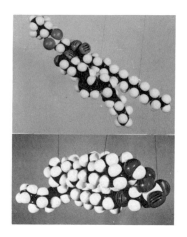

Fig. 10. Advanced biolyotropics and parts of their regulation instrumentation (top to bottom): Lecithin
molecule. Cholesterol fitting a phospholipid
tail for regulation of membrane fluidity.22

oligomer and -monomer recognition and interaction patterns.
They act by a well-balanced interplay of more ordered or
disordered, that is, more rigid or flexible parts, where
domain transitions within biopolymeric organizations
achieve new levels of interspecies and interdomain cooperation, partially simulated by our rather primitive approaches in handling low temperature devices by special mixing
techniques of simple thermotropics.

Within those interacting networks, thermotropic and
lyotropic aspects (Fig. 11) contribute to a more general
method of intelligently handling dynamic order-disorder
patterns, suitably assisted by corresponding energetic interaction facilities. By this, water, as well as all the
other more flexible solvent-like parts of interacting mesogens, would appear only as a special expression of a more
comprehensive solvent characteristic, attributable to solvent species that display transient order patterns within
movable order-disorder interplays and might thus intimately
respond to and follow up the individual statics and dynamics of the solutes. Within the realms of proteins, nucleic acids and membrane components, dynamic parts of
those entities will perform assisting solvent roles in
solubilizing and mobilizing the rather rigid solute elements that must be built up in order to secure the dynami-
cally mediated and newly gained qualities in recognition,
function, collective operation modes and organizational
behavior.

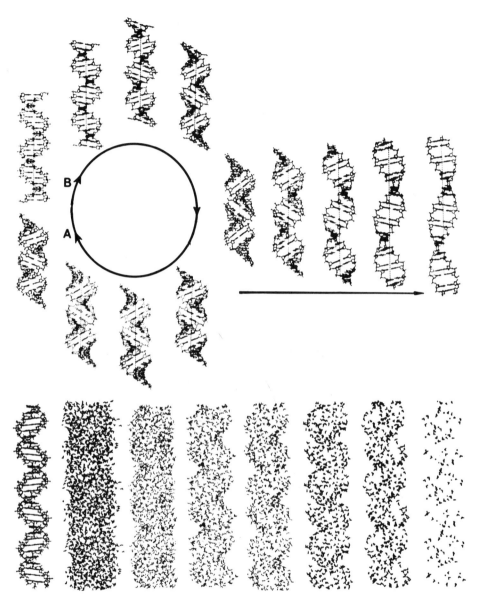

Fig. 11. Nucleic acids B-DNA domain operation modes: Backbone flexibilities mediating A-B transitions and
intercalation geometries - an instrumentation assumed to be partially based on nonlinear excita-
tions.[31] Modified from refs. 32 and 33 (top).
Water molecules connectivity pathways and Na+
counterion orientations from low to high humidi-
ty as resulting from extended Monte Carlo simula-
tions. Modified from ref. 34 (bottom).

It was the rather transparent structural design of nucleic acids that first allowed for deeper insights into the operational strategies in recognition and action processes between nucleic acids alone, nucleic acids in combination with small and large molecule effectors, and nucleic acids in cooperation with the partner of the "grand game": the proteins. The store of DNA information appears today as a remarkably plastic macromolecule varying greatly in base composition.[35,36] RNAs in the form of tRNAs give insight into both their more rigid, ordered parts as the different defined arrangements of anticodon loops,[37] and their more flexible, operational instrumentation mainly represented by the sensitive hinge regions.[38] These complex prodynamic domains seem to be capable of synergetic-directed hysteresis regulation modes.[19,39] The interaction patterns of nucleic acids with low- and high-molecular-weight effectors, that are, in principle, elucidable by electrostatic potential and field calculations (Fig. 12),[40,41] inspired research strategies for alienating macromolecular design[2,33] and, by this, provoking eukaryotic self non-recognition and defense strategies (Figs. 13 and 14).

Though proteins are the real functional entities, their considerably higher degrees of complexity in dynamic action modes have limited and delayed our approaches. Static disorder patterns in a number of functional proteins as, for instance, trypsin-trypsinogen, citrate synthase and immuno-globulins[42] are well known. We are not only aware of the precise matrix fits of substrates within their binding sites, but have learned much about the dynamic interplays of protein matrices and flexible ligands within dynamic recognition modes.[22,43] Carboxypepdidase, hemo- and myoglobin, cytochromes and many other "small" protein structures have advanced our understanding.[22,44] We try to elucidate a dynamic hierarchy of growing complexity when we follow the cooperative dynamics of side chains, peptide segments and whole domain areas, and endeavor to include mediating solvent and counterion environments into the picture.[45] We hope to decode regulation languages and have learned, in the meantime, something of synergetics, competing subsystems, slaving order parameters and nonlinear dynamics far from thermal equilibria.[46]

At this point in a structure inquiry into interdependences and complexities, we begin to understand how proteins might mediate transmembrane passages, for instance, and how the lipid veil could assist these efforts.[48] General nucleic acid-protein recognition and action modes might be determined from the first insights into the sensitive re-

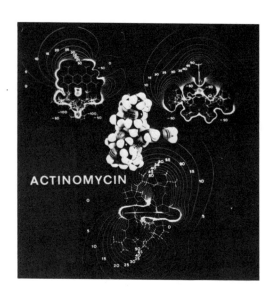

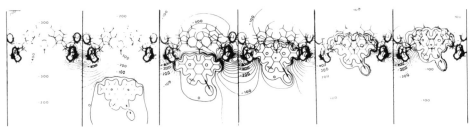

Fig. 12. CPK-model and electrostatic potentials of action-
mycin, a $[d(G-C]_n$ -specific intercalator that
simulates by its twofold symmetry recognition
strategies of repressor proteins. Modified from
ref. 40 (top).
Sequence of electrostatic contour maps through
the intercalation plane of the nucleic acid re-
ceptor in mediating the receipt of ethidium bro-
mide. Modified from ref. 41 (bottom).

gions of gen-expression regulations, cro-repressor-operator[49] interactions serving as a starting-point for applyi
mesogenic strategies to our old structural pictures.

     How well even a rather simple biomacromolecule emplo
all these strategies in handling energetic and entropic e
ficiency, might be demonstrated by the antibiotic transmen
brane ion-carrier valinomycin.[22,55] A well-balanced inte
play of energetic and entropic elements, working on a rather small but highly sophisticated polydepsipeptide, un-
ravels mechanisms of recognition, handling and manipula-

Fig. 13. Base-pair and strand-analogs: low and high
molecular-weight effectors for alienating nucleic
acid design.[50]

tion of "guest" ions by operating on differently inter-
changed domain arrangements and interfaces, that match im-
pressively our own intelligence (Figs. 16 and 17). Evolution might have so favored this quasi-enzyme "small"-mole-
cule that similar strategies built up carrier subunit arrangements[56] (Fig. 18), and later on transferred the whole
action to the level of eukaryotic organism behavior.

As is the case with valinomycin and even more sophisticated interaction patterns of large biomesogen organizations, energetic patterns are rather well accounted for by
our computer calculated possibilities and increasingly well
visualized by developing output facilities. The dynamic
programs imprinted into their molecular interaction patterns and dynamic interplays, however, remain less obvious,
both in interpretation and visualization.

Thus, if one tries to conceive this mesogenic world
just from the viewpoint of entropy (Fig. 19), old cate-
gories submerge into a more general amphiphilic pattern
based on amphiphilicity in terms of order-disorder prefer-
ences rather than on various geometric and polar-apolar
characteristics. As pointed out earlier, it is from this
viewpoint that thermotropics apparently resemble lyotropics in that the more flexible disordered segments will
function as a solvent for the more rigid and ordered entities. The lyotropics, on the other hand, display within
their solvent=solute distributions more interactive coherence than had originally been anticipated from a classic
structural view of the molecular species involved. Both
the rather primitive thermotropic alkoxybenzoic acid and
the more advanced lyotropic Olson-RNA, for instance, display common mesogenic concepts of order-disorder distributions. Their hydrogen-bonded cores are "solubilized"

Fig. 14. Alienating nucleic acid structural and functional
design by low and high molecular-weight effectors
(top to bottom): Intercalation of a tilorone-
type cytokine inducer into a B-DNA, the bibasic
terminals of the intercalator imitating the twofold recognition symmetries of repressor proteins.[51] Base-pair match of a semiplastic interferon inducer.[33,52] Base-pair fit of an artificial strand analog - built up for inhibition of
reverse transcriptases[53] - in a triple stranded
arrangement with two mate polynucleotides.

by long chained hydrocarbon terminals in the case of thermotropic and by a well balanced movable backbone design
with adjustable water shells and counterion patterns in
that of the lyotropic. The same also hold true even more
so if the "phase" - or the "domain"-transitions are more
accurately mediated by nucleic acid-protein-water interaction modes (Fig. 20). While the specific characteris-

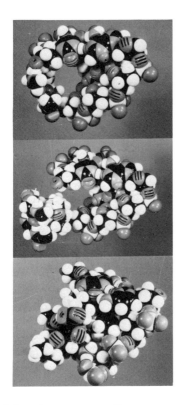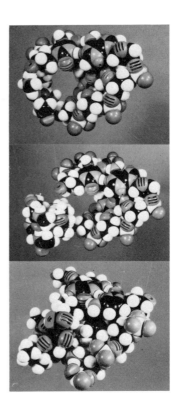

Fig. 15. Aspects of cro-repressor-operator recognition
strategies[49] (top to bottom): Fit of an antiparallel protein β-sheet in the minor groove of
B-DNA.[54] Suggestions for derived steroid hormone
operation modes.[19,30,33] Recognition fit of helical parts of the repressor molecule in the B-DNA
major groove. Flexible parts of the protein that
mediate initial stages of approach and recognition are omitted because of limitations of the
"static" photograph.

tics of the solvent partners provide a spectrum of entropy-
driven organizational forces for the lyotropics, the pronounced order-disorder gradients of a majority of thermotropics should offer a similar basis for entropy to work on
in an effort to minimize the more constrained and ordered
interfaces between the rigid cores and the flexible terminals.

Especially for the biologically important hydration
and counterion sites, these interrelationships might also

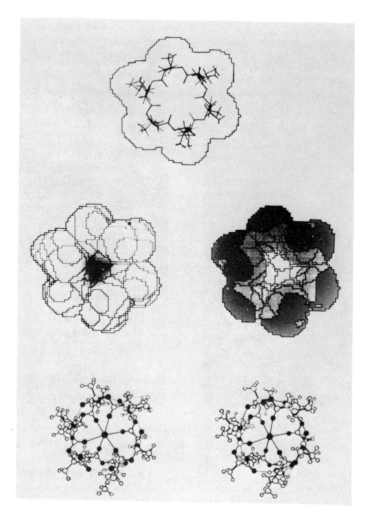

Fig. 16. Valinomycin (top to bottom): Schematic representation of the ionophore. Surface electrostatic
potentials of the free and the K[+]- complexed form
(negative values: dark; positive values: light).
Skeletal view of the valinomycin-K[+]-complex.[57],
[55,20]

be regarded as being caused by special order parameters,
imposed by the transient geometry and dynamics of the bio-
macromolecule solute upon the transient hydrogen bond network of the water solvent. Hydrophilic parts of the solute
imprint consecutive connectivity pathways with enhanced
lifetimes, while hydrophobic areas tend to limit interaction possibilities and, by this, favor intra-solvent arrangements.[59] The statics and dynamics of solvent-solute
line connectivity and closed-surface connectivity sites[34]

Fig. 17. Valinomycin (top to bottom): K⁺ facing the flower-like invitation pattern. The beginning for the successive "striptease." The carrier displaying a polar hydration-shell simulation for the guest ion and an apolar gliding pattern towards the membrane interior: the transmembrane passage. K⁺ exposed by the carrier "sees" the post-membrane water phase. Anchored by its hydrophobic domain within the lipid phase, the carrier dismisses the K⁺ guest into the new water-cover.[20]

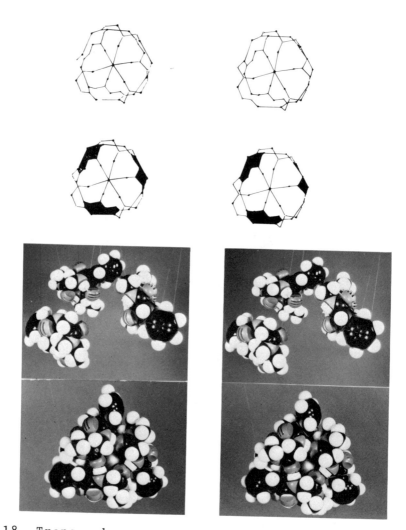

Fig. 18. Transmembrane carrier relationships (top to bottom):
Backbone abstractions of valinomycin, indicating
in the second row the threefold symmetry of the
complex cyclo(D-Val-L-Lac-L-Val-D-Hyi)$_3$-ionophore.
Modified from ref. 57.
Trimer subunit pattern of the phytotoxin tentoxin:
cyclo(L-MeAla-L-Leu-MePhe $\left[(Z)\Delta\right]$-Gly) displaying
conformations for building up a hydrogen bond network between three cyclotetrapeptide moieties, providing a cavity for K$^+$-ion complexation.
Structural proposal for a (tentoxin)$_3$-K$^+$-complex
that resembles in geometry and dynamics the domain
operation modes of valinomycin.56

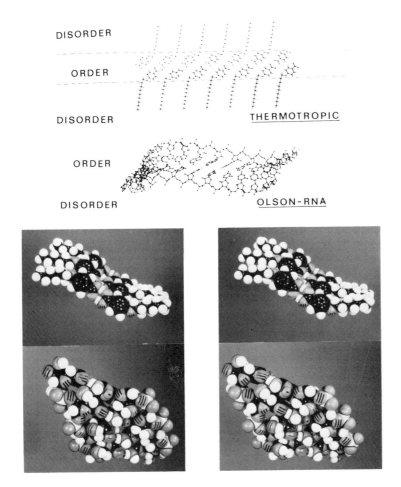

Fig. 19. Order-disorder distributions suggested for an artificial thermotropic (n-alkoxy-benzoic acid dimer)
and a biolyotropic (Olson-RNA).[27,58] Skeletal
drawings (top) and CPK-representations (bottom).

contribute significantly to the overall characteristics of
biomesogen organization. The synergism between the two
subsystems "solvent" and "solute", coupled with crossed
feedbacks, constitutes new aspects of nonlinear dynamics
of this new functional system that might exhibit re-entrancy
and hysteresis, for instance.

The entropy-modulated dynamic and creative interfaces
of biomesogen organizations provide essentials for nonlinear behavior patterns with spatial-temporal coherence.

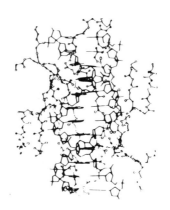

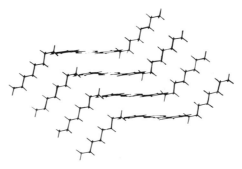

Fig. 20. Protein-nucleic order-disorder patterns resem-
bling primitive thermotropic action (top to bottom): Hypothetical B-DNA-prealbumin complex in
skeletal presentation, viewed along and perpendicular to the B-DNA helix axis.[61] Thermotropic
n-alkoxybenzoic acid dimer.[58]

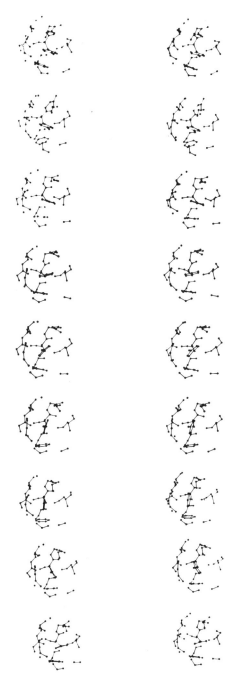

Fig. 21. Rotational isomerization of Tyr-35 of bovine pan-
creatic trypsin inhibitor mediated by structural
changes in the environing protein matrix.   Adapted from ref. 60.

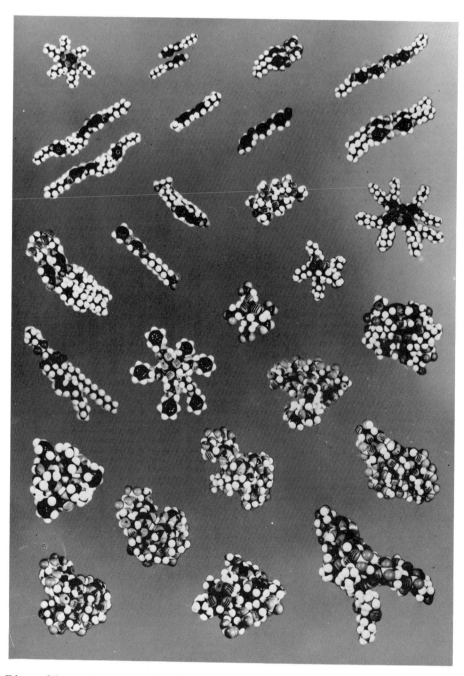

Fig. 22. Survey of the introduced (bio)mesogneic species
in CPK-presentation and - (continued Fig. 23)

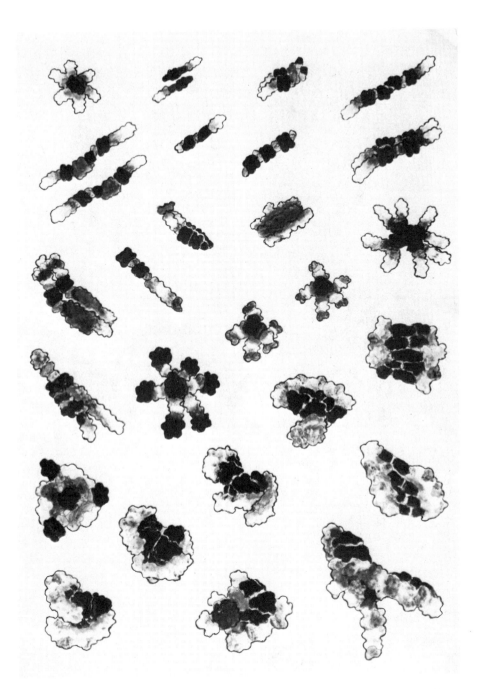

Fig. 23. -- their suggested simplified order-disorder pat-
        terns (ordered parts: dark; disordered parts:
        light).

445

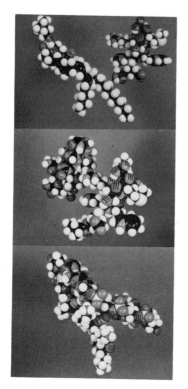
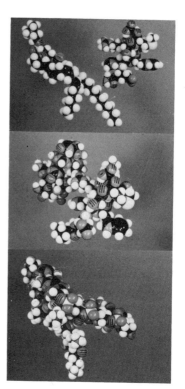

Fig. 24. Biomesogenic backbone geometries providing structural prerequisites for dynamic order-disorder
patterns (left to right and top to bottom): phospholipid and helical protein, nucleic acid strand
and protein helix, protein-nucleic acid backbone
interplays, presumably engaged in the induction
of interferon.[33,52]

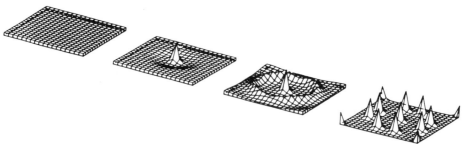

Fig. 25. Synergetics in pattern formation. Adapted ref. 62.

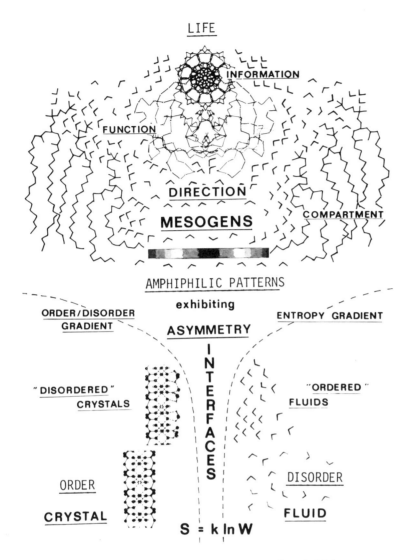

Fig. 26. The evolution of the amphiphilic patterns.

The complexity of the picture when extending this special view into the general action of mesogenic order-disorder distributions is, although insufficiently and only within a small facet, reflected by the computer sequence of Fig. 21, which displays mesogen characteristics of a protein matrix that mediates the rotational isomerization of an aromatic side-chain.[60] Ranging from simple thermotropic and lyotropic synthetics to the highly evolved biomesogenic species of life today (Figs. 22-26), (bio)-mesogens display unique order-disorder distributions in addition to their highly sophisticated energetic instrumentation. In close analogy to the order-disorder patterns of thermotropic synthetics - rod and disclike cores fitted with differently shaped flexible alkyl-chain terminals - nucleic acids, as well as proteins and membrane components, exhibit more or less rigid cores in a wide variety of suitably designed backbone geometries, partially fluidized and designed for an infinity of specialization purposes by variously extended flexible and dynamic side-chain patterns including solvation shell and counterion cloud continuations (Figs. 22-24). By differently valuing the specific order-disorder gradients of mesogenic individuals, entropy just in its tendency to minimize order and to maximize disorder should at least partially act as a driving force in building up cooperative and dynamic amphiphilic patterns with a pronounced tendency for self-organizations.

By reflecting within their individual molecular designs not only the polar-apolar distributions but also the order-disorder patterns of their zones in which they are formed, the mesogenic constituents of the "grand" amphiphilic pattern of life provided decisive prerequisites for the projection of individual molecular facilities into the structural and functional amplifications of cooperative and dynamic mesogen domain ensembles that gained -far from thermal equilibria - directionality and linked together, subjected to slaving synergetics (Fig. 25), entropy and information, in a delicate mesogen balance (Fig. 26).

CONCLUSIONS

Born within the tension between the realms of crystalline order and fluid disorder, molecularly imprinted by the very thesis and antithesis of their origin and later in the fluctuations of the conquered areas of their increasingly complex world, mesogens managed to escape from the sterile extrema-futilities that bounded their extended interface-birth zones. By intelligently reflecting and handling the order-disorder thesis and antithesis of their origin within their individual molecular designs and their cooperative interaction patterns, the mesogenic constituents of the developing amphiphilic patterns of life succeeded in obtaining a fertile and creative synthesis, avoiding statics and hyperdynamics - the disadvantages of the extreme states that tried to govern their origin -

and optimizing the promising advantages of the "meso" positions of ongoing ordered dynamics. Compensating for the loss of order-disorder tension, entropy endowed them with the gifts of self-organization, cooperation and a new-born directioned dynamic order within enlarging molecular ensembles and growing organizations, strategies and means that were later to be followed by the artificial mesogenic creations of the "grand" process.

REFERENCES

1. S. Hoffmann, "Evolution of the Amphiphilic Pattern" in Darwin Today, E. Geissler and W. Scheler, Eds., Akademie-Verlag, Berlin (1983).
2. S. Hoffmann and W. Witkowski, "Biomesogens and their Models" in Mesomorphic Order in Polymers and Polymerization in Liquid Crystalline Media, A. Blumstein, Ed., ACS Symp. Ser. 74, 178 (1978).
3. A. Blumstein, Ed., Mesomorphic Order in Polymers and Polymerization in Liquid Crystalline Media, ACS Symp. Ser. 74 (1978).
4. O. Lehmann, Die Scheinbar Lebenden Kristalle, Schreiber Verlag, Esslingen (1907).
5. C. Destrade, N.H. Tinh, H. Gasparoux, J. Malthete, and A.M. Levelut, Mol. Cryst. Liq. Cryst., 71, 111 (1981).
6. G.W. Gray and A.J. Leadbetter, Phys. Bull. 28, (1977).
7. S. Hoffmann, H. Schubert, A. Kolbe, H. Krause, Z. Palacz, Ch. Neitzsch and M. Herrmann, Z. Chem. 18, 93 (1978).
8. S. Hoffmann, W. Weissflog, W. Kumpf, W. Brandt, W. Witkowski, H. Schubert, G. Brezesinski, and H.D. Dörfler, Z. Chem. 19, 62 (1979).
9. C. Piechocki, J. Simon, A. Skoulios, D. Guillon, and P. Weber, J. Am. Chem. Soc., 104, 5245 (1982).
10. W.G. Miller, Ann. Rev. Phys. Chem., 29, 519 (1978).
11. D.L. Patel and R.D. Gilbert, J. Polym. Sci., Polym. Phys. Ed., 20, 877 (1982).
12. A. Blumstein, S. Vilasagar, S. Ponrathnam, S.B. Clough, R.B. Blumstein and G. Maret, J. Polym. Sci., Polym. Phys. Ed., 20, 877 (1982).
13. S.G. Kostromin, R.V. Talroze, V.P. Shibaev and N.A. Plate, Makromol. Chem. Rapid. Commun. 3, 803 (1982).
14. S. Hoffmann and U. Kullack, unpublished results.
15. H. Finkelmann, H. Benthak, and G. Rehage, J. Chim. Phys., 80, 163 (1983).
16. D.B. DuPrè and E.T. Samulski, "Polypeptide Liquid Crystals" in Liquid Crystals - the Fourth Estate of Matter, F.D. Saeva, Ed., Marcel Dekker Inc., (1979).

17. N. Ise, T. Okubo, K. Yamamoto, M. Matsuoka, H. Kawai, T. Hashimoto and M. Fujimura, J. Chem. Phys., 78, 541 (1983).
18. E. Iizuka and J.T. Yang, "Formation of the Liquid Crystals of Polyribonucleotide Complexes" in Liquid Crystals and Ordered Fluids, J.F. Johnson and R.S. Porter, Eds., Plenum Publ. Co., N.Y. (1978).
19. S. Hoffmann and W. Witkowski, "Biomesogen-Regulationen" in Wirkstofforschung 1981, H. Possin, Ed., Wissenschafts Publ. MLU, Halle (1983).
20. R.J.P. Williams, Biochim. Biophys. Acta, 416, 237 (1975
21. C.J. Alden and S.H. Kim, "Accessible Surface Areas of Nucleic Acids and Their Relation to Folding, Conformational Transition, and Protein Recognition" in Nucleic Acid Geometry and Dynamics, R.H. Sarma, Ed., Pergamon Press, N.Y. (1980).
22. S. Hoffmann, Molekulare Matrizen (I Evolution, II Proteine, III Nucleinsäuren, IV Membranen),Akademie-Verlag, Berlin (1978).
23. R. Lewin, Science, 218, 872 (1983).
24. S. Arnott and R. Chandrasekaran, "Fibrous Polynucleotide Duplexes Have Very Polymorphic Secondary Structure" in Biomolecular Stereodynamics, R.H. Sarma, Ed., Adenine Press, N.Y. (1981).
25. R. Chandrasekaran and A.K. Mitra, "Conformational Flexibilities of Some Polypeptide Helices and Disulfide Bridged Peptides" in Conformation in Biology, R. Srinavasan and R.H. Sarma, Eds., Adenine Press, N.Y. (1983).
26. D.W. Urry, C.M. Venkatachalam, M.M. Long and K.U. Prasad, "Dynamic β-Spirals and a Librational Entropy Mechanism of Elasticity" in Conformation in Biology, R. Srinavasan and R.H. Sarma, Eds., Adenine Press, N.Y. (1983).
27. W.K. Olson, "Spatial Configuration of Ordered Polynucleotide Chain: A Novel Double Helix," Proc. Natl. Acad. Sci. USA, 74, 1775 (1977).
28. R.H. Sarma, C.K. Mitra and M.H. Sarma, "Structure of the DNA Double Helix in Solution, Crystals and Theory. Coming of Age of Z-DNA" in Biomolecular Stereodynamics, R.H. Sarma, Ed., Adenine Press, N.Y. (1981).
29. K.C. Chou, M. Pottle, G. Némethy, Y. Ueda and H.A. Scheraga, J. Mol Biol., 162, 89 (1982).
30. S. Hoffmann, "Steroid/DNA-Strukturkomplementaritäten" in Wirkstofforschung 1980, H. Possin and G. Barysh, Eds., Wissenschafts Publ. MLU, Halle (1981).

31. J.A. Krumhansl and D. Alexander, "Nonlinear Dynamics and Significant Conformational Changes in Biomolecular Structures" in Conversation in Biomolecular Stereodynamics III. Program and Collected Abstracts, R.H. Sarma, Ed., Adenine Press, N.Y. (1983).
32. H.M. Sobell, "Structural and Dynamic Aspects of Drug Intercalation into DNA and RNA" in Nucleic Acid Geometry and Dynamics, R.H. Sarma, Ed., Pergamon Press, N.Y. (1980).
33. S. Hoffmann, Wiss. Z. Univ. Halle, 32, 51 (1983).
34. E. Clementi, "Structure of Water and Counterions for Nucleic Acids in Solution" in Structure and Dynamics: Nucleic Acids and Proteins, E. Clementi and R.H. Sarma, Eds., Adenine Press, N.Y. (1983).
35. R.E. Dickerson, H.R. Drew, B.N. Conner, R.M. Wing, A.V. Fratini, and M.L. Kopka, Science, 216, 475 (1982).
36. D.J. Patel, A. Pardi and K. Itakura, Science, 216, 581 (1982).
37. J.L. Sussman, S.R. Holbrook, R.W. Warrant, G.M. Church, and S.H. Kim, J. Mol. Biol., 123, 607 (1978).
38. S.C. Harvey and J.A. McCammon, Nature, 294, 286 (1981).
39. S. Micciancio and G. Vassallo, N. Chim.(D), 1, 627 (1982).
40. P.M. Dean and L.P.G. Wakelin, "Electrostatic Components of Drug-Receptor Recognition II. The DNA-Binding Antibiotic Actinomycin", Proc. R. Soc. London B, 209, 473 (1980).
41. P.M. Dean and L.P.G. Wakelin, Phil. Transact. R. Soc. London, 287, 571 (1979).
42. R. Huber and W.S. Bennett, Pure Appl. Chem, 54, 2489, (1982).
43. R.J.P. Williams, Angew. Chem., 89, 805 (1977).
44. P.G. Debrunner and H. Frauenfelder, Ann. Rev. Phys. Chem., 33, 283 (1982).
45. E. Clementi, G. Corongiu, M. Gratarola, P. Habitz, C. Lupo, P. Otto and D. Vercanteren, Int. J. Quant. Chem. Quant. Chem. Symp., 16, 409 (1982).
46. H. Haken, Naturwissenschaften, 67, 121 (1980).
47. Y.A. Ovchinnikov, Biochem. Soc. Symp., 46, 103 (1982).
48. D. Marsh, "Biomembranes" in Supramolecular Structure and Function, G. Pifat and J.N. Herak, Eds., Plenum Press Publ., N.Y. (1983).
49. D.H. Ohlendorf, W.F. Anderson, R.G. Fisher, Y. Takeda, and B.W. Matthews, Nature, 298, 718 (1982).
50. S. Hoffmann and W. Witkowski, "Alienating Nucleic Acids by Distant Mono-, Olio- and Polymeric Analogs" Nucleic Acids Symp. Ser., 4, s221 (1978).
51. S. Hoffmann, W. Witkowski and R. Skölziger, unpublished results.
52. S. Hoffmann, Z. Chem, 22, 357 (1982).

53. S. Hoffmann, W. Witkowski, P. Venker and H. Rössler, Z. Chem., <u>16</u>, 402 (1976).
54. G.M. Church, J.L. Sussman and S.H. Kim, "Secondary Structural Complementarity Between DNA and Proteins", Proc. Natl. Acad. Sci. USA, <u>74</u>, 1458 (1977).
55. Y.A. Ovchinnikov and V.T. Ivanov, Tetrahedron, <u>31</u>, 2177, (1975).
56. S. Hoffmann and E. Müller, unpublished results.
57. C. Etchebest, R. Lavery and B. Pullman, Stud. Biophys., <u>90</u>, 7 (1982).
58. R.F. Bryan, P. Hartley, R.W. Miller, and M.S. Shen, Mol. Cryst. Liq. Cryst., <u>62</u>, 281 (1980).
59. M.U. Palma, "Internal Dynamics of Biomolecules in Solution: An Introduction and Some Comments on Structural Defect Formation and Annealing in Duplex Helical Structures" in <u>Structure and Dynamics: Nucleic Acids and Proteins</u>, E. Clementi and R.H. Sarma, Eds., Adenine Press, N.Y. (1983).
60. J.A. McCammon, C.Y. Lee and S.H. Northrup, J. Am. Chem. Soc., <u>105</u>, 2232 (1983).
61. C.C.F. Blake, Endeavour, <u>2</u>, 137 (1978).
62. A. Gierer, <u>Die Physik und das Verstandnis des Lebendigen</u>, Festvortrag - Hauptversammlung der Max Planck Gesellschaft (1982).

INDEX